New National Framework

MATHEMATICS

7

M. J. Tipler   K. M. Vickers

J. Douglas

OXFORD
UNIVERSITY PRESS

Great Clarendon Street, Oxford, OX2 6DP, United Kingdom

Oxford University Press is a department of the University of Oxford.
It furthers the University's objective of excellence in research, scholarship,
and education by publishing worldwide. Oxford is a registered trade mark of
Oxford University Press in the UK and in certain other countries

First published by Nelson Thornes Ltd in 2002
This edition first published by Oxford University Press in 2014

British Library Cataloguing in Publication Data
Data available

978-0-7487-6751-9

20 19 18 17 16

Printed in Great Britain by CPI Group (UK) Ltd., Croydon CR0 4YY

Acknowledgements

The authors and publishers would like to thank Jocelyn Douglas for her contribution to
the development of this book.

The publishers thank the following for permission to reproduce copyright material.
Casio: 5, 51, 99; Corel (NT): 2, 23, 34, 102, 135, 141, 285, 364, 367, 374, 377, 385, 403, 409;
Digital Stock (NT): 148, 308; Digital Vision (NT): 176, 237, 329, 366; Image 100 (NT): 327
The publishers have made every effort to contact copyright holders but apologise if any
have been overlooked.

Although we have made every effort to trace and contact all
copyright holders before publication this has not been possible in all
cases. If notified, the publisher will rectify any errors or omissions at
the earliest opportunity.

Contents

Contents

Introduction

We hope that you enjoy using this book. There are some characters you will see in the chapters that are designed to help you work through the materials.

These are

 This is used when you are working with information.

 This is used where there are hints and tips for particular exercises.

 This is used where there are cross references.

 This is used where it is useful for you to remember something.

 These are blue in the section on number.

 These are green in the section on algebra.

 These are red in the section on shape, space and measures.

 These are yellow in the section on handling data.

Number Support

Place value – whole numbers

Millions	Hundreds of thousands	Tens of thousands	Thousands	Hundreds	Tens	Units
5	6	3	2	4	7	9

This is a place value chart.

This chart shows **place value**.

In 5 632 479   the value of the 5 is 5 millions or 5 000 000
               the value of the 6 is 6 hundreds of thousands or 600 000
               the value of the 3 is 3 tens of thousands or 30 000
               the value of the 2 is 2 thousands or 2000
               the value of the 4 is 4 hundreds or 400
               the value of the 7 is 7 tens or 70
               the value of the 9 is 9 units or 9.

Practice Questions 1, 2, 9, 26, 29, 33

Reading and writing whole numbers

Large numbers are **read** and **written** in groups of three.

MILLIONS			THOUSANDS					
Hundreds	Tens	Units	Hundreds	Tens	Units	Hundreds	Tens	Units
		9	6	7	4	1	0	8

The number 9 674 108 is read as 'nine million, six hundred and seventy-four thousand, one hundred and eight'.

Practice Questions 6, 8

Multiplying and dividing whole numbers by 10, 100 and 1000

We use **place value** to **multiply and divide by 10, 100 and 1000**.

Examples

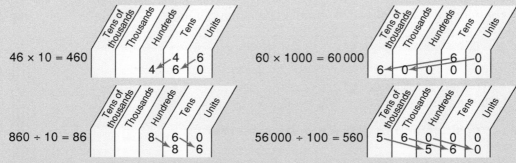

$46 \times 10 = 460$

$60 \times 1000 = 60\,000$

$860 \div 10 = 86$

$56\,000 \div 100 = 560$

When we **multiply by 10, 100** or **1000** the digits move to the **left** by one place for 10, two places for 100 and three places for 1000.

When we **divide by 10, 100** or **1000** the digits move to the **right** by one place for 10, two places for 100 and three places for 1000.

Practice Question 21

Putting whole numbers in order

To **put numbers in order** we compare digits with the same place value. We start at the left.

Example   Put   8356   8563   8536   in order from largest to smallest.
Start at the left and compare the thousands digits. They are all the same.
Look at the hundreds digits. 8563 and 8536 have the biggest hundreds digit.
Look at the tens digits in these two numbers.
8563 has more tens than 8536.   8563 is bigger.
In order the numbers are 8563, 8536, 8356.

< means '**is less than**'.   > means '**is greater than**'.
≤ means '**is less than or equal to**'.   ≥ means '**is greater than or equal to**'.

Examples   4 < 7   3 > 0   □ ≤ 2   □ ≥ 3

What numbers could go in the boxes?

Practice Questions 14, 44, 57, 71

Rounding to the nearest 10 or 100

27 cm is closer to 30 cm than to 20 cm.
27 cm **rounded** to the nearest 10 cm is 30 cm.

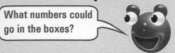

327 mm is closer to 300 mm than to 400 mm.
327 mm **rounded** to the nearest 100 mm is 300 mm.

Numbers exactly halfway between are rounded up.
465 to the nearest 10 is 470.
250 to the nearest 100 is 300.

Practice Questions 17, 27, 47

Negative numbers

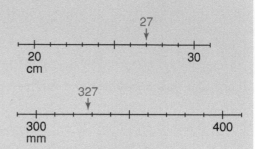
Its cold here.

A temperature of ⁻5 °C means 5 °C below zero.
Numbers less than zero are called **negative numbers**.
The number ⁻4 is read as 'negative 4' and means '4 less than zero'.

We can show positive and negative numbers on a number line.

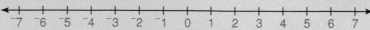

Example   The temperature is ⁻7 °C.
If the temperature rises 5 °C it will then be ⁻2 °C.

We can put **positive and negative numbers in order**
by comparing them.

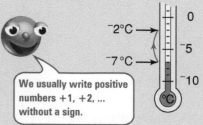
We usually write positive numbers +1, +2, ... without a sign.

Example   These temperatures are in order from coldest to
warmest.
⁻10 °C   ⁻2 °C   4 °C

Practice Questions 16, 23, 24, 28, 53

Multiples, factors, primes, squares

The **multiples** of a number are found by multiplying the number by 1, 2, 3, 4, 5, 6, ...

Example   The multiples of 6 are 6, 12, 18, 24, 30, 36, ...

The **factors** of a number are all of the numbers that will divide into that number leaving no remainder. We usually list them as pairs.

Example   The factor pairs of 24 are 1 and 24, 2 and 12, 3 and 8, 4 and 6.

A **prime number** has exactly two factors, itself and 1.
1 is not a prime number.

Example   17 can only be divided by 17 and 1 so it is a prime number.

> 1 is not a prime because it has only got one factor.

A whole number, when multiplied by itself, gives a **square number**.

Example $4 \times 4 = 16$.   16 is a square number.

The first 10 square numbers are 1, 4, 9, 16, 25, 36, 49, 64, 81, 100.

Practice Questions 7, 15, 25, 34

Mental calculation

You should be able to quickly recall the **addition and subtraction facts to 20**.

You should be able to quickly recall the **multiplication facts up to 10×10**.

> Knowing these will help you a lot.

Addition and subtraction are **inverse operations**. One undoes the other.
If $56 + 8 = 64$ then we know $8 + 56 = 64$ ⟵ commutative law
$64 - 8 = 56$ ⟵
$64 - 56 = 8$. ⟵ inverse operations.

Multiplication and division are also **inverse operations**.
If $16 \times 5 = 80$ then we know $5 \times 16 = 80$ ⟵ commutative law
$80 \div 5 = 16$ ⟵
$80 \div 16 = 5$. ⟵ inverse operations.

When two numbers add to 10 they are **complements in 10**.
When two numbers add to 100 they are **complements in 100**.

Examples   7 and 3 are complements in 10. $7 + 3 = 10$
53 and 47 are complements in 100. $53 + 47 = 100$
250 and 750 are complements in 1000. $250 + 750 = 1000$.

For mental calculation we often **partition** a number.

$68 = 60 + 8$ $352 = 300 + 50 + 2$
tens units   hundreds tens units

Practice Questions 3, 4, 5, 10, 11, 13, 18, 19, 22, 40, 43, 68, 69

Estimating

Always **estimate** the answer to a calculation first.

Example $23 + 89$
23 is about 20.   89 is about 90.
$20 + 90 = 110$.
We estimate the answer to $23 + 89$ is about 110. The actual answer to $23 + 89$ is 112.

Practice Questions 30, 37

3

Written calculation

Examples

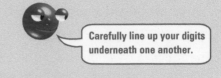

```
  386          4 11 1        1·8          8 1
+ 427          5̶2̶6         +7·9         9̶·7
─────         ─387         ─────        −3·8
  813          139          9·7         ─────
  11          ─────         ─────        5·9
                            1↖           ↗
```

Line up the decimal points

Carefully line up your digits underneath one another.

Examples

1. 527 × 2

527 × 2 is approximately equal to 500 × 2 = 1000.

This is called the grid method.

	500	20	7
2	1000	40	14

= 1054   **or**

```
  527
×   2
─────
 1054
   1
```

2. 56 × 24

56 × 24 is approximately equal to 60 × 20 = 1200.

	50	6
20	1000	120
4	200	24

1120
 224
─────
1344

or

```
   56
  ×24
 ─────
 1120   ← 56 × 20
  224   ← 56 × 4
 ─────
 1344
```

3. 8·9 × 7 is approximately equal to 9 × 7 = 63.

8·0 × 7 = 56·0
0·9 × 7 = 6·3
          ─────
          62·3
            1

or

	8	0.9
7	56	6.3

5.6 + 6.3 = 62.3

4. 342 ÷ 7

```
  342
−  70    10 × 7
─────
  272
− 140    20 × 7
─────
  132
−  70    10 × 7
─────
   62
−  56     8 × 7
─────
    6
```

Answer   48 R6 or 48$\frac{6}{7}$ ← remainder
                          ← number you are dividing by

Practice Questions 32, 35, 38, 41, 42, 48, 49, 50, 52, 55, 59, 61, 64, 66, 67, 70, 72, 73

Fractions

$\frac{2}{5}$ is read as 'two-fifths' and means 2 out of every 5.

$\frac{2}{5}$ ← numerator
    ← denominator

Example

3 out of 7 or $\frac{3}{7}$ of these horses are red.

Some fractions are **equivalent**. Equivalent means 'equal in value'.

Examples $\frac{1}{2} = \frac{2}{4} = \frac{3}{6} = \frac{4}{8} = \frac{5}{10} = \frac{50}{100}$ $\frac{1}{4} = \frac{2}{8} = \frac{3}{12} = \frac{4}{16} = \frac{5}{20} = \frac{25}{100}$

Sometimes a division is written as a fraction.

Example $\frac{53}{4}$ means $53 \div 4$

We can find a **fraction of a quantity**.

Examples $\frac{1}{10}$ of $40 = 40 \div 10$
                          $= 4$
              $\frac{1}{5}$ of $35 = 35 \div 5$
                          $= 7$

> We find $\frac{1}{10}$ by dividing by 10, and $\frac{1}{5}$ by dividing by 5.

Practice Questions 12, 31, 36, 54, 65

Decimals

In 0·25, the digit 2 means 2 tenths
        the digit 5 means 5 hundredths.

We can write **fractions as decimals**.
You should know these.

$\frac{1}{2} = 0.5$ $\frac{1}{4} = 0.25$ $\frac{3}{4} = 0.75$ $\frac{1}{5} = 0.2$

$\frac{1}{10} = 0.1$ $\frac{1}{100} = 0.01$

$\frac{2}{10} = 0.2$ $\frac{2}{100} = 0.02$

$\frac{3}{10} = 0.3$ and so on $\frac{53}{100} = 0.53$ and so on

Practice Question 58

Percentages

7% is read as 'seven percent' and means 7 out of 100.
Rachel got 69 out of 100 in her maths test. This is 69%.

You should know these.
100% = 1   50% = $\frac{1}{2}$ = 0·5   25% = $\frac{1}{4}$ = 0·25   75% = $\frac{3}{4}$ = 0·75   10% = $\frac{1}{10}$ = 0·1

Practice Questions 46, 56, 60, 62, 63

Ratio and proportion

There is one red square for every two blue squares.

Jack gets 3 apples for every 2 apples that Mary gets.
If Jack gets 12 apples (3 × 4) then Mary gets 8 apples (2 × 4).

Practice Questions 20, 39, 45, 51

Using the calculator

We use a calculator to find the answers to
difficult calculations.
**Always try to use a mental or written
method first.**

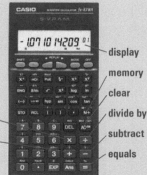

display
memory
clear
divide by
subtract
equals
multiply by
add

> Always ask yourself if you could do it using a mental or written method instead.

Practice Questions 74, 75, 76, 77, 78, 79

Practice Questions

1 What is the value of the 8 in these?
  a 34 859   **b** 584 392   **c** 123 468   **d** 4 348 213   **e** 38 050 712
  f 1 800 904   **g** 7 342 814   **h** 8 234 620   **i** 85 263 047

2 Which digit in each of the numbers in question **1** has a place value of tens of thousands?

3 Partition these numbers.
  a 72   **b** 86   **c** 420   **d** 571   **e** 967

4 Find the complements in 100 of these.
  a 37   **b** 81   **c** 56   **d** 49   **e** 62   **f** 78

5 Calculate the answers mentally.
  a Carly threw two darts. Write down all the ways she could win a prize.
  b Craig threw three darts. Write down all the ways he could win a prize.

WIN A PRIZE IF
• 2 darts add to 10
• difference between two darts is 5
• 3 darts add to 20

② ③ ④ ⑥
⑦ ⑧ ⑨ ⑩
⑪ ⑭ ⑮ ⑯

6 Write in figures
  a four hundred and eighty
  b sixty-two thousand, five hundred and four
  c five hundred and two thousand, four hundred and twenty
  d three million, four hundred thousand, eight hundred
  e six hundred and thirty thousand and twelve.

7 Write down the first 6 multiples of
  a 7   **b** 9

8 Write these in words.
  a 7521   **b** 12 045   **c** 250 018   **d** 25 482 134   **e** 5 062 011

9 Which is bigger, 5 thousands or 51 hundreds?

10 Do these multiplications mentally.
  a 9×7 **b** 8×5 **c** 9×6 **d** 8×6 **e** 7×8
  f 5×7 **g** 60×2 **h** 80×2 **i** 40×3 **j** 70×5
  k 40×9 **l** 60×8 **m** 300×5 **n** 600×4 **o** 900×8
  p 600×7 **q** 400×6 **r** 700×4 **s** 800×9

11 What numbers could go in the empty boxes?
  a ☐ + ☐ − 10 = 24
  b 3 × ☐ × ☐ = 30

Find at least 2 ways for each.

12 What fraction of each of these is coloured?
  a   **b**   **c**
  d   **e** 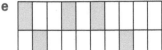   **f**

13 Find the answer to these mentally.

   a   42 ÷ 7   **b**   80 ÷ 10   **c**   45 ÷ 9   **d**   72 ÷ 8   **e**   48 ÷ 6

   f   36 ÷ 4   **g**   320 ÷ 8   **h**   360 ÷ 6   **i**   280 ÷ 7   **j**   180 ÷ 3

   k   240 ÷ 6   **l**   540 ÷ 9   **m**   630 ÷ 7   **n**   560 ÷ 8

14 Which of these statements are true?

   a   596 > 634   **b**   407 032 < 407 049   **c**   10 000 > 9999

15 Write down the factor pairs of

   a   8   **b**   20   **c**   45   **d**   60.

16

$$\overset{\text{°C}}{\underset{\substack{-90 \quad -80 \quad -70 \quad -60 \quad -50 \quad -40 \quad -30 \quad -20 \quad -10 \quad 0 \quad 10 \quad 20 \quad 30 \quad 40}}{\vert\ \ \vert\ \ \vert\ \ \vert\ \ \vert\ \ \vert\ \ \vert\ \ \vert\ \ \vert\ \ \vert\ \ \vert\ \ \vert\ \ \vert\ \ \vert}}$$

Use a copy of this number line and mark these temperatures.

   a   ⁻89°C. The coldest temperature ever measured (at a place in Antarctica).

   b   ⁻27°C. The lowest temperature ever measured in Britain (on 11 February 1985 at Braemar).

   c   32°C. 13 days in a row were hotter than this in the heatwave in 1976.

17 This list gives the number of people at an art gallery. Round them

   a   to the nearest 10   **b**   to the nearest 100.

     Monday   89   Tuesday   124   Wednesday   189   Thursday   173

     Friday   364   Saturday   757   Sunday   650

18 **a**   23 + 69 = 92

     Write down 3 more facts that are true using the numbers 23, 69 and 92.

   b   17 × 5 = 85

     Write down 3 more facts that are true using the numbers 17, 5 and 85.

   c   If 17 × 26 = 442, find the answers to   **i**   442 ÷ 26   **ii**   442 ÷ 17

19   1 × 25 = 25

    2 × 25 = 50

    4 × 25 = 100

    Continue this pattern to find the answer to 16 × 25.

20 Lamb should be cooked 80 minutes for each kg.

    How long does it take to cook a 2 kg piece of lamb?

21 Calculate

   a   52 × 10   **b**   750 ÷ 10   **c**   2 × 100   **d**   46 000 ÷ 100   **e**   820 × 100

   f   73 × 1000   **g**   56 000 ÷ 1000   **h**   50 × 1000   **i**   72 000 ÷ 10.

22 Find the answers to these mentally.

   a   Find three numbers next to each other in a row on the triangle, which add to

      i 18   **ii** 24   **iii** 33   **iv** 42   **v** 69   **vi** 72.

   b   Find a pair of numbers one above the other that have a

     difference of   **i** 4   **ii** 6   **iii** 10. For instance .

   c   Which numbers from 5 to 20 can you *not* make by adding only digits in the blue row? For instance 15 = 4 + 3 + 2 + 3 + 3.

   d   Copy this diagram. Put the digits from the red, blue and green rows in the circles so that each line of numbers adds to 15.

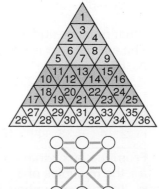

23 A scuba diver is 20 m below sea level.
Her position is given as ⁻20 m.
Give her new position as a negative number if she
   a goes down a further 10 m
   b comes up 10 m.

24 Andy has £25 in his bank account.
Write his new bank balance as a positive or negative number if
   a he withdraws £30   **b** he deposits £30.

25 Write down the first seven prime numbers.

26 What is
   a 100 more than 5824   **b** 10 less than 7269   **c** 1000 less than 60 427
   d 1000 more than 29 842   **e** 100 less than 6080   **f** 10 less than 14 007?

27 It takes Raylene an hour, give or take 10 minutes, to travel to school.
   a What is the shortest time it takes Raylene to travel to school?
   b What is the longest time?

28 The temperature was ⁻15 °C. How much has it risen if it is now
   a 0 °C   **b** ⁻8 °C   **c** 10 °C   **d** ⁻5°C   **e** ⁻3°C?

29 Find the number halfway between
   a 38 and 40   **b** 45 and 55   **c** 260 and 280   **d** 3500 and 4500
   e 3560 and 3570   **f** 27 400 and 27 800   **g** 45 670 and 45 680.

30 There are 884 men and 703 women at a conference.
Rory estimated there were about 1400 people altogether.
Explain why Rory is wrong.

31 Find these.
   a $\frac{1}{4}$ of £16   **b** $\frac{1}{2}$ of £10   **c** $\frac{1}{10}$ of 80 m   **d** $\frac{1}{5}$ of 40 ℓ   **e** $\frac{1}{3}$ of 27 g

32

a	**b**	**c**	**d**	**e**
79 + 5	63 − 8	47 +36	82 −49	364 +123

f	**g**	**h**	**i**	**j**
567 +863	71 −36	296 −159	564 −328	581 +639

   k 56 + 83 + 27   **l** 92 + 41 + 83   **m** 79 + 89 − 64   **n** 84 − 52 + 78
   o 96 − 35 − 29   **p** 73 + 96 − 29   **q** 83 − 27 − 35   **r** 36 + 46 − 39

33 What must be added or subtracted to the first number to get the second number?
The answer to **a** is **add 40**.
   a 528, 568   **b** 8325, 8125   **c** 7432, 7932   **d** 59 624, 51 624
   e 609 231, 409 231   **f** 724 164, 720 164   **g** 80 326, 78 326   **h** 87 541, 92 541
   i 27·68, 37·68   **j** 53·71, 52·71   **k** 59·86, 89.86

34

| 2 | 4 | 6 | 8 | 11 | 16 | 18 | 19 | 21 | 24 | 25 | 30 | 49 | 60 | 64 | 290 |

From the box, write down
   a the multiples of 3   **b** the numbers that can be divided by 10
   c the prime numbers less than 20   **d** the factors of 16
   e the square numbers.

35 Calculate

a	21	b	32	c	37	d	46	e	52
	× 3		× 4		× 5		× 6		× 8

f 37 × 4   g 36 × 5   h 321 × 9   i 864 × 7   j 392 × 8

36 Copy this number line. Mark these fractions on the line.

a $\frac{1}{5}, \frac{3}{5}$

0   1

b $\frac{1}{6}, \frac{5}{6}, \frac{1}{3}, 1\frac{2}{3}$

0   1   2

37 Estimate the answers to these.

a 86 + 29   b 23 + 94   c 85 + 89   d 96 − 51   e 98 − 37
f 52 × 6   g 89 × 4   h 37 × 7   i 66 ÷ 8   j 71 ÷ 9

38 Find the answers to these.

a	4·2	b	7·2	c	9·5	d	8·8	e	9·6
	+3·1		+8·4		+2·6		−3·4		−5·4

f	7·3	g	9·3	h	4·0	i	8·7
	+8·9		−2·8		−2·6		+9·9

39 For every 1 kg of strawberries, Mrs Thorn makes four jars of jam.
How many jars does 3 kg of strawberries make?

40 Find the missing numbers mentally.

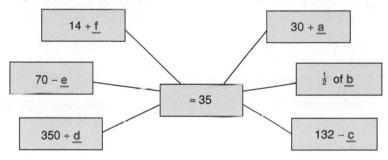

14 + **f**   30 + **a**   70 − **e** $\frac{1}{2}$ of **b**   = 35   350 ÷ **d**   132 − **c**

41 Calculate

a 90 ÷ 9   b 5)‾125‾   c 8)‾128‾   d 4)‾144‾
e $\frac{720}{9}$ f $\frac{189}{7}$ g 162 ÷ 6   h 234 ÷ 9

42 Find the remainder in these divisions.

a 64 ÷ 3   b 58 ÷ 4   c 76 ÷ 8   d 83 ÷ 4   e 64 ÷ 7
f $\frac{47}{2}$ g $\frac{85}{9}$ h $\frac{53}{5}$

43 Find the answers to these mentally.

a Gerri got 36 and 9 with two darts.
   She threw three darts altogether. Her total was 57.
   What did she score with her third dart?

b Gerri threw 4 darts and hit the same number each time.
   Her total was 64.
   What number did she hit each time?

44 Write the numbers in these lists in order from the smallest to the largest.
    a 3041, 3401, 3104, 3410   b 1431, 1354, 1541, 1435, 1543
    c 82 345, 82 352, 82 435, 82 325   d 614 248, 624 149, 614 420, 621 447

45 What goes in the gap?
    a There is one blue square for every ___ white squares.

    b There are ___ blue squares for every one white square.

46 What goes in the gaps?
    a About ___% of the garden is trees.
    b About ___% of the garden is flowers.
    c About ___% of the garden is lawn.

47 Jan's bank balance to the nearest £10 is £870.
Which of these amounts might Jan have in her bank account?
    £871 £875 £865 £868 £862

48 a Find the sum of 336 and 527.
   b Find the difference between 873 and 495.
   c Increase 652 by 389.
   d What is 474 less than 691?

49 Eight people paid a total of £128 for a mini bus tour.
    a How much change did they get from £150?
    b What was the cost for each person?

50 A parking building charges 80p for the first hour and 50p for each half hour or part half hour after that. Rob parked there from 11:15 a.m. until 2:55 p.m. How much did he pay?

51 In Lisa's class there are 2 boys for every 3 girls.
    a If there are 4 boys, how many girls are there in Lisa's class?
    b If there are 10 boys, how many girls are there?

52 Alison had three pieces of ribbon. One was 1·3 m and the other two were 4·7 m each. What total length of ribbon did she have?

53 Write these temperatures in order, coldest first.
    a 3 °C ⁻10 °C 0 °C   b 5 °C ⁻2 °C ⁻4 °C

54 Which fractions from the box are equivalent to
    a $\frac{1}{2}$ b $\frac{1}{4}$ c $\frac{1}{3}$?

$\frac{4}{8}$ $\frac{3}{12}$ $\frac{5}{20}$ $\frac{10}{20}$ $\frac{4}{12}$ $\frac{5}{15}$ $\frac{50}{100}$ $\frac{25}{100}$

55 Calculate.
    a $3·2 \times 4$ b $4·2 \times 3$ c $5·1 \times 6$ d $2·9 \times 3$ e $4·8 \times 4$
    f $7·6 \times 3$ g $8·1 \times 9$ h $3·7 \times 5$ i $6·4 \times 7$

56 Out of every 100 students, 19 have a Saturday job.
What percentage of students is this?

57 a If $4160 < \square < 4180$, write down three whole numbers that could go in the box.
   b If $18\ 560 \leqslant \square \leqslant 18\ 562$, write down three whole numbers that could go in the box.

58 Write these as decimals.
   a $\frac{1}{2}$ b $\frac{1}{4}$ c $\frac{3}{4}$ d $\frac{1}{10}$ e $\frac{1}{100}$ f $2\frac{1}{2}$
   g $4\frac{3}{4}$ h $\frac{3}{10}$ i $\frac{42}{100}$ j $\frac{39}{100}$ k $\frac{6}{100}$ l $\frac{4}{100}$

59 Copy these.
   Find the answers.

   a 52   b 36   c 82   d 36   e 27
      ×12   ×19   ×14   ×21   ×24

   f 52   g 39   h 72   i 83   j 93
      ×36   ×58   ×27   ×54   ×72

   k 8)361   l 4)593   m 5)437   n 7)632   o 8)507
   p 3)862   q 6)421   r 9)815   s 4)702   t 7)383

60 Which of these is the same as 100%?
   A 100   B $\frac{1}{100}$ C 1   D $\frac{1}{10}$

61 Cass bought 36 of these small cakes.
   How much did she pay?

64p

62 Write these as decimals.
   a 50%   b 25%   c 75%   d 10%

63 Write these as fractions.
   a 50%   b 25%   c 75%   d 10%

64 There are 7 boxes of books.
   3 of the boxes have 48 books each.
   The other boxes have 36 books each.
   How many books are there altogether?

65 Wasim had 30 bars of chocolate.
   He gave $\frac{1}{5}$ of these to his sister.
   He gave $\frac{1}{4}$ of the rest to his brother.
   How many did he have left?

66 What operation does ∗ stand for?
   a 14 ∗ 24 = 336   b 253 ∗ 11 = 23   c 8258 ∗ 379 = 8637

67 Pete has three 35p stamps and five 25p stamps.
   Find all the different amounts you could stick on a parcel.

68 Melanie has these coins.

Write down all the different ways she could pay for these.
   a   b   c

50p
per slice

45p
each

38p
each

69 Copy and complete this to make the answer 45.
You may use any of these operations: + − × ÷
24___2___33 = 45

70 Find two consecutive numbers with a product of 156.

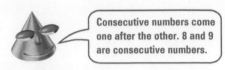

Consecutive numbers come one after the other. 8 and 9 are consecutive numbers.

71 5   7   4   3   6   9
   a What is the smallest even number that can be made using three of these digits?
   b What is the largest odd number that can be made?

72 Use all the digits 2, 3, 4, 7, 8 and 9 to make two 3-digit numbers so that
   a the difference between these numbers is as small as possible
   b the sum of these numbers is 1032.
Is there more than one answer to each of these?

73 Choose any four numbers from the grid.
Add them up.
How many sets of four numbers add to 1000?

59	550	189	9
99	184	723	29
56	236	385	270
211	403	305	177

This question takes a long time.

Use your calculator to answer questions 74–79.

74 Calculate   **a** 25 × 36   **b** 189 × 17   **c** 1045.1 ÷ 7   **d** 83.7 × 5.4
                e 219.18 ÷ 39   **f** 56.72 × 8.3   **g** £2.65 × 14   **h** £440 ÷ 16

75 **a** How many diaries can be bought with £20?
     How much change will there be?
   b How much change is there from £20 if two
     calculators and three diaries are bought?

Calculators	£4.95
Diaries	£1.89

76 **a** Fourteen people ate a meal at this cafe.
     How much did it cost altogether?
   b A group all had the special. It cost £92.65
     altogether. How many were in the group?
   c One morning 27 people ate breakfast at the
     cafe. How much did this cost altogether?

CAFE

Meals £9·70

Breakfast £6·70

Special £5·45

77 Each ● is a different missing digit.
   ●● × 5 ● = 4056
What is the calculation?

78 **a** Find different ways of completing ■ ■ × ■ = 252.
   b Each ● represents one of the digits 0, 2, 3, 4, 5, 6.
     Each digit is used only once.
     Replace each ● to make this true.
     ●● × ● = ● ● ●

79 Two whole numbers, when divided, give the answer 2.72727272 ...
What are the two numbers?

1 Place Value, Ordering and Rounding

You need to know

✓ place value — whole numbers .... page 1

✓ reading and writing whole numbers .... page 1

✓ multiplying and dividing whole numbers by 10, 100 and 1000 ... page 1

✓ putting whole numbers in order .... page 2

✓ rounding to the nearest 10 or 100 .... page 2

⋯ Key vocabulary ⋯

between, compare, decimal number, decimal place, digit, greater than (>), greatest value, hundredth, least value, less than (<), most significant digit, nearest, one decimal place (1 d.p.), order, place holder, place value, round, tenth, thousandth, value

▶▶ In Days Gone By

Many ancient civilisations had number systems that used symbols.

The Egyptian number system had symbols for 1, 10, 100, 1000, 10 000, 100 000, 1 000 000.

The symbol for 1 was a stroke |

10 was a heel bone ∩

100 was a coiled rope ⌒

1000 was a lotus flower ⚱

10 000 was a bent finger ⌐

100 000 was a tadpole ⌇

1 000 000 was an excited man ⚟

The order in which the symbols were written was not important.

For instance, 23 could be written as ∩ ∩ ||| or | ∩ ∩ || or || ∩ ∩ | etc.

9 could be written as |||||/||| or |||/||||| or ||||||||| etc.

Write these numbers using the Egyptian system.

    8   16   33   58   386   4900   56 423

Write other numbers using the Egyptian system.

Decimal place value

The **place** of a digit tells you its **value**.

Tens of thousands	Thousands	Hundreds	Tens	Units	•	Tenths	Hundredths	Thousandths
	7	2	0	0	•	5	0	3

This is a *place value* chart.

In 7200·503 the 7 means 7 thousands
     the 2 means 2 hundreds
     the 5 means 5 tenths
     the 3 means 3 thousandths.

7200·503 is a *decimal number* with three *decimal places*.

The zeros are **place holders**.

We use zeros as place holders rather than spaces.

Worked Example
What is the value of the 7 in   a 2786·03   b 5·637?
Answer
a   The 7 is in the hundreds place so the value of the 7 is **seven hundreds** or **700**.
b   The 7 is in the thousandths place so the value of the 7 is **seven thousandths** or **0·007**.

Exercise 1

1   What is the value of the digit 4 in these?
   a   0·459   **A** 4 tenths   **B** 4 hundredths   **C** 4 thousandths
   b   8·542   **A** 4 tenths   **B** 4 hundredths   **C** 4 thousandths
   c   0·04   **A** 4 tenths   **B** 4 hundredths   **C** 4 thousandths
   d   60·314   **A** 4 tenths   **B** 4 hundredths   **C** 4 thousandths
   e   32·143   **A** 4 tenths   **B** 4 hundredths   **C** 4 thousandths
   f   856·004   **A** 4 tenths   **B** 4 hundredths   **C** 4 thousandths

2   What is the hundredths digit in these?
   a   4·063   b   568·230   c   42·807   d   30·049   e   4·500

3   What is the thousandths digit in the numbers in **question 2**?

4   What is the value of the digit 4 in these?
   a   7·462   b   410·057   c   9·734   d   4672·30   e   7105·04

5   What is the value of the digit 7 in the numbers in **question 4**?

6   a   Make the largest number you can with Mark's cards.
    There must be at least one digit after the decimal
    point and the last digit must not be zero.
   b   Make the smallest number you can with Mark's
    cards.
    There must be at least one digit before the
    decimal point and the first digit must not be zero.
   c   How many different decimal numbers, each with
    two decimal places, can you make using all of
    Mark's cards? The last digit must not be zero.
    Write them down.

Make the largest number you can with your cards.

MARK

7 Nathan cut three pieces of rope. Each was between 1·1 m and 1·2 m. Write down how long each piece might be.

*8 How many times larger is the first 8 than the second 8 in these?
The answer to **a** is **100 times**.
a 818·3	**b** 188·05	**c** 18·83	**d** 0·88	**e** 0·838
f 885·6	**g** 846·82	**h** 983·08	**i** 58·804	

*9 i Repeat question **6a** and **b** with the cards
ii With these cards, make the third largest number with one decimal place. The last digit must not be zero.

`3` `8` `2` `2` `1` `6` `0` `0` `9` `·`

Review 1 What is the tenths digit in these?
a 0·368 **b** 0·502 **c** 7·92 **d** 9·091 **e** 56·00

Review 2 What is the hundredths digit in the numbers in review **1**?

Review 3 What is the value of the digit 8 in these?
a 6·84	**b** 0·08	**c** 3·048	**d** 8·02	**e** 5800·9
f 4·832	**g** 83 241·7	**h** 194·83	**i** 18 057·6	

Review 4 `4` `9` `0` `3` `·`

a Make the largest number you can with these cards.
There must be at least one digit after the decimal point and the last digit must not be zero.
b Make the smallest number you can with these cards.
There must be at least one digit before the decimal point and the first digit must not be zero.

Review 5 How many times larger is the first 4 than the second 4 in these?
*a 454·63 **b** 164·42 **c** 0·44 **d** 453·42 **e** 504·354

Review 6
*i Repeat **Review 4** with the cards `8` `3` `6` `4` `3` `1` `1` `7` `0` `·`

ii With these cards, make the second largest number with one decimal place. The last digit must not be zero.

 Puzzle

1 I am a decimal number. Just two of my digits are the same.
I have two digits before my decimal point.
I have two digits after my decimal point.
My tens digit is smaller than my units digit.
My units digit is the same as my tenths digit.
My tens digit is one less than my hundredths digit.
My digits add up to 11.
What number am I?

2 A number has five digits, two of which are the same.
The number is between 20 and 30.
The difference between the tenths digit and the units digit is the same as the hundredths digit.
The tens digit is the same as the thousandths digit.
The tenths digit is three times greater than the tens digit.
The sum of all the digits is 16.
There are two numbers it might be. What are they?

Reading and writing decimals

Discussion

Kieran: 3·72 km is three point seven two kilometres.

Amber: 3·72 km is three kilometres, seven hundred and twenty metres.

Kieran   Amber

Are Kieran and Amber both correct? **Discuss**.

How do we usually say these? **Discuss**.

£3·84   £6·05   £0·60   8·04 km   7·2 m   8·5 hours

At the last Winter Olympics Georg Hackl of Germany won the Luge event in 3 minutes 18·436 seconds.
18·436 is said as 'eighteen point four three six' not 'eighteen point four hundred and thirty-six.'

Are any other Olympic events timed to a thousandth of a second?

Worked Example
Write 17·384 in words.
Answer
Seventeen point three eight four.

Worked Example
Write one hundred and two thousand, six hundred and eight point seven zero two in figures.
Answer
102 608·702

This small space shows where the thousands end.

Worked Example
Write as a decimal number, sixteen thousand and two and thirty-five hundredths.
Answer
16 002·35

Worked Example
Write these in words and then as a decimal number.
a $6\frac{3}{10}$ b $14\frac{37}{100}$ c $3\frac{189}{1000}$
Answer
a **six and three tenths, 6·3**
b **fourteen and thirty-seven hundredths, 14·37**
c **three and one hundred and eighty-nine thousandths, 3·189**

Exercise 2

1 Some Olympic times are given in red. Write them in words.
  a Yordanka Donkova, Hurdles, **12·38** sec
  b Marie-Jose Perec, 400 m, **48·25** sec
  c Shiva Keshavan, Luge, **52·315** sec

2 Write these in figures.
  a eight point six
  c three hundred and seventy point zero two
  e four hundred and twenty thousand and sixty-four point zero zero seven
  g three million, six hundred and fourteen thousand point five two
  b seventeen point four one
  d five thousand and two point nine
  f two hundred and eleven thousand point six eight

3 Write these as decimal numbers.
  a three hundred and five and four tenths
  c seventeen and three tenths
  e sixty-five and seven hundredths
  g eighty-three hundredths
  i three hundred and two and twelve hundredths
  k four hundred and three and and twenty-five thousandths
  b twenty-five and six tenths
  d two thousand, five hundred and six and five hundredths
  f seven thousand and forty-one and eight thousandths
  h seventy-eight thousandths
  j nine thousand and fifty-two hundredths
  l twenty and fifty thousandths

4 Write these in words and then as a decimal number.
  a $4\frac{7}{10}$
  b $216\frac{1}{10}$
  c $12\frac{59}{100}$
  d $3\frac{8}{100}$
  e $114\frac{77}{100}$
  f $6\frac{451}{1000}$
  g $42\frac{367}{1000}$
  h $805\frac{29}{1000}$
  i $617\frac{2}{1000}$

Review 1 Write these in words. a 7·726   b 5 842·04

Review 2 Write as a decimal number.
a two hundred and six point eight three two
b five hundred and eighteen thousand and sixteen point zero two eight

Review 3 Write these as decimals.
a five hundred and six and seven hundredths
b three hundred and two and four thousandths
c forty-six and forty-two thousandths
d twenty and eleven hundredths
e six thousand and ten and thirty-five thousandths

Review 4 Write these in words and then as a decimal number.
a $9\frac{3}{10}$ b $8\frac{63}{100}$ c $19\frac{4}{100}$ d $24\frac{362}{1000}$ e $28\frac{19}{1000}$

Discussion

What is the purpose of the zeros in each of these? **Discuss.**

0·86,   8·06,   8·60,   0·086,   8·006

Adding and subtracting 0·1 and 0·01

We can use our understanding of place value to **add or subtract 0·1 and 0·01**.
0·1 is one tenth and 0·01 is one hundredth.

Worked Example
a What length is 0·1 m less than 4·36 m?
b What number is 0·01 more than 4·09?
Answer
a 4·36 m – 0·1 m = **4·26** m   subtract one tenth from the tenths

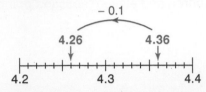

b 4·09 + 0·01 = **4·10**   add one hundredth to the hundredths

Exercise 3

1 Write down the first six numbers if you
   a count on in 0·1s from 3·5   b count back from 12·6 in 0·1s
   c count on in 0·01s from 4·08   d count back from 2·13 in 0·01s
   e count back from 12 in 0·01s.

2 What length is 0·1 m less than   a 27·68 m   b 4 m   c 1·03 m?

3 What length is 0·01 m greater than   a 5·72 m   b 8·69 m   c 3·09 m?

4 a What number is 0·01 less than 26·59?
   b Find the sum of 18·642 and 0·01.
   c Find the difference between 13·056 km and 0·1 km.
   d The difference between two lengths is 0·1. One length is 8·64 m.
      What might the other be?

5 What must be added or subtracted to the first number to get the second?
   The answer to **a** is **add 0·1**.
   a 8·6, 8·7   b 9·73, 9·63   c 0·892, 0·902   d 62·037, 61·937

*6 Graham wanted a length of wood 1·83 m long. He measured a piece he had in the garage.
   It was 10 cm too short.
   How long was the piece in the garage?

Review 1
a What mass is 0·1 kg less than 28·352 kg?
b Find the sum of 8·647 km and 0·01 km.
c Find the difference between 18·072 ℓ and 0·1 ℓ.

Review 2   What must be added or subtracted to the first number to get the second number?
a 18·65, 18·55   b 204·831, 204·841   c 3·99, 4   d 5·903, 6·003

Multiplying and dividing by 10, 100 and 1000

T

Investigation

10, 100 and 1000

You will need a calculator or a spreadsheet package or a place value board and a copy of the table.

1 Fill in the table.

Number	×10	×100	×1000	÷10	÷100	÷1000
856						
43						
8·7						
2·95						
17·04						
0·042						

You could use a calculator, spreadsheet package or place value board to help you.

2 Explain what happens to the digits when you multiply a number by 10, 100 or 1000. Does the number get bigger or smaller?

3 Explain what happens to the digits when you divide a number by 10, 100 or 1000. Does the number get bigger or smaller?

T

If you use a spreadsheet to fill in the table, change the numbers in column A and see if you get the same answers to questions 2 and 3.

	A	B	C	D	E	F	G
1	Number	×10	×100	×1000	÷10	÷100	÷1000
2	856	=A2*10	=A2*100	=A2*1000	=A2/10	=A2/100	=A2/1000
3	43						
4	8.7						

Use the fill down feature on your spreadsheet.

Examples

$4·63 \times 10 = 46·3$

$4·63 \times 1000 = 4630$

To multiply by 10, move each digit one place to the left.
To multiply by 100, move each digit two places to the left.
To multiply by 1000, move each digit three places to the left.

We use a zero rather than a space to show there are no units. The zero is called a place holder.

Number

Examples

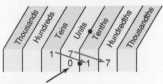

$17 \div 10 = 1.7$ $17 \div 100 = 0.17$

This zero stops the decimal point from getting lost.

To divide by 10, move each digit one place to the right.
To divide by 100, move each digit two places to the right.
To divide by 1000, move each digit three places to the right.

Worked Example

a 13×10 b 203.6×100 c 2.7×1000
d $86 \div 10$ e $462 \div 100$ f $2.7 \div 1000$

Answer

a **130**   move each digit one place to the left
b **20 360**   move each digit two places to the left
c **2700**   move each digit three places to the left
d **8·6**   move each digit one place to the right
e **4·62**   move each digit two places to the right
f **0·0027**   move each digit three places to the right

Exercise 4

1 Does × or ÷ go in the box?
   a 6 ☐ 10 = 60   b 42 ☐ 10 = 420   c 5200 ☐ 100 = 52
   d 5 ☐ 10 = 0·5   e 23·01 ☐ 100 = 2301

2 Which of 10, 100 or 1000 goes in the box?
   a 76 × ☐ = 760   b 4080 ÷ ☐ = 408   c 7 ÷ ☐ = 0·7
   d 31 ÷ ☐ = 3·1   e 8·26 × ☐ = 8260   f 4·6 × ☐ = 460
   g 894 ÷ ☐ = 8·94   h 6·05 × ☐ = 605   i 0·81 × ☐ = 810
   j 6 ÷ ☐ = 0·006   k 0·4 × ☐ = 40   l 7·2 ÷ ☐ = 0·072
   m 0·5 ÷ ☐ = 0·005   n 0·36 × ☐ = 360

3 Find the answers to these.
   a 59 × 10   b 59 × 100   c 59 ÷ 10   d 59 ÷ 1000
   e 59 × 1000   f 59 ÷ 100   g 1·23 × 100   h 10·8 ÷ 10
   i 351 ÷ 1000   j 724·3 × 100   k 0·02 ÷ 100   l 3·04 × 10
   m 48·24 × 100   n 2 ÷ 100   o 67·9 × 1000   p 841·05 ÷ 100
   q 67·03 ÷ 100   r 8 ÷ 1000   s 0·06 × 100

4 What number goes in the box?
   a ☐ ÷ 10 = 0·5   b ☐ × 10 = 600   c ☐ ÷ 100 = 0·07   d ☐ ÷ 1000 = 0·8

5 Nick bought plastic to cover his books. He needed 41·5 cm for each book.
   How much plastic did he need for 10 books?

6 a Dianne measured the thickness of 1000 sheets of paper as
      264 mm.
      How thick is each of these sheets?
  *b Dianne measured a stack of 500 sheets of card as 502 mm.
      How thick is each sheet?

Review

		A						**A**							
1·2	0·12	620	14	12	1·2			620	14	6200			6·2	0·12	6200

												A	
0·062	62	140	1·4		0·0012	0·62	1·2	0·12		6·2	0·12	620	6·2

	A							
0·14	620	62		120	140	0·62	62	12

Use a copy of this box.

Find the answers to these in the box. Write the letter above the answer. The first one is done.

A 62 × 10 = 620 **E** 62 × 100 **O** 0·62 ÷ 10 **T** 6200 ÷ 1000 **N** 0·062 × 1000
I 62 ÷ 100 **L** 1·4 × 100 **Y** 14 ÷ 10 **C** 140 ÷ 1000 **R** 0·014 × 1000
F 1·2 ÷ 1000 **H** 12 ÷ 100 **S** 0·012 × 100 **K** 0·012 × 1000 **B** 1·2 × 100

Putting decimals in order

To put decimals in order, **compare digits in the same position**, starting from the left.

Example

Remember: < means *is less than* and > means *is greater than*.

8·73 > 8·69

Starting at the left the units digit in both numbers is 8. The tenths digits are 7 and 6 and 7 > 6.

0·417 < 0·419

Starting at the left, the first digits that are not the same are the thousandths. 7 < 9.

 We start at the left because this is the *most significant digit*.

Worked Example
Put these numbers in order, from largest to smallest.
5·06   5·6   5·065   3·76   5·142   6·72

Answer
Compare the units digits first, then the tenths, then the hundredths and then the thousandths.
The numbers from largest to smallest are
6·72, 5·6, 5·142, 5·065, 5·06, 3·76

Exercise 5

1 Are these true or false?
  a 0·8 > 0·7   b 3·4 < 3·5   c 5·7 > 5·9   d 3·42 > 3·402
  e 0·23 > 0·32   f 8·6 ≠ 8·600   g 4·6 ≠ 4·06   h 0·56 < 0·48
  i 4·704 < 4·74   j 7·19 > 7·189   k 0·043 < 0·034

 ≠ means is not equal to.

2 Does < or > go in the box?
  a 16·75 ☐ 16·57   b 7·526 ☐ 7·52   c 8·07 ☐ 8·071   d 0·643 ☐ 0·634
  e 0·012 ☐ 0·02

3 Put these lists in order from smallest to largest.
  a 0·82, 0·28, 0·2, 0·8   b 0·7 m, 0·69 m, 0·72 m, 0·68 m
  c 13·4 ℓ, 14·05 ℓ, 14·5 ℓ, 13·38 ℓ   d 8·52 g, 8·052 g, 8·5 g, 8·025 g
  e 0·369 m, 0·693 m, 0·6 m, 0·03 m, 0·963 m

4 Which number is halfway between these?
The answer to **a** is **0·45**.

0·4   0·45   0·5

a 0·4 and 0·5	**b** 4 and 5	**c** 6 and 7	**d** 2 and 4
e 9 and 13	**f** 7·6 and 7·7	**g** 9·3 and 9·4	**h** 0·8 and 0·9
i 11·4 and 11·5	**j** 9·4 and 9·6	***k** 8·3 and 8·6	***l** 5·7 and 6·0

5 This table gives the weights of five of Rochelle's kittens.

Cheeky	Wolf	Tricks	Panda	Lucy
0·563 kg	589 g	0·613 kg	500 g	0·6 kg

a Which kitten weighed the most?
b Which kitten weighed the least?
c Put the weights of the kittens in order from lightest to heaviest.
d Rochelle has a sixth kitten, Missy. Missy's weight is exactly halfway between Panda's and Lucy's weights. How much does Missy weigh?

Remember:
When comparing measurements they must be in the same units.
1 kg = 1000 g
1 ℓ = 1000 mℓ
1 km = 1000 m

6 Which of <, > or = goes in the box?
a 850 mℓ ☐ 8 ℓ
b 5·5 ℓ ☐ 550 mℓ
c 3500 m ☐ 3·5 km
d 8350 g ☐ 8·305 kg

7 $32·4 \leqslant \square \leqslant 32·6$
What number could go in the box if the number has
a one decimal place   **b** two decimal places?

≤ means *is less than or equal to.*

8 Estimate the decimal number that goes in each box.

3·0   a   b   c   4·0   d   e   5·0

*9 7·5 lies halfway between two other numbers.
Write down four possible pairs of numbers that could go in the boxes.

☐   7·5   ☐

*10 a Fran has these cards. Write down all the numbers with two decimal places she could make with them. Do not have zero as the first or last digit.
Put the numbers in order from largest to smallest.

8 7 0 2 .

b Repeat **a** for these cards.

0 3 4 8 0 .

T ***11** Use a calculator to generate 10 random decimal numbers.
Write them down. Put them in order.

You *could* use a graphical calculator.

Review 1 Which of <, > or = goes in the box?
a 12·83 ☐ 12·38
b 4·06 ☐ 4·062
c 0·63 ☐ 0·629
d 4200 mℓ ☐ 4·2 ℓ
e 5505 m ☐ 5·5 km

Review 2 Put these in order from smallest to largest.
a 4·1, 4·12, 4·214, 3·99   **b** 0·864, 0·87, 0·86, 0·684, 0·846

Review 3 Which number is halfway between
a 14 and 15   **b** 8 and 13   **c** 0.3 and 0.4   **d** 8.6 and 8.7?

Review 4

	Vault	Bars	Beam	Floor
Angela	7·85	6·05	8·15	7·58
Maddison	8·30	8·10	8·05	8·13

a Who got the higher score on the beam?
b On which piece of equipment did Angela score lowest?
c Put Maddison's scores in order from highest to lowest.

*Review 5   2.4 is halfway between two numbers.
What might they be?

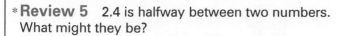

2·4

Decimal Yes/No – a game for a group

To play

● Choose a leader.
  The leader chooses a decimal number less than 30 with no more than 5 digits.
● The other students in the group have to work out what the number is.
  They do this by taking turns to ask the leader questions.
  The leader may *only* answer Yes or No to each question.
● The student who works out the number is the leader for the next round.

Example
The leader wrote down 13·053

Student	Question	Answer
Zara	Are there 2 digits after the point?	No
Max	Are there 3 digits after the point?	Yes
Shanna	Is the number less than 10?	No
Irina	Is the number less than 20?	Yes
Zara	Is the second digit less than 5?	Yes
Max	Is the second digit bigger than 2?	Yes
Shanna	Is the second digit 3?	Yes
Irina	Is there another 3?	Yes
Zara	Is the last digit 3?	Yes
Max	Do the digits add up to more than 13?	No
Shanna	Is one of the digits 0?	Yes
Irina	Is the tenths digit 0?	Yes
Zara	Is the hundredths digit less than 5?	No
Max	Is the number 13·053?	Yes

Max is leader for the next game.

Rounding to the nearest 10, 100 or 1000

Ben's father bought a new car for £5863.
He wanted to insure it.
He was asked to round the price he paid to the nearest £100.
£5863 is between £5000 and £5900.
£5863 to the nearest £100 is £5900.

£5863

5800   5900

£5,863 is closer to £5,900 than to £5,800.

Number

4372 is between 4000 and 5000.
4372 is closer to 4000 than 5000.
4372 rounded to the nearest 1000 is 4000.

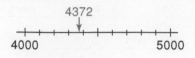

When a number is exactly halfway between two numbers we round up.
45 is halfway between 40 and 50.
45 rounded to the nearest 10 is 50.

A number rounded to the nearest 100 is 300.
The greatest value it could have been is 349.
The least value it could have been is 250.

Try and explain why.

Exercise 6

1 The table gives the ten longest rivers in the world in alphabetical order.
   a Round each length to the nearest 100 km.
   b Put the rivers in order from longest to shortest.
   c Which five rivers have the same length to the nearest 1000 km?

	Km		Km
Amur	4416	Ob-Irtysh	5410
Amazon	6570	Paraná	4500
Huang He	4840	Yangtze	6380
Mississippi	6020	Yenisei	5310
Nile	6670	Zaire(Congo)	4630

2 This table gives the number of fans at some football games.
   Round these numbers
   a to the nearest 10   b to the nearest 100   c to the nearest 1000.

Manchester United	61 864
Arsenal	33 575
Liverpool	39 504
Southampton	27 398

Find out how many fans were at ten FA cup matches and round the numbers.

3 This table shows the diameters of the planets in kilometres.

Mercury	Venus	Earth	Mars	Jupiter	Saturn	Uranus	Neptune	Pluto
4878	12 104	12 756	6794	142 800	120 000	52 000	48 400	2300

   a Round each diameter to the nearest thousand kilometres.
   b Put the planets in order from smallest to largest.

4 There were 8723 people at a cricket match.
   a The radio report said that there were about 9000 people at the match.
      Was this rounded to the nearest thousand or nearest hundred?
   b A newspaper report said that nearly 10 000 people were at the match.
      Was the newspaper correct?
      Explain.

5 The population of Hereford to the nearest thousand is 48 000.
   a What is the smallest number of people who might live there?
   b What is the greatest number of people?

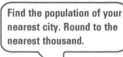
Find the population of your nearest city. Round to the nearest thousand.

6 How long is this string to the nearest
   a 10 cm   b cm   c 100 cm   d mm?

123   124   125   126   127 cm

Review 1 This table shows the five highest mountains in the world.

a Round each height to the nearest 10 m.
b Put the mountains in order, highest first.
c Which four mountains are the same height to the nearest thousand metres?

	metres
K2	8611
Everest	8863
Kanchenjunga	8586
Makalu	8463
Lhotse	8511

Review 2 £15 386 was raised at a charity evening.

a How much, to the nearest £100, was raised?
b How much, to the nearest £1000, was raised?
c Was the headline in the 'The Mail' correct? Explain.
d At another charity evening, £15 000 to the nearest thousand pounds was raised. What are the greatest and least amounts, to the nearest pound, that could have been raised?

THE MAIL
Nearly £20,000 raised at charity evening

Rounding to the nearest whole number or to one decimal place

To **round** to the **nearest whole number** we look at the digit in the first decimal place.

If this digit is 5 or more we round up. Otherwise the whole number is left unchanged.

Examples 3·7 kg to the nearest whole number is **4 kg**.
↑
this digit is bigger than 5 so we round 3·7 up to 4.

13·484 ℓ to the nearest whole number is 13 ℓ.
2.199 km to the nearest whole number is 2 km.

To **round** to **one decimal place** we look at the digit in the second decimal place.

If this digit is 5 or more we round up. Otherwise the number in the first decimal place stays the same.

Examples 4·682 to one decimal place is **4·7**.
↑
5 or more so round the first decimal place up

18·349 to one decimal place is **18·3**.
↑
less than 5 so the first decimal place stays the same

4·97 to one decimal place is **5·0**.

Decimal place is often written as **d.p.**

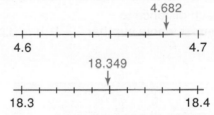
We put zero to show we have rounded to one decimal place and not to the nearest whole number.

Example 8·69 to 1 d.p. is 8·7.

Exercise 7   **Except for question 6**

1 Round these to the nearest whole number.

a 14·71   b 8·34   c 17·205   d 19.684   e 62·867
f 13·50   g 26·285   h 633·851   i 526·4 g   j 82.913 kg
k 904·56 m   l 350·67 m   m 1086·71 mm   n 1492·03 ℓ

2 Round the numbers in question 1 to one decimal place.

Look at the second decimal place.

3 This is the answer to a calculation.
Rob rounded this to the nearest whole number.
He said 'the 8 is 5 or more so 3·48 becomes 3·5.
Now 3·5 to the nearest whole number is 4'.
Explain why Rob is wrong.

$$3.48$$

4 What goes in the gaps?
   a 4·98 rounded to the nearest whole number is _____ and to 1 d.p. is _____ .
   b 0·96 rounded to the nearest whole number is _____ and to 1 d.p. is _____ .

5 This table shows the points given to the five dogs in the final of the 'most lovable pooch' competition.

Name	Misty	Dougal	Katie	Hound Dog	Muss
Points	93·42	86·87	91·95	88·60	92·00

Round each to
a the nearest whole number   **b** one decimal place.

*6 Round the answers to these to
   i the nearest whole number   **ii** one d.p.
   a $56 \div 17$ **b** $824 \div 26$ **c** $9174 \div 824$ **d** $96·8 \div 3·2$

Review 1 Round these to the nearest whole number.
a 15·83   **b** 7·04   **c** 19·55   **d** 36·147   **e** 43·996

Review 2 Round the numbers in **Review 1** to one decimal place.

Review 3 This table shows the distances, in metres, five people threw in a 'welly throwing' competition.

Name	B.Smith	R.Johnson	M.Carter	S.Patel	M.O'Reagan
Distance (m)	17·23	20·49	16·58	19·98	18·00

Round each distance to
a the nearest whole number   **b** one d.p.

Rounding up or down

In some practical problems we must decide whether it is more sensible to **round up** or **round down**.

Worked Example
How many 8 seater mini-buses are needed to take 107 people to the theatre?
Answer
$\frac{107}{8} = 13·375$
To round down is not sensible because after 13 mini-buses have been filled there are still some people left over.
This is shown by the decimal part of the answer.
14 mini-buses are needed.

$$13.375$$

Exercise 8

1 Would the answers to each of these be rounded up or rounded down?
   Find the answers.
   a 100 fans of a football team hired mini-buses.
      Each minibus could hold 16 people.
      How many mini-buses were needed?
   b There are 100 students at a tennis coaching school.
      They are put into teams of 6.
      How many full teams can be made?

2 'Day Tripper' coaches can carry only 52 people.
   How many coaches are needed for
   a 162 people   b 217 people   c 380 people?

3 Jason had a holiday job packing chocolates.
   A box holds 48 chocolate cremes.
   How many boxes can be filled with
   a 72 chocolate cremes   b 186 chocolate cremes?

4 A lift can hold 64 large boxes.
   How many trips are needed for
   a 187 boxes   b 271 boxes   c 385 boxes?

5 Trees at 'Tall Trees' are all £12.
   a Tony had £100 to buy trees for his new garden. How many trees could he buy?
   b How many trees could be bought with £236?
   c How many trees could be bought with £472?

Review 1 A mini-bus can carry 15 people. How many mini-buses are needed for
a 84 people   b 107 people   c 186 people?

Review 2 Iced buns cost 35p. How many can be bought for
a 90p   b £2   c £4·25?

Summary of Key Points

 A This chart shows **place value**.

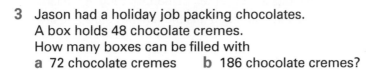

The zeros are place holders.

The place of a digit tells us its value.
This chart shows the number five thousand, six hundred and four point one zero three.

8·163 is said as 'eight point one six three'.
Four and seventeen hundredths is written as 4·17.

 We use place value to **add and subtract 0.1 and 0.01**
Examples   8·04 − 0·1 = 7·94   subtract 1 tenth from the tenths
6·89 + 0·01 = 6·90   add 1 hundredth to the hundredths

 To **multiply by 10, 100 or 1000** move each digit one place to the left for each zero in 10, 100 or 1000, ...
Examples   46 × 10 = 460   0·8 × 100 = 80   0·520 × 1000 = 520

To **divide by 10, 100 or 1000** move each digit one place to the right for each zero in 10, 100 or 1000, ...
Examples   530 ÷ 10 = 53   4·8 ÷ 100 = 0·048   834 ÷ 1000 = 0·834

 To **put decimals in order**, compare digits in the same position, starting at the left.
Example   4·63 > 4·628   Starting at the left, the first digits that are not the same are the hundredths. 3 > 2.

 Rounding to the nearest 10, 100 or 1000
Example   6562 is closer to 7000 than to 6000.
6562 to the nearest thousand is 7000.

When a number is **halfway between** two numbers we **round up**.
Example   185 to the nearest ten is 190.

 To **round to the nearest whole number** we look at the tenths digit.
If the tenths digit is 5 or more we round up. Otherwise the whole number stays the same.
Example   16·54 to the nearest whole number is 17.
↑
The tenths digit is 5 or more so 16 becomes 17.

To **round to one decimal place** we look at the digit in the second decimal place.
If this digit is 5 or more we round up. Otherwise the digit in the first decimal place stays the same.
Examples   4·58 to 1 d.p. is 4·6.
4·82 to 1 d.p. is 4·8.
5·97 to 1 d.p. is 6·0.

 In some practical problems we must decide whether it is more sensible to **round up** or **round down**.

Test yourself

1 What is the value of the 8 in these?
   a 43 825   **b** 508·72   **c** 396·84   **d** 13·08   **e** 2·918

2 Make the largest number you can with these cards. You must have at least one digit
   after the decimal point. The last digit must not be zero.

3 Write these in words.
   a 168·03   **b** 8·964   **c** 96·086   **d** $4\frac{273}{1000}$

4 Write these as decimal numbers.
   a eighty-nine thousand and sixty-two and seven hundredths
   b eleven and sixty-four hundredths
   c four hundred and two and thirty-nine thousandths

5 Find the answers to these.
   a 8·7 + 0·1   **b** 14·06 − 0·1   **c** 3·98 + 0·01   **d** 5·00 − 0·01

6 Calculate
   a 48 × 1000   **b** 96 ÷ 10   **c** 540 ÷ 100   **d** 83 ÷ 1000   **e** 5·7 × 10
   f 0·89 × 100   **g** 0·52 × 1000   **h** 6·37 ÷ 1000   **i** 52 ÷ 1000

7 Ten friends won £36·50 in a raffle. It was to be divided equally.
   How much did each get?

8 Are these statements true or false?
   a 8·6 < 8·06   **b** 7·257 > 7·527   **c** 4·23 = 4·230   **d** 8·36 > 8·036

9 Put these in order from smallest to largest.
   a 4·27, 4·72, 4·7, 4·074   **b** 0·869, 0·689, 0·06, 0·078, 0·798, 0·85

10 Find the number halfway between these.
   a 18 and 19   **b** 8 and 12   **c** 5 and 10   **d** 0·4 and 0·5
   e 0·8 and 0·9   **f** 1·9 and 2·0   **g** 6·8 and 6·9   **h** 8·4 and 8·5

11 Six friends were bored one Saturday.
   They decided to see who could throw a heavy can
   the furthest.
   a Who threw the can the greatest distance?
   b Who threw it the next greatest distance?
   c Who threw it the fourth greatest distance?
   d Who threw it the shortest distance?

Name	Distance
Jake	865 cm
Brian	8·284 m
Aled	8·732 m
Toby	870 cm
Rob	8·928 m
Samuel	8·099 m

12 Look at the table in **question 11**.
   a Round the distance Jake threw to the nearest ten centimetres.
   b Round the distance Toby threw to the nearest hundred centimetres.
   c Round the distances Brian, Aled, Rob and Samuel threw to the nearest whole number.
   d Round the distances Brian, Aled, Rob and Samuel threw to one decimal place.

13 Mike needed 61 m of wallpaper.
   How many rolls, 8 m long, would he need?

2 Special Numbers

You need to know

✓ negative numbers   page 2

✓ multiples, factors primes, squares   page 3

· Key vocabulary ·

common factor, consecutive, divisible, divisibility, factor, highest common factor (HCF), integer, least common multiple (LCM), minus, multiple, negative, plus, prime, prime factor, properties, sign, square number, squared, square root, triangular number

All in a Spin

T

				14	13			
		5	4	3	12			
		6	1	2	11			
		7	8	9	10			

- Barry began to make a spiral of numbers from 1 to 100. Use a copy of this. Finish the spiral of numbers.

- Colour all the multiples of 2. What do you notice?

- Use another copy of the spiral. Colour all the square numbers. What do you notice?

- Use more copies of the spiral. Find other number patterns.

30

Putting integers in order

Remember
⁻3 °C means 3° below zero.
A temperature of ⁻18 °C is colder than
⁻10 °C.   ⁻18 °C < ⁻10 °C

The numbers ..., ⁻4, ⁻3, ⁻2, ⁻1, 0, 1, 2, 3, ... are called the **integers**.

Worked Example
Put the integers ⁻1, ⁻4, 2, 0, ⁻3, in order from smallest to largest.

Answer
We can show these integers on a number line.

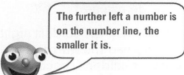

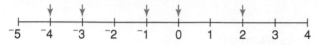

In order, these integers are **⁻4, ⁻3, ⁻1, 0, 2**.

The further left a number is on the number line, the smaller it is.

Exercise 1

1 Are these true or false?
a ⁻3 °C > 0 °C   b ⁻2 °C < 2 °C   c ⁻2 °C > ⁻5 °C   d ⁻2 °C < ⁻3 °C   e ⁻4 °C > ⁻5 °C

2 Which of > or < goes between these?
a 2 ⁻2   b ⁻1 1   c ⁻8 0   d 0 ⁻5
e ⁻9 ⁻4   f ⁻6 ⁻7   g ⁻14 ⁻7   h ⁻5 ⁻15

3

One of each pair of numbers is at A and the other at B.
Which is at A?
a ⁻7, ⁻3   b ⁻20, ⁻4   c ⁻3, ⁻10   d ⁻25, ⁻35
e ⁻8, ⁻40   f ⁻13, ⁻14   g ⁻24, ⁻23

4 Which of the three integers is closest to zero?
a ⁻4, ⁻3, ⁻5   b ⁻6, ⁻8, ⁻3   c ⁻2, ⁻1, ⁻8   d ⁻15, 16, ⁻17   e ⁻3, 4, ⁻11

5

	6 am	midday	6 pm	midnight
York	⁻7 °C	⁻3 °C	⁻3 °C	⁻4 °C
Manchester	⁻5 °C	0 °C	4 °C	⁻3 °C
Gloucester	2 °C	4 °C	0 °C	2 °C
Bristol	5 °C	6 °C	1 °C	1 °C
Glasgow	⁻4 °C	⁻1 °C	⁻2 °C	⁻1 °C

Put the temperatures at these times in order from coldest to warmest.
a 6 am   b midday   c 6 pm   d midnight

6 Put the integers ⁻7, ⁻3, ⁻6, ⁻2, ⁻1, 4, 3, ⁻4 on a number line.
Which number is halfway between these?

   a ⁻7, ⁻3   **b** ⁻6, ⁻2   **c** ⁻1, ⁻7   **d** ⁻2, 4
   e ⁻7, 3   **f** ⁻7, ⁻2   **g** ⁻2, 3   **h** ⁻4, ⁻1

*7 Use a graphical calculator to generate ten random numbers between – 30 and 30. Put them in order.

Review 1   Are these true or false?

a 7 > ⁻7   **b** 0 < ⁻3   **c** 1 < ⁻1   **d** ⁻2 > ⁻1   **e** ⁻5 < ⁻2
f ⁻5 > ⁻6   **g** ⁻52 > ⁻32   **h** ⁻29 < ⁻30   **i** ⁻50 < ⁻51

Review 2   These were the temperatures at midnight, local time, on February 12th.

Beijing	⁻5 °C	Kiev	⁻2 °C	New York	⁻6 °C
Calgary	⁻22 °C	Montreal	⁻24 °C	Warsaw	⁻3 °C
Jerusalem	⁻1 °C	Moscow	⁻8 °C		

a Which city had the coldest night?
b Which city had the warmest night?
c Which was colder, Moscow or New York?
d Write the temperatures in order, from coldest to warmest

Review 3   Put the integers ⁻5, ⁻9, ⁻3, 1, 2, ⁻6 on a number line.
Which number is halfway between these?

a ⁻5, ⁻9   **b** ⁻3, 1   **c** 2, ⁻6   ***d** ⁻6, ⁻3   ***e** ⁻6, ⁻9

You'll meet negative numbers again in the graphs chapter on page 220.

? Puzzle

Use a copy of this.

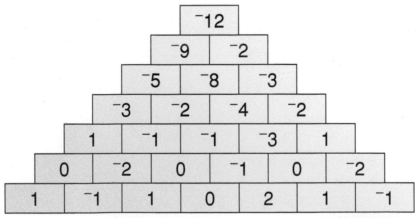

Find a path from the bottom of this pyramid to the top.
Always move into a square which has a smaller integer.
You may not move sideways.

Find as many paths as possible.

10 paths	Excellent
8 paths	Very good
6 paths	Good
4 paths	Keep trying

Using integers

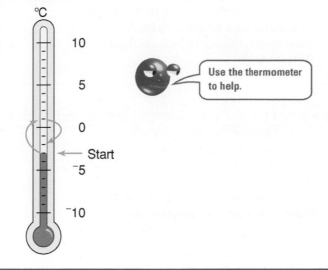

Worked Example

The temperature in York was ⁻3 °C.
It rose 4 °C, then fell 3 °C.
What was the temperature then?

Answer

The temperature started at ⁻3 °C.
It rose 4 °C to 1 °C.
It fell 3 °C to ⁻2 °C.
The temperature would be **⁻2 °C**.

Use the thermometer to help.

Exercise 2

1   Use the thermometer to help.
   a   The temperature was 4 °C. It rose 6°. What was the temperature then?
   b   The temperature was 4 °C. It fell 6°. What was the temperature then?
   c   The temperature was ⁻4 °C. It rose 6°. What was the temperature then?
   d   The temperature was ⁻4 °C. It fell 6°, then rose 2°. What was the temperature then?
   e   The temperature was ⁻6 °C. It rose 4°, then fell 1°. What was the temperature then?
   f   The temperature was ⁻6 °C. It fell 4°, then rose 5°. What was the temperature then?
   g   The temperature was 1 °C. It fell 5°, then rose 3°. What was the temperature then?
   h   The temperature was ⁻4 °C. It rose 3°, then fell 1°. What was the temperature then?
   i   The temperature was ⁻1 °C. It fell 4°, then rose 7°. What was the temperature then?

2   Use the thermometer to help.
   What is the difference between these night and day temperatures?
   a   night ⁻4 °C, day 2 °C   **b**   night ⁻4 °C, day 0 °C   **c**   night ⁻10 °C, day 2 °C
   d   night ⁻3 °C, day ⁻1 °C   **e**   night ⁻9 °C, day ⁻2 °C   **f**   night ⁻14 °C, day ⁻7 °C

3   A lift starts at floor ⁻2 and goes
       up 2 floors
       down 3 floors
       up 5 floors
       down ___ floors
   It ends up at floor ⁻2. What is the missing number?

Third floor	3
Second floor	2
First floor	1
Ground	0
Basement	⁻1
Garage	⁻2
Store	⁻3

4   In a board game Victoria has scored 5 points.
   What will Victoria's score be, if on the next 3 turns she
   a   gains 2 points, loses 4 points and gains 1 point
   b   loses 8 points, gains 2 points and gains 4 points
   c   gains 1 point, loses 9 points and loses 3 points
   d   loses 4 points, gains 3 points and loses 7 points?

5   The height above sea level of some places is shown.

    a   What is the difference in height between London and the Caspian Sea?

    b   What is the difference in height between The Dead Sea and Ayers Rock?

    c   How much higher is Nairobi than Rotterdam?

Place	Height above sea level (m)
Ayers Rock	900
Caspian Sea	⁻30
London	30
Nairobi	1800
Rotterdam	⁻5
The Dead Sea	⁻400

6   In a test, +1 is given for a correct answer and ⁻1 for a wrong answer.

    a   Emma's score came to ⁻4 + 6.
       How many wrong answers did she give?

    b   Sari's score came to ⁻5 + 7.
       What was her total score?

    c   Pete's total score for three questions was ⁻1.
       How many wrong answers did he give?

7   Priya's mother played nine holes of golf.
The table shows how far above or below par she was for each hole.

Hole Number	1	2	3	4	5	6	7	8	9
Above/Below Par	+1	⁻2	+3	⁻1	+4	⁻3	+2	⁻2	+1

Work out how far above or below par Priya's mother was at the end of the nine holes.
You could use this number line to help.

*8   The temperature is below zero.
It rises 3 degrees then falls 5 degrees.
What could the temperature be now?

Review 1   What will the new temperature be if the temperature
a   is ⁻6 °C then rises 3 °C
b   is ⁻6 °C then drops 5 °C, then rises 2 °C
c   is ⁻6 °C then drops 4 °C, then rises 7 °C?

Review 2   The midnight temperature was ⁻8 °C and the midday temperature was 3 °C.
What is the difference between these temperatures?

Review 3   In a maths quiz, a correct answer gets 2 points and a wrong answer loses 1 point.
There are 20 questions.
Part way through this quiz Sandy has a score of 2.
What will Sandy's new score be if he
a   gets the next 5 questions correct
b   gets the next 5 questions wrong
c   gets 2 of the next 5 questions correct and 3 wrong
d   gets 4 of the next 5 questions wrong and 1 correct?

Adding and subtracting integers

Discussion

- 4 + 1 = 5   ⁻3 + 5 = 2   2 – 2 = 0   ⁻3 – 2 = ⁻5
  4 + 0 = 4   ⁻3 + 4 = 1   2 – 1 = 1   ⁻3 – 1 = ⁻4
  4 + ⁻1 = 3   ⁻3 + 3 = 0   2 – 0 = 2   ⁻3 – 0 = ⁻3
  4 + ⁻2 = 2   ⁻3 + 2 = ⁻1   2 – ⁻1 = 3   ⁻3 – ⁻1 = ⁻2
  4 + ⁻3 = 1   ⁻3 + 1 = ⁻2   2 – ⁻2 = 4   ⁻3 – ⁻2 = ⁻1
  4 + ⁻4 = 0   ⁻3 + 0 = ⁻3   2 – ⁻3 = 5   ⁻3 – ⁻3 = 0
  4 + ⁻5 = ⁻1   ⁻3 + ⁻1 = ⁻4   2 – ⁻4 = 6   ⁻3 – ⁻4 = 1

⁻3 + ⁻1 is said as negative 3 plus negative 1.

⁻3 – ⁻2 is said as negative 3 minus negative 2.

What are the next three lines of each of these number patterns? **Discuss.**
Use your patterns to find the answers to these.

4 + ⁻9   ⁻3 + ⁻7   2 – ⁻10   ⁻3 – ⁻10

- How could you use the number line to help you find the answers to these? **Discuss.**

4 + ⁻3   ⁻3 + 4   ⁻3 + ⁻2   3 – 4   ⁻2 – 1   2 – ⁻3   ⁻3 – ⁻4

$$\text{⁻7 ⁻6 ⁻5 ⁻4 ⁻3 ⁻2 ⁻1   0   1   2   3   4   5   6   7}$$

Beth wrote this.

Adding and subtracting are inverse operations.
To add a positive number move →. To subtract a positive number move ←.
To add a negative number move ←. To subtract a negative number move →.

What does Beth mean? **Discuss.**

Exercise 3

1 Write down the next three lines of these patterns.

	a	b	c	d	e
	3 + 1 = 4	2 + 1 = 3	5 + ⁻1 = 4	2 – 1 = 1	⁻2 – 1 = ⁻3
	3 + 0 = 3	2 + 0 = 2	5 + ⁻2 = 3	2 – 0 = 2	⁻2 – 0 = ⁻2
	3 + ⁻1 = 2	2 + ⁻1 = 1	5 + ⁻3 = 2	2 – ⁻1 = 3	⁻2 – ⁻1 = ⁻1
	3 + ⁻2 = 1	2 + ⁻2 = 0	5 + ⁻4 = 1	2 – ⁻2 = 4	⁻2 – ⁻2 = 0
	3 + ⁻3 = 0	2 + ⁻3 = ⁻1	5 + ⁻5 = 0	2 – ⁻3 = 5	⁻2 – ⁻3 = 1
	3 + ⁻4 = ⁻1	2 + ⁻4 = ⁻2	5 + ⁻6 = ⁻1		

2 Use the number patterns in question 1 to find the answers to these.

a 3 + ⁻8	b 3 + ⁻9	c 2 + ⁻7	d 2 + ⁻9	e 5 + ⁻9
f 2 – ⁻7	g 2 – ⁻9	h 3 – ⁻5	i 3 – ⁻7	j ⁻2 – ⁻5
k ⁻2 – ⁻8	l ⁻2 – 2	m ⁻2 – 4		

Number

T

3 Use a copy of this table.
Fill in the green shaded section first.
Use the pattern to fill in the rest of the table.

second number

+	⁻3	⁻2	⁻1	0	1	2	3
3						5	6
2							5
1							
0							
⁻1							
⁻2							
⁻3							

first number

4 Use the table you completed in question **3** to help answer these.

a $1 + {}^-2$	**b** $2 + {}^-1$	**c** $3 + {}^-2$	**d** $2 + {}^-3$	**e** $1 + {}^-3$
f ${}^-2 + 3$	**g** ${}^-3 + 3$	**h** ${}^-1 + 3$	**i** ${}^-3 + 2$	**j** ${}^-2 + {}^-1$
k ${}^-3 + {}^-1$	**l** $0 + {}^-2$	**m** ${}^-2 + {}^-3$	**n** ${}^-3 + {}^-3$	

Use the number line to help you find the answers to questions **5, 6** and **7**.

5

a ${}^-3 + 4$	**b** ${}^-5 + 2$	**c** ${}^-1 + 4$	**d** ${}^-2 + 6$	**e** $2 + {}^-5$
f $3 + {}^-2$	**g** $1 + {}^-3$	**h** $6 + {}^-4$	**i** ${}^-2 + {}^-3$	**j** ${}^-3 + {}^-1$
k ${}^-1 + {}^-2$	**l** ${}^-1 + {}^-4$	**m** ${}^-5 + {}^-1$	**n** ${}^-2 + {}^-3$	

6

a ${}^-2 - 5$	**b** ${}^-1 - 5$	**c** ${}^-3 - 1$	**d** $2 - 4$	**e** $3 - 4$
f $2 - {}^-1$	**g** $3 - {}^-2$	**h** $4 - {}^-3$	**i** $0 - 3$	**j** $0 - {}^-3$
k ${}^-2 - {}^-3$	**l** ${}^-3 - {}^-2$	**m** ${}^-1 - {}^-4$	**n** ${}^-1 - {}^-2$	**o** ${}^-2 - {}^-1$

7

a $3 + 6 - 4 - 10$	**b** $4 + 5 - 10 - 7$	**c** $4 + 6 - 11 + 3$	**d** $3 + 5 - 4 - 7$
e $6 - 4 + 2 - 10$	**f** $5 + 3 - 6 - 8$	**g** $7 + 6 - 3 - 9 - 4$	

8

a Choose a number card to give the answer 5. $\boxed{{}^-6} + \boxed{3} + \boxed{} = 5$

b Choose a card to give the lowest possible total.
What is this total? $\boxed{{}^-4} + \boxed{} =$

9 Here is a list of numbers.   ⁻5, ⁻4, ⁻3, ⁻1, 0, 1, 2, 3, 4, 5.
a Which of these numbers goes in the gaps?

$$^-3 + \underline{} = 2 \qquad 2 + \underline{} = {}^-3$$

b Choose two of the numbers for these gaps. $\underline{} + \underline{} = {}^-4.$
c Choose another two numbers which have a total of ⁻4. $\underline{} + \underline{} = {}^-4.$
d What is the total of all nine numbers in the list?
e Choose three different numbers from the list $\underline{} + \underline{} + \underline{} =$
so that the total is as small as possible.

T

10 Use a copy of this magic square.
Remember   Each row, each column and each diagonal must add to the same number.

	⁻6	1
		⁻1
⁻3		

   a   What does the diagonal 1, ⁻1, ⁻3 add up to?
   b   Complete the magic square.

* **11**   ⁻5   0   1   ⁻9   5   9   ⁻4   ⁻1

   a   Choose a number card to give the answer ⁻5.   ⁻4 – ☐ = ⁻5

   a   Choose a number card to give the answer 5.   ⁻4 – ☐ = 5

* **12**   Here is a list of numbers. ⁻5, ⁻3, ⁻2, ⁻1, 0, 1, 3, 4, 5
   a   Which of the numbers goes in the gaps?

   1 – ⁻2 = ___   ⁻5 – ___ = ⁻2

   b   Choose two numbers for these gaps.   ___ – ___ = ⁻4
   c   Choose another two numbers for the gaps in **b**.
   d   Choose three different numbers from the list so this total is as large as possible.
   ___ – ___ – ___ =

* **13**   What numbers are Tania and Gareth thinking of?
   a   **b**

I am thinking of a negative number. I add 12, then another 10. My answer is −3.

I am thinking of a negative number. I subtract 6, then add 12. My answer is −17.

* **14**   **a**   Two numbers are added.
           The answer is ⁻2.
           What might the numbers be?

           b   The difference between
           two numbers is ⁻1.
           What might the numbers be?

Review 1
a ⁻2 + 3   **b** ⁻4 + 2   **c** 5 + ⁻2   **d** 3 + ⁻4
e ⁻2 + ⁻4   **f** ⁻1 + ⁻5   **g** 3 + 5 – 6 – 2 **h** 4 + 6 – 10 – 8

Review 2   Here is a list of numbers. ⁻6, ⁻3, ⁻2, ⁻1, 0, 1, 3, 5
a   Which of the numbers goes in the gap?   3 + ___ = 2
b   Choose two of the numbers for these gaps.   ___ + ___ = ⁻5
c   Choose another two which have a total of ⁻5.
d   Choose three different numbers from the list so that this total is a small as possible.

___ + ___ + ___ =

Review 3

⁻4	2
3	⁻1

Two counters are dropped onto this board.
The numbers are added to get the score.
Write down all the possible scores.

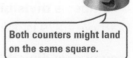
Both counters might land
on the same square.

37

*Review 4

a Choose a number card to give the answer 6.

b Choose a number card to give the answer ‾2.

$\boxed{^-5}-\boxed{}={}^-2$

c Choose two number cards to give the answer ‾3.
d Choose two more numbers to give the answer ‾3.

$\boxed{}-\boxed{}={}^-3$

e Choose two number cards to give the largest possible answer. What is your answer?

$\boxed{}-\boxed{}={}$

 Integers – a game for a group or class

You will need to draw a diagram like this for each turn.

To play
- Choose a leader.
- The leader calls out 4 integers between ‾10 and 10.
- As each one is called, put it in one of your blue boxes.
- Whoever gets the biggest total is the leader for the next turn.

 Practical

Make up a game that uses negative numbers. Write clear instructions for your game. Play your game to make sure it works.

Divisibility

A number is **divisible by 2** if it is an even number.
A number is **divisible by 3** if the sum of its digits is divisible by 3.
Example 198 is divisible by 3 since $1 + 9 + 8 = 18$ and 18 is divisible by 3.

A number is **divisible by 4** if the last two digits are divisible by 4.
Example 248 is divisible by 4 since 48 is.

A number is **divisible by 5** if its last digit is 0 or 5.
A number is **divisible by 6** if it is divisible by both 2 and 3.
A number is **divisible by 8** if half of it is divisible by 4.
Example Half of $432 = 216$.   216 is divisible by 4 so 432 is divisible by 8.

A number is **divisible by 9** if the sum of its digits is divisible by 9.
Example 279 is divisible by 9 since $2 + 7 + 9 = 18$ and 18 is divisible by 9.

A number is **divisible by 10** if the last digit is 0.

Exercise 4

T   1

222	259	528	341	205	306	244	473	543	833	157	526	192	341	803
376	882	536	454	653	508	746	101	531	514	926	391	124	233	674
625	359	411	538	344	177	712	377	685	369	632	407	356	416	642

I'm not over it yet!

Use a copy of this table.
Shade the squares with numbers that are
a   divisible by 5   b   divisible by 4   c   divisible by 3.
What word does the shading spell?

2   a   Which of 96, 154, 267, 432, are divisible by both 3 and 4?
     b   Which of 108, 162, 234, 378, are divisible by both 2 and 9?

3   Which of 116, 175, 216, 224, 360, 413 are divisible by
     a 2   b 3   c 4   d 5   e 6   f 8   g 9   h 10?

*4   Choose a number from 168, 174, 177, 200, 388, 418, 423 for each gap.
     Use each number only once.
     a   ___ is divisible by 2.   b   ___ is divisible by 3.   c   ___ is divisible by 4.
     d   ___ is divisible by 5.   e   ___ is divisible by 6.   f   ___ is divisible by 8.
     g   ___ is divisible by 9.

Review   Which of 75, 150, 188, 220, 235, 240, 336, 522 are divisible by
a 5   b 3   c 4   d 10   e 2   f 6   g 8   h 9?

*Investigation

Divisibility by 4
The sum of four even numbers is divisible by 4.
When is this true? When is it false? **Investigate**.

Multiples

Remember
The numbers 4, 20, 28, 36 are all multiples of 4.
They are all divisible by 4.
To find the **multiples** of a number we multiply by 1, 2, 3, 4, 5, 6, ...

Worked Example
Find the first five multiples of 3.

Answer
We find 3 × 1, 3 × 2, 3 × 3, 3 × 4, 3 × 5.
The first five multiples of 3 are **3, 6, 9, 12, 15**.

Worked Example
Find the smallest multiple of both 6 and 8.

Answer
The multiples of 6 are 6, 12, 18, **24**, 30, 36, 42, 48, ...
The multiples of 8 are 8, 16, **24**, 32, 40, 48, 56, ...
24 is the smallest multiple of both.

24 is called the **Lowest Common Multiple (LCM)** of 6 and 8.
It is the smallest number that is a multiple of both 6 and 8.

Exercise 5

1 Write down the first 6 multiples of   **a** 5   **b** 8.

You could use a counting stick to help.

2 **a** Write down the first 10 multiples of 3.
    b Write down the first 10 multiples of 4.
    c What is the Lowest Common Multiple of 3 and 4?

3 Find the LCM of these
    a 2 and 5   **b** 3 and 8   **c** 3 and 7   **d** 7 and 10   **e** 4 and 5
    f 4 and 6   **g** 6 and 5   **h** 6 and 9   **i** 8 and 10

T

4 Use a copy of this grid.
    Shade the multiples of 5.
    Shade the numbers that are divisible by 3.
    Which number is left?

30	3	18	25
35	21	5	6
12	15	22	9
24	33	27	45

5 **a** The numbers 246, 264, 426 all have the digits 2, 4 and 6.
      What other numbers can be made with these three digits?
    b What is the largest multiple of 8 you can make from the digits 2, 4 and 6?
    c What is the largest multiple of 7 you can make from these digits?

*6 John and Lara are nurses.
    John works 3 days, then has a day off.
    Lara works 2 days, then has a day off.
    They both have a day off on July 4th.
    When do they next have a day off together?

Review 1 Find the LCM of these.
a 2 and 9   **b** 4 and 7   **c** 3 and 5   **d** 4 and 10

Review 2 What is the largest multiple of 6 you can make using the three digits 4, 5 and 6?

Review 3 Julia paints her toenails every 5 days. Diana paints hers every 4 days.
Both girls paint their toenails on Christmas Day.
On what day in January do they both paint them again?

Prime numbers

Remember A prime number is divisible by only two numbers, itself and 1.

Examples 5, 11, 19, 37
1 is not a prime number.

The **prime numbers less than 30** are 2, 3, 5, 7, 11, 13, 17, 19, 23, 29.

A prime number can only be drawn as a rectangle that is a straight line of dots.

Example 11   • • • • • • • • • • •
11 can only be divided by 1 and 11. 11 is a prime number.

35   • • • • • • •
      • • • • • • •   35 can be divided by 5 and 7
      • • • • • • •   as well as 1 and 35.
      • • • • • • •   35 is not a prime number.
      • • • • • • •

To **test if a number is prime**, try to divide it by each of the prime numbers in turn.

Examples To test if 149 is prime, try dividing it by 2, 3, 5, 7 and 11.

Exercise 6

1 a Use a copy of this grid.
Colour 1.
Colour all the multiples of 2 that are bigger than 2.
Colour all the multiples of 3 that are bigger than 3.
Colour all the multiples of 5 that are bigger than 5.
Colour all the multiples of 7 that are bigger than 7.

1	2	3	4	5	6	7	8	9	10
11	12	13	14	15	16	17	18	19	20
21	22	23	24	25	26	27	28	29	30
31	32	33	34	35	36	37	38	39	40
41	42	43	44	45	46	47	48	49	50
51	52	53	54	55	56	57	58	59	60
61	62	63	64	65	66	67	68	69	70
71	72	73	74	75	76	77	78	79	80
81	82	83	84	85	86	87	88	89	90
91	92	93	94	95	96	97	98	99	100

b What is the missing word?
'The numbers not coloured are all _____ numbers.'

∗c Why was there no need to colour the multiples of 11?

2 For which of these numbers can you only draw a rectangle of dots that is just a straight line?
5   7   9   12   13   17   19   21   23   27   29

3 In each list there is one number which is not prime. Which number?
a 1, 2, 3, 5, 7   **b** 2, 3, 5, 7, 9, 11   **c** 7, 11, 13, 17, 27
d 2, 5, 11, 21, 23   **e** 7, 11, 17, 23, 25, 29

4 Test whether these numbers are prime.
a 33   **b** 41   **c** 45   **d** 47   **e** 51
f 52   **g** 57   **h** 63   **i** 71   **j** 78
k 81   **l** 87   **m** 91

5 A lawn has an area of 41 m². What must its length and breadth be if they are both whole numbers.

6 Explain why 2 is the only even prime number.

7 Two prime numbers are added.
The answer is 18.
What could the numbers be?

Find both sets of answers.

*8 Find pairs of primes with a difference of 4.

Review 1 Write down all the prime numbers between 8 and 24.

Review 2 Test whether these numbers are prime.
a 36 **b** 37 **c** 49 **d** 55 **e** 67 **f** 83

Review 3 Which of these lists has the most primes? How many does it have?
A 1, 2, 3, 11, 12, 13, 21, 22, 33 **B** 1, 4, 5, 9, 14, 19, 24, 29
C 2, 7, 13, 15, 19, 21, 27, 28 **D** 1, 5, 17, 19, 23, 25, 28, 29

*Review 4 Two prime numbers are added.
The answer is 36.
What could the numbers be?

Find all the possible answers.

Practical

There is a lot of information about prime numbers on the Internet.
Use the Internet to write a project on primes.

You can find out about
● primes and codes
● history of primes
● largest known primes
● formulae for finding primes and their exceptions

*Investigation

Primes

1 Investigate whether these are true or false.
   a Every even number greater than 4 is the sum of two prime numbers.
   b Between every number greater than 1 and its double, there is at least one prime number.

2 Primes with a difference of two are called *Twin Primes*.
Use a copy of this table and finish filling it in.

Twin Primes	Sum	Product
3 and 5	8	15
5 and 7	12	35
11 and 13	24	143

Hint: 15 is 16 – 1, 35 is 36 – 1 and 143 is 144 – 1

What do you notice about the sums? **Investigate**.
What do you notice about the products? **Investigate**.

Factors

Remember
The **factors** of 8 are 1, 2, 4 and 8.   8 is divisible by each of these factors.
A **factor** divides exactly into the given number.
A factor that is a prime number is a **prime factor**.

Example   To find the factors of 48 check which of the numbers 1, 2, 3, 4, 5, ... divides into 48.
        48 can be divided by 1, 2, 3, 4, 6, 8, 12, 16, 24 and 48.
        The factors of 48 are 1, 2, 3, 4, 6, 8, 12, 16, 24, 48.

1 and 48, 2 and 24, 3 and 16, 4 and 12, 6 and 8 are all the **factor pairs** of 48.
The prime factors of 48 are 2 and 3.

Worked Example
What is the biggest factor of both 12 and 30?

Answer
The factors of 12 are 1, 2, 3, 4, **6**, 12.
The factors of 30 are 1, 2, 3, 5, **6**, 10, 15, 30.
The **common factors** of 12 and 30 are 1, 2, 3 and 6.
The biggest common factor is **6**.

6 is called the **Highest Common Factor (HCF)** of 12 and 30.

Exercise 7

1 Find all the factors of these.
    a   16   b   18   c   27   d   35   e   60
    f   42   g   64   h   75   i   90

2   What are the prime factors of the numbers in **question 1**?

3   Use a copy of this diagram. Colour all the parts with
    a   factors of 15
    b   prime numbers
    c   multiples of 9.
    Which part is not coloured?

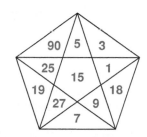

4   a   8 has four factors: 1, 2, 4, 8.
        What other 1-digit number has 4 factors?
    b   20 has six factors: 1, 2, 4, 5, 10, 20.
        There are 2 numbers less than 20 that have six factors.
        What numbers are they?
    c   What is the smallest number that has three factors?

Find all the answers.

5   a   What two numbers multiply to give 72?
    b   What three numbers other than 1 multiply to give 72?

6   a   Write down the factors of 16.
    b   Write down the factors of 40.
    c   4 is a common factor of 16 and 40. Write down all the other common factors of 16 and 40.
    d   What is the highest common factor of 16 and 40?

7 Find the HCF of these.
   a 18 and 24   **b** 16 and 36   **c** 40 and 25   **d** 6 and 18
   e 20 and 36   **f** 32 and 80   **g** 11 and 17

*8 i The number in each square is the product of the numbers in the circles on either side.

$56 = 8 \times 7$
$35 = 7 \times 5$
$40 = 5 \times 8$

How can you use these factor pairs of 56, 35 and 40 to find the numbers in the circles?

   ii Find the numbers in these circles.

a 28, 12, 21   **b** 18, 15, 30   **c** 42, 48, 56

*9 We can use factors to help us calculate.

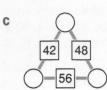

$45 \times 12 = 45 \times 2 \times 6$ $\frac{108}{36} = \frac{12 \times 9}{12 \times 3}$
        $= 90 \times 6$ $= \frac{9}{3}$
        $= 540$ $= 3$

There is more about calculating using factors on page 64.

Calculate these using factors.
   a 15×6 **b** 28×4 **c** $\frac{72}{18}$ **d** 35×12 **e** 15×24
   f 35×18 **g** $\frac{120}{24}$ **h** $\frac{144}{36}$

*10 a What is the HCF of 16 and 56?
    b Copy and complete $\frac{16}{56} = \frac{8 \times 2}{8 \times \_}$

This is called cancelling. There is more about cancelling on page 110.

    c Use **b** to write $\frac{16}{56}$ in its lowest terms.

*11 Cancel these fractions by first finding the HCF.
   a $\frac{4}{12}$ **b** $\frac{14}{35}$ **c** $\frac{16}{24}$ **d** $\frac{32}{72}$ **e** $\frac{40}{64}$ **f** $\frac{27}{45}$

*12 A storeroom is 12 feet by 25 feet by 16 feet. How many boxes 3 feet by 5 feet by 2 feet will fit in it if it is stacked full?

T

Review 1 Use a copy of this grid.
Shade
a the prime numbers
b the numbers that have 3 as a factor
c the factors of 56
d the multiples of 7.
What shape does the shading make?

65	100	76	82	92	44
87	29	15	8	42	99
35	6	28	60	11	56
300	21	17	7	30	33
77	4	1	63	72	14
110	55	74	40	200	76

Review 2 Find all the factor pairs of **a** 18   **b** 32.

Review 3 Find the HCF of **a** 16 and 24   **b** 20 and 45   **c** 9 and 19.

Review 4 Calculate these using factors. **a** 21×6 **b** $\frac{144}{48}$

* Review 5 The number in each square is the product of the numbers in the circles on either side. Use factors to find the numbers in the circles.

*Review 6 Cancel these fractions. **a** $\frac{3}{15}$ **b** $\frac{12}{30}$ **c** $\frac{35}{50}$

Factor Fun – a game for two players (A and B)

You will need a copy of this grid.

To play

- Player A crosses out any number. This is A's score.

- Player B then crosses out as many factors as possible of this number and adds them together. This is B's score.

- Player B then crosses out a number that is still left and adds it to his or her score.

1	2	3	4	5	6
7	8	9	10	11	12
13	14	15	16	17	18
19	20	21	22	23	24
25	26	27	28	29	30
31	32	33	34	35	36

- Player A then crosses out as many factors as possible of this number and adds them to his or her score. Then A crosses out a number that is still left and adds this to his or her score.

- The game continues in this way until all the numbers have been crossed out.
- The winner is the player with the highest score.

Example   If A crosses out 28 the game starts as shown.

A crosses out 28.
A's score = 28.

B crosses out 1, 2, 4, 7, 14.
B's score = 1 + 2 + 4 + 7 + 14
= 28

B crosses out 32.
B's score = 28 + 32
= 60

A crosses out 8, 16.
A's score = 28 + 8 + 16
= 52

A crosses out 17.
A's score = 52 + 17
= 69

B cannot cross out any factors of 17.
B crosses out 35.
B's score is now 60 + 35 = 95.
It is now A's turn.

These are the only factors of 32 left.

? Puzzle

1 I am a prime number.
I am a factor of 21.
I am not a factor of 12.
What number am I?

2 I am a factor of 36 and also of 48.
I am not a factor of 15 or of 20.
What numbers could I be?

*3 A 30 year old mother has two children at school. The age of each child is a factor of the mother's age. The sum of their ages is also a factor of the mother's age. How old are the children?

Investigation

Factors

1 What is the smallest number with exactly 4 factors?   5 factors?
Choose any number of factors. Try and find a number with this many factors. Have you found the smallest number with this many factors? **Investigate**.

2 Which 2-digit number has the greatest number of factors?
Is there more than one answer? **Investigate**.

3 What sort of number has an odd number of factors? **Investigate**.

Triangular numbers

 Practical

Begin with 2 students
    then 3 students
    then 4 students
    then 5 students
    and so on.

Count the number of handshakes if every student shakes hands with every other student.
Put your answers on a table like this.

Number of Students	Number of Handshakes
2	1
3	
4	
5	
⋮	

What do you notice about the numbers you get?

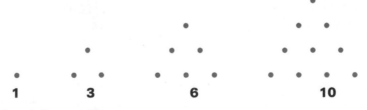

| 1 | 3 | 6 | 10 |

1, 3, 6, 10 are **triangular numbers**.

Exercise 8 **You will need triangular and square dotty paper.**

1 1, 3, 6, 10 are the first four triangular numbers.
Draw the next triangle in the pattern shown on the previous page.
What is the next triangular number after 10?

2

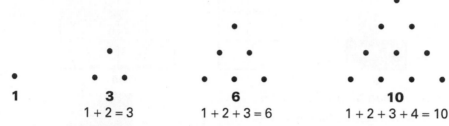

1	**3**	**6**	**10**
	1 + 2 = 3	1 + 2 + 3 = 6	1 + 2 + 3 + 4 = 10

Continue this pattern to find the next three triangular numbers.

3 1   3   6   10   15
    2   3   4   5

$$3 - 1 = 2$$
$$6 - 3 = 3$$
$$10 - 6 = 4$$
$$15 - 10 = 5$$

The difference between 15 and the next triangular number will be 6.
15 + **6** = 21. The next triangular number must be 21.
Continue this pattern to show that 28, 36, 45 and 55 are triangular numbers.

4 Some triangular numbers can be written as the sum of two other triangular numbers.
Find the three smallest triangular numbers that can be written like this.

Review

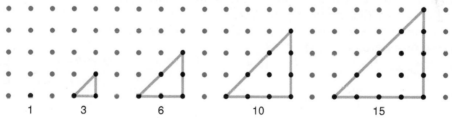

| 1 | 3 | 6 | 10 | 15 |

The first 5 triangular numbers can be shown like this.
Draw the next triangle in this pattern.
What is the next triangular number after 15?

Square numbers

Remember

1, 4, 9, 16 are **square numbers**.
1 = 1 × 1   4 = 2 × 2
9 = 3 × 3   16 = 4 × 4

We can write 3 × 3 as 3^2.
We read 3^2 as '3 squared'.

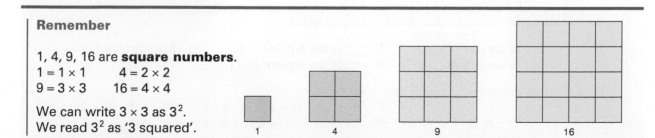

| 1 | 4 | 9 | 16 |

Number

Investigation

Triangular and Square Numbers

1   3   6   10

The triangular numbers can be drawn like this.

We can put the first two together like this.
Can you put the red and yellow ones together to get a square?
Can you put any of the other diagrams together to get a square?
Investigate.
Draw diagrams for some of the bigger triangular numbers like 15, 21, 28, ...
Do any of these fit together to make a square?

What can you say about the sum of any two
consecutive triangular numbers?

> Consecutive means one after the other.

Exercise 9

1 There are $5 \times 5 = 25$ small squares in this large square.
This shows that 25 is a square number.
   a Draw diagrams to show that 36 and 49 are square numbers.
   b What numbers go in the gaps?
      $8 \times 8 =$ ___ $9 \times 9 =$ ___ $10 \times 10 =$ ___ $11 \times 11 =$ ___ $12 \times 12 =$ ___

2 1, 4, 9 ___, 25, 36, 49, ___, 81, ___, 121, 144.
This is a list of the first 12 square numbers.
Which numbers are missing?

3 $3 \times 3 = 3^2$. Write these the same way.
   a 6×6 b 4×4 c 5×5 d 2×2 e 8×8 f 12×12

4 Copy and complete this table.

1^2	2^2	3^2	4^2	5^2	6^2	7^2	8^2	9^2	10^2	11^2	12^2
1	4										

5 a 16 is the square of ___.   b ___ is the square of 6.   c 5 squared equals ___.
  d 81 is the square of ___.   e ___ is the square of 11.   f 144 is the square of ___.

6 Calculate mentally.
   a $3^2 + 6$ b $5^2 + 7$ c $4^2 + 9$ d $(2+4)^2$ e $(3+4)^2$ f $3^2 + 2^2$
   g $6^2 + 8^2$ h $8^2 - 9$ i $6^2 - 11$ j $10^2 - 5^2$ k $9^2 - 4^2$

> In **d** and **e** work out the bracket first.

48

7 a Find the next three lines of this number pattern.

$1^2 = 1$
$2^2 = 1 + 3$
$3^2 = 1 + 3 + 5$

b What goes in the gap?
27^2 is equal to the sum of the first ___ odd numbers.

8 What is the total area of a courtyard made of two squares, one of side 4 m and the other of side 5 m?

9 Which is bigger, the sum of the squares of 3 and 5 or the square of the sum of 3 and 5?

*** 10 a** 100 can be written as the sum of two prime numbers in 6 different ways.
One way is $100 = 59 + 41$.
Find the other 5 ways.

b Which of the square numbers from 1 to 144 *cannot* be written as the sum of two prime numbers?

*** 11** Find the squares of these using the grid method.

a 16   **b** 23   **c** 18   **d** 1·4

> See page 4 for the grid method.

Review 1

a 8 squared equals ___
b ___ is the square of 9
c 1 is the square of ___
d 11 squared equals ___
e ___ is the square of 7.
f ___ is the square of 12.

Review 2
$3 \times 3 = 3^2$. Write these the same way.

a 1×1 **b** 7×7 **c** 9×9 **d** 11×11

Review 3
Calculate mentally.   **a** $4^2 + 8$ **b** $(3 + 5)^2$ **c** $2^2 + 4^2$ **d** $7^2 - 6$ **e** $8^2 - 5^2$

Review 4
Which is smaller, the sum of the squares of 5 and 6 or the square of the sum of 5 and 6?

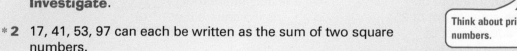

Investigation

Squares, Factors, Primes

1 The square number 4 has three factors 1, 2 and 4.
Which other square numbers, less than 100, have exactly 3 factors?
How are these square numbers different from the other square numbers?
Investigate.

> Think about prime numbers.

*** 2** 17, 41, 53, 97 can each be written as the sum of two square numbers.

$17 = 1 + 16$ $41 = 16 + 25$ $53 = 4 + 49$ $97 = 16 + 81$

Another six 2-digit primes can be written in the same way.
Investigate to find these 6 numbers.

Square roots

The answer to 'What number squared gives 9?' is 3.   3 is called the **square root** of 9.
The sign for a square root is $\sqrt{\ }$. $\sqrt{9}$ means 'the square root of 9'.

Squaring and finding the square root are inverse operations. One 'undoes' the other.

Example $3^2 = 9$ and $\sqrt{9} = 3$.

$3 \longrightarrow$ | square | $\longrightarrow$ 9

$3 \longleftarrow$ | square root | $\longleftarrow$ 9

Exercise 10

1 Copy and complete this table.

$\sqrt{1}$	$\sqrt{4}$	$\sqrt{9}$	$\sqrt{16}$	$\sqrt{25}$	$\sqrt{36}$	$\sqrt{49}$	$\sqrt{64}$	$\sqrt{81}$	$\sqrt{100}$	$\sqrt{121}$	$\sqrt{144}$
1	2										

2 **a** the square root of 36 is ___
    b 4 is the square root of ___
    c the square root of ___ is 7
    d ___ is the square root of 9
    e 12 is the square root of ___
    f ___ is the square root of 1
    g the square root of 100 is ___
    h 5 is the square root of ___
    i ___ is the square root of 4

3 **a** $4^2 = 16$ and $\sqrt{16} =$ ___
    b $5^2 =$ ___ and $\sqrt{25} = 5$
    c $3^2 =$ ___ and $\sqrt{9} = 3$
    d $7^2 = 49$ and ___ $= 7$
    e $6^2 = 36$ and ___ $= 6$
    f ___ $= 81$ and $\sqrt{81} = 9$
    g ___ $= 1$ and $\sqrt{1} = 1$
    h $8^2 = 64$ and ___ $= 8$
    i $10^2 = 100$ and $\sqrt{100} =$ ___
    j $2^2 = 4$ and $\sqrt{4} =$ ___

4 Calculate mentally.
    a $\sqrt{1+3}$ **b** $\sqrt{4+5}$ **c** $\sqrt{9+7}$ **d** $\sqrt{24+12}$ **e** $\sqrt{20-4^2}$ **f** $\sqrt{40-2^2}$ **g** $\sqrt{34-3^2}$

* **5** Which is bigger, the square root of the sum of 9 and 16 or the sum of the square roots of 9 and 16?

Review 1
a 10 is the square root of ___
b 9 is the square root of ___
c ___ is the square root of 25
d ___ is the square root of 16
e 2 is the square root of ___
f ___ is the square root of 64
g the square root of 1 is ___

Review 2
a $5^2 = 25$ and $\sqrt{25} =$ ___
b $4^2 = 16$ and ___ $= 4$
c $11^2 = 121$ and $\sqrt{121} =$ ___
d ___$^2 = 64$ and $\sqrt{64} = 8$
e ___$^2 = 4$ and $\sqrt{4} = 2$

Review 3   Calculate mentally.   **a** $\sqrt{19+6}$ **b** $\sqrt{73+8}$ **c** $\sqrt{50-1^2}$ **d** $\sqrt{73-3^2}$

Investigation

Happy Numbers

Is 13 a happy number?
Step 1 Square each digit and find the sum. $1^2 + 3^2 = 1 + 9 = 10$
Step 2 Keep squaring the digits and finding the sum until you get a single digit.
$1^2 + 0^2 = 1 + 0 = 1$.
If the single digit is 1, the number is a happy number. 13 is a happy number.

Is 865 a happy number?
Step 1 $8^2 + 6^2 + 5^2 = 64 + 36 + 25 = 125$
Step 2 $1^2 + 2^2 + 5^2 = 1 + 4 + 25 = 30$
        $3^2 + 0^2 = 9 + 0 = 9$
865 is not a happy number.

Is your house number a happy number?
Is your birthdate (MM DD YY) a happy number?
Is your telephone number a happy number? **Investigate**

You could use a spread sheet to help.

Calculator squares and square roots

Squares on the calculator

To find 7^2 **Key** ⑦ x^2 ⑤   to get 49.

To find $6\cdot3^2$ **Key** ⑥ • ③ x^2 ⑤   to get 39·69.

Square Roots on the calculator

To find $\sqrt{81}$ **Key** √ ⑧ ① ⊖   to get 9.

To find $\sqrt{5\cdot29}$ **Key** √ ⑤ • ② ⑨ ⊖   to get 2·3.

To find $\sqrt{29}$ **Key** √ ② ⑨ ⊖   to get   **5.385164807**
        $\sqrt{29} = 5\cdot4$ (1 d.p.)

Exercise 11

Estimate the answers first. See page 74 for more on estimating.

1 Calculate
   a 11^2 **b** 9^2 **c** 5^2 **d** 15^2 **e** 18^2 **f** 24^2
   g 91^2 **h** 87^2 **i** 113^2 **j** 302^2 **k** $7\cdot9^2$ **l** $9\cdot3^2$
   m $16\cdot4^2$ **n** $8\cdot3^2$ **o** $31\cdot6^2$ **p** $56\cdot1^2$ **q** $0\cdot13^2$ **r** $0\cdot47^2$
   s $0\cdot85^2$ **t** $0\cdot12^2$.

2 Calculate

a $\sqrt{196}$ b $\sqrt{484}$ c $\sqrt{784}$ d $\sqrt{1156}$

e $\sqrt{1024}$ f $\sqrt{2\cdot25}$ g $\sqrt{2\cdot89}$ h $\sqrt{21\cdot16}$

i $\sqrt{62\cdot41}$ j $\sqrt{0\cdot16}$ k $\sqrt{0\cdot01}$ l $\sqrt{0\cdot04}$

m $\sqrt{0\cdot49}$ n $\sqrt{0\cdot81}$.

3 Give the answers to these to 1 d.p.

a $\sqrt{54}$ b $\sqrt{32}$ c $\sqrt{94}$ d $\sqrt{104}$

e $\sqrt{811}$ f $\sqrt{964}$ g $\sqrt{2\cdot6}$ h $\sqrt{5\cdot9}$

i $\sqrt{16\cdot9}$ j $\sqrt{27\cdot1}$ k $\sqrt{64\cdot7}$.

4 Give the answer to these to the nearest tenth.

a $\sqrt{19}$ b $\sqrt{27}$ e $\sqrt{86}$ d $\sqrt{134}$

e $\sqrt{341}$ f $\sqrt{34\cdot3}$ g $\sqrt{71\cdot4}$ h $\sqrt{92\cdot8}$

i $\sqrt{105\cdot2}$.

5      $1^2 = 1$
       $11^2 = 121$
       $111^2 = 12321$

a Use your calculator to find the next line of this pattern.

b Without using your calculator, write down the answers to 11111^2 and 111111^2.

6 a Find the answer to $20^2 - 19^2$.
     Find two consecutive numbers that add to this total.

  b Calculate $15^2 - 14^2 + 13^2 - 12^2$.
     Find four consecutive numbers that add to this total.

  c Which 4 consecutive numbers do you think $32^2 - 31^2 + 30^2 - 29^2$ is the sum of?

  d What about $66^2 - 65^2 + 64^2 - 63^2$?

Review 1

T

| $\overline{20\cdot1}$ | $\overline{3\cdot24}$ | $\overline{31}$ | $\overset{A}{\overline{169}}$ | $\overline{2\cdot3}$ | $\overline{1\cdot96}$ | | $\overline{2\cdot1}$ | $\overset{A}{\overline{169}}$ | $\overline{0\cdot3}$ | | $\overline{21\cdot3}$ | $\overline{1\cdot5}$ | $\overline{2\cdot3}$ | $\overline{1\cdot3}$ | $\overline{8\cdot6}$ | $\overline{6724}$ |

| $\overline{1\cdot3}$ | $\overline{729}$ | | $\overset{A}{\overline{169}}$ | $\overline{729}$ | | $\overline{1\cdot5}$ | $\overline{729}$ | $\overline{20\cdot1}$ | $\overset{A}{\overline{169}}$ | $\overline{2\cdot3}$ | $\overline{0\cdot36}$ | $\overline{8\cdot6}$ | $\overline{6724}$ |

| $\overline{0\cdot16}$ | $\overline{2\cdot3}$ | $\overset{A}{\overline{169}}$ | $\overline{196}$ | $\overline{8\cdot6}$ |

Use a copy of this box.
Use your calculator to find these. Put the letter that is beside each above its answer.

A $13^2 = 169$ V 14^2 N 27^2 D 82^2 O $1\cdot8^2$ Z $\sqrt{961}$

W $\sqrt{4\cdot41}$ S $\sqrt{0\cdot09}$ E $\sqrt{73\cdot96}$ K $0\cdot6^2$ G $0\cdot4^2$ T $1\cdot4^2$

U $\sqrt{2\cdot25}$ I $\sqrt{1\cdot69}$ B $\sqrt{453\cdot69}$ R $\sqrt{5\cdot29}$ M $\sqrt{404\cdot01}$

Review 2   Give the answers to these to 1 d.p.

a $\sqrt{8}$ b $\sqrt{21}$ c $\sqrt{129}$ d $\sqrt{6\cdot8}$ e $\sqrt{17\cdot4}$

Investigation

Last Digits

1 $4^2 = 16$. The last digit of 4^2 is 6.
   $3^2 = 9$. The last digit of 3^2 is 9.
   Show that 1, 4, 9, 6, 5 are the last digits of $1^2, 2^2, 3^2, 4^2, 5^2$.

   Copy and complete this list for the last digits of $1^2, 2^2, 3^2, ..., 9^2$.
   1, 4, 9, 6, 5, ___, ___, ___, ___
   Describe the pattern.

   0, 1, 4, 9, 6, 5 are the last digits of $10^2, 11^2, 12^2, 13^2, 14^2, 15^2$.
   Predict the last digits of the square numbers $16^2, 17^2, 18^2, 19^2$ and 20^2.
   Use your calculator to check your prediction.

   Predict the last digits of the square numbers 20^2 to 40^2.

   Explain why 713, 258, 422 and 597 could not be square numbers.

> Look at the last digits.

2 9604 is a square number. ▲■^{2} = 9604. ▲■ is a 2-digit number.
   ■ could be 8. What other number could ■ be?
   What could ⬤ be for these square numbers?
   ▲⬤^{2} = 2209   ▲⬤^{2} = 3025   ▲⬤^{2} = 1681   ▲⬤^{2} = 4356

   Khalid was asked to find what number multiplied by itself gives the answer 4489. His calculator was broken.
   He wrote $30 \times 30 = 900$
   $40 \times 40 = 1600$
   $50 \times 50 = 2500$
   $60 \times 60 = 3600$
   $70 \times 70 = 4900$
   Khalid now knew the number must be either 63 or 67. Explain how he knew this.
   Is the number 63 or 67?

   Use Khalid's method to find what number multiplied by itself gives these.
   2209, 3025, 1681, 4356, 1024
   Do not use a calculator.

More squares and square roots

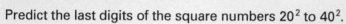

$4 \times 4 = 16$ and $0.4 \times 0.4 = 0.16$ and $\sqrt{0.16} = 0.4$

$0.1^2 = 0.01$ and $\sqrt{0.01} = 0.1$
$0.2^2 = 0.04$ and $\sqrt{0.04} = 0.2$
$0.3^2 = 0.09$ and $\sqrt{0.09} = 0.3$
$\vdots$
$0.9^2 = 0.81$ and $\sqrt{0.81} = 0.9$

*Worked Example
Find   a 40^2 b 400^2

Answer

a $40^2 = (4 \times 10)^2$
$= 4^2 \times 10^2$
$= 16 \times 100$
$= \mathbf{1600}$

b $400^2 = (4 \times 100)^2$
$= 4^2 \times 100^2$
$= 16 \times 100 \times 100$
$= \mathbf{160\ 000}$

Worked Example

Find **a** $\sqrt{6400}$ **b** $\sqrt{640\ 000}$

Answer

a $\sqrt{6400} = \sqrt{64 \times 100}$
$= \sqrt{64} \times \sqrt{100}$
$= 8 \times 10$
$= \mathbf{80}$

b $\sqrt{640\ 000} = \sqrt{64 \times 10\ 000}$
$= \sqrt{64 \times 100 \times 100}$
$= \sqrt{64} \times \sqrt{100} \times \sqrt{100}$
$= 8 \times 10 \times 10$
$= \mathbf{800}$

Notice that the numbers 6400 and 640 000 have an *even* number of zeros.

Exercise 12

1 Copy and complete these tables.

a

0.1^2	0.2^2	0.3^2	0.4^2	0.5^2	0.6^2	0.7^2	0.8^2	0.9^2
0·01	0·04	0·09						0·81

b

$\sqrt{0.01}$	$\sqrt{0.04}$	$\sqrt{0.09}$	$\sqrt{0.16}$	$\sqrt{0.25}$	$\sqrt{0.36}$	$\sqrt{0.49}$	$\sqrt{0.64}$	$\sqrt{0.81}$
	0·2			0·5				

*∗**2** Find **a** 20^2 **b** 50^2 **c** 80^2 **d** 100^2

*∗**3** Find **a** 300^2 **b** 500^2 **c** 700^2 **d** 900^2

*∗**4** Find **a** $\sqrt{400}$ **b** $\sqrt{900}$ **c** $\sqrt{2500}$ **d** $\sqrt{4900}$

*∗**5** Find **a** $\sqrt{90\ 000}$ **b** $\sqrt{360\ 000}$ **c** $\sqrt{640\ 000}$

Review 1 Find **a** 0.2^2 **b** 0.8^2 **c** $\sqrt{0.09}$ **d** $\sqrt{0.49}$

*∗**Review 2** Find **a** 30^2 **b** 90^2 **c** 600^2 **d** 800^2

*∗**Review 3** Find **a** $\sqrt{1600}$ **b** $\sqrt{8100}$ **c** $\sqrt{490\ 000}$ **d** $\sqrt{810\ 000}$

Summary of key points

A The numbers ... $^-3$, $^-2$, $^-1$, 0, 1, 2, 3, 4, ... are the **integers**.

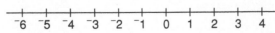

On the number line the further to the left an integer is the smaller it is.

Example $^-5 < ^-3$

B We can **add and subtract integers**. Sometimes we use a number line to help.

Examples $4 + {}^-2 = 2$ $3 + {}^-5 = {}^-2$ ${}^-2 + {}^-5 = {}^-7$ ${}^-3 + 7 = 4$

$3 - {}^-4 = 7$ ${}^-6 - {}^-8 = 2$ ${}^-7 - {}^-3 = {}^-4$

C A number is **divisible** by 2 if it is an even number,

by 3 if the sum of its digits is divisible by 3,

by 4 if the last two digits are divisible by 4,

by 5 if the last digit is 0 or 5,

by 6 if it is divisible by both 2 and 3,

by 8 if half of it is divisible by 4,

by 9 if the sum of its digits is divisible by 9,

by 10 if the last digit is 0.

D The **multiples** of a number are found by multiplying the number by each of 1, 2, 3, 4, ...

Example The multiples of 7 are 7, 14, 21, 28, 35, ...

The **Lowest Common Multiple (LCM)** of 4 and 5 is 20. This is the smallest number that is a multiple of both 4 and 5.

E **Prime numbers** are only divisible by themselves and 1.   1 is not a prime number. The prime numbers less than 30 are 2, 3, 5, 7, 11, 13, 17, 19, 23, 29. We can use divisibility tests to test if other numbers are prime.

F A **factor** divides exactly into a given number.

Example The factors of 44 are 1, 2, 4, 11, 22, 44.

The **factor pairs** of 44 are 1 and 44   2 and 22   4 and 11

The **Highest Common Factor (HCF)** of 18 and 27 is 9. This is the biggest number that is a factor of both 18 and 27.

G The first four **triangular numbers** are 1, 3, 6, 10.

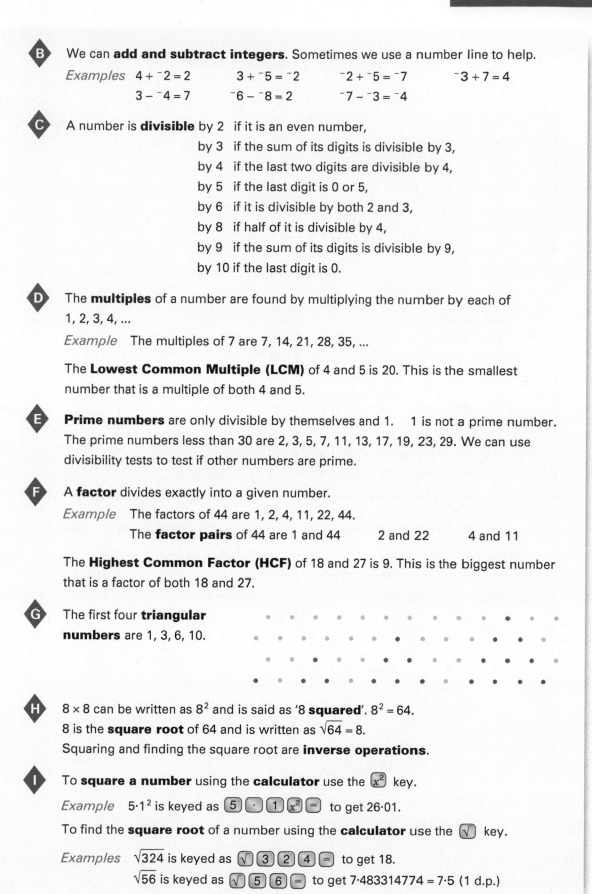

H 8×8 can be written as 8^2 and is said as '8 **squared**'. $8^2 = 64$.

8 is the **square root** of 64 and is written as $\sqrt{64} = 8$.

Squaring and finding the square root are **inverse operations**.

I To **square a number** using the **calculator** use the $\boxed{x^2}$ key.

Example $5 \cdot 1^2$ is keyed as $\boxed{5}\boxed{\cdot}\boxed{1}\boxed{x^2}\boxed{=}$ to get 26·01.

To find the **square root** of a number using the **calculator** use the $\boxed{\sqrt{}}$ key.

Examples $\sqrt{324}$ is keyed as $\boxed{\sqrt{}}\boxed{3}\boxed{2}\boxed{4}\boxed{=}$ to get 18.

$\sqrt{56}$ is keyed as $\boxed{\sqrt{}}\boxed{5}\boxed{6}\boxed{=}$ to get 7·483314774 = 7·5 (1 d.p.)

Number

Test yourself *(calculator not allowed)* **except for questions 13, 14**

1 3   ⁻1   ⁻4   0   1   ⁻6
   Put these integers in order from smallest to largest.

   Use a number line.   **A**

2 Which number is halfway between ⁻2 and ⁻5?   **A**

3 The noon temperature at Bristol was 8 °C. The midnight temperature was ⁻3 °C.
   What is the difference between these temperatures?   **B**

4 ⎡2⎤ ⎡⁻1⎤ ⎡0⎤ ⎡5⎤ ⎡⁻3⎤ ⎡1⎤ ⎡⁻2⎤ ⎡3⎤ ⎡⁻6⎤

   a Choose number cards that make the answer true. Is there more than one way?
   i ⎡ ⎤ + ⎡2⎤ = ⁻1   **ii** ⎡ ⎤ + ⎡ ⎤ = ⁻4

   iii ⎡⁻3⎤ − ⎡ ⎤ = 3   **iv** ⎡ ⎤ − ⎡ ⎤ = ⁻4

   ⎡ ⎤ + ⎡ ⎤ + ⎡ ⎤ = __
   ⎡ ⎤ − ⎡ ⎤ = __

   b Choose three number cards so the answer is a small as possible.
   c Choose two number cards to give the largest possible answer.   **C**

5 Which of 108, 117, 121, 216, 387, 600, 640, 702 are divisible by
   a 5   **b** 3   **c** 4   **d** 6   **e** 8   **f** 9?   **C**

6 Write down three properties of each of these numbers.
   1   3   4   8
   For example, three properties of the number 6 might be;
   it is a triangular number, it is a factor of 12, it is a multiple of 3.   **C D E F G H**

7 **a** The number 613 has the digits 6, 1 and 3. What other numbers can you make with   **C**
        these three digits?
   b What is the largest multiple of 4 you can make using the three digits 6, 1 and 3?   **D**

8 **a** What is the LCM of 4 and 6?   **b** What is the HCF of 24 and 40?   **D F**
   c Write down all the pairs of factors of 72.

9 Two prime numbers are added.
   The answer is 24.
   What might the numbers be?   *Find all three possible sets of answers.*   **E**

10 1, 3, 6, 10 are the first four triangular numbers.   **G**
    1 + **2** = 3   3 + **3** = 6   6 + **4** = 10   10 + **?** = **?**
    Explain how you can use this to find the fifth triangular number. What is it?

11 **a** ___ squared is 64.   **b** 25 is the square of ___.   **H**
    c The square of ___ is 81.   **d** 4 is the square root of ___.
    e The square root of 9 is ___.   **f** The square root of ___ is 6.

12 **a** 4^2 **b** $\sqrt{100}$ **c** 7^2 **d** $\sqrt{49}$ **e** 0.5^2 **f** $\sqrt{0.36}$ **H**

13 **a** 6.8^2 **b** 29^2 **c** $\sqrt{8.41}$ **d** $\sqrt{39.69}$ **I**

14 Give the answers to 1 decimal place.   **a** $\sqrt{21}$ **b** $\sqrt{33}$ **c** $\sqrt{109}$ **I**

*15 **a** 60^2 **b** 200^2 **c** $\sqrt{3600}$ **d** $\sqrt{160\,000}$ **H**

*16 13 and 31 are both prime numbers.   **E**
    Which other 2-digit primes are also prime when the digits are reversed?

3 Mental Calculation – Whole Numbers and Decimals

You need to know

Key vocabulary

add, addition, approximate, approximately, calculation, column, commutative complements, decrease, difference, divide, division, double, estimate, factor, halve, increase, multiplication, multiply, nearly, order of operations, partition, product, row, subtract, subtraction, sum

Much Ado About Something

This is how Jade found the answer to 6×8.

Step 1 Subtract 6 from 10 and 8 from 10 to get 4 and 2.

Step 2 Multiply the two answers together.

       This gives the units digit. $4 \times 2 = \mathbf{8}$

Step 3 Write down the original numbers, one under the other as shown in black.

       Write down the answers from step 1, one under the other as shown in red.

6	4
8	2

       Subtract diagonally to get the tens digit. $6 - 2 = \mathbf{4}$ or $8 - 4 = \mathbf{4}$

Step 4 Write down the answer. **48**

Does this work for other 1-digit numbers?

Adding and subtracting whole numbers

Discussion

Jack made 46 runs in the first innings of cricket and 29 in the second innings.
Jack used a number line to explain how he worked out the total mentally.

Discuss Jack's way of adding 46 and 29.
How else could Jack add 46 and 29 mentally? **Discuss**.

Here are some ways to **add and subtract mentally**.

Example   47 + 68

47 + 68 = 40 + 7 + 60 + 8 = 40 + 60 + 7 + 8 = 100 + 15 **= 115** 40 and 60 are complements in 100.	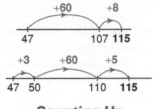	+70 diagram 47 + 68 = 47 + 70 – 2 = 117 – 2 **= 115**
Partitioning	**Counting Up**	**Adding too much then taking some away (compensation)**

Example   417 + 388

417 + 388 = 400 + 400 + 17 – 12 = 800 + 17 – 12 = 817 – 12 **= 805** **Nearly Doubles**	417 + 388 = 420 + 390 – 3 – 2 = 810 – 3 – 2 **= 805** **Nearly Numbers**

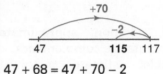

Mentally means 'in your head'. These boxes explain ways you could do it.

Example   605 – 293

605 – 293 = 605 – 200 – 90 – 3 = 405 – 90 – 3 = 315 – 3 **= 312**	+7 +300 +5 diagram 293 300 600 605 7 + 300 + 5 = **312**	–300 diagram 605 – 293 = 605 – 300 + 7 = 305 + 7 **= 312**
Partitioning	**Counting Up**	**Subtracting too much then adding some back (compensation)**

Look for **complements in 10 or 100** to help you to add.

Examples   8 + 6 + 4 = 8 + 10   45 + 8 + 55 + 22 = 45 + 55 + 8 + 22
                        = 18   = 100 + 30
                                                                                = 130

Exercise 1 **This exercise is to be done mentally.**

1 **a** 4 + 8 + 12 + 6 + 13   **b** 9 + 7 + 13 + 11 + 15   **c** 16 – 7 + 4 – 2 + 17
    d 19 – 6 + 11 – 12 + 8   **e** 15 + 12 – 8 + 16 – 17

2 **a** 30 + 10 + 10   **b** 50 + 20 – 10   **c** 80 – 20 – 10   **d** 80 + 40 – 40
    e 40 + 50 + 60   **f** 80 + 30 + 20   **g** 100 – 60 + 40   **h** 90 – 30 + 50
    i 60 + 50 – 30   **j** 160 – 50   **k** 130 + 70 + 40

3 **a** 36 + 20   **b** 86 – 30   **c** 57 + 40   **d** 93 – 50   **e** 37 + 80
    f 36 + 23   **g** 89 – 35   **h** 24 + 66   **i** 58 – 33   **j** 87 + 19
    k 96 – 8   **l** 57 + 14   **m** 86 – 19   **n** 82 – 29

4 **a** 48 + 52 + 9   **b** 63 + 37 + 16 + 4   **c** 59 + 7 + 41 + 9
    d 83 + 5 + 17 + 10   **e** 7 + 72 + 18 – 3   **f** 21 + 36 – 19 + 64
    g 56 + 12 + 54 + 88   **h** 147 + 9 + 53 – 4

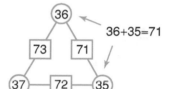

Look for complements.

T

5 This is an arithmagon.
    The number in each square is the sum of the numbers in the two
    circles on either side of the square.
    Use a copy of these.
    Fill them in.

    a   **b**   **c**

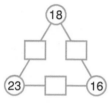   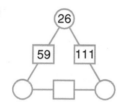   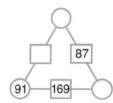

6 **a** Find the total of the numbers in the
        i orange row   * **ii** green row   * **iii** yellow row.
    b Subtract the total of the blue row from the total of the orange
        row.
    c Choose any triangle of four numbers.
        Add together the 'top' number and the
        number 'below' it.   **25 + 17 = 42**
        Add the other two numbers
        together.   **16 + 18 = 34**
        Find the difference between the two totals.   **42 – 34 = 8**
        Try this for other triangles of four numbers. What do you notice?

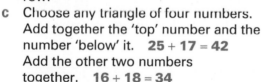

7 **a** 174 + 36   **b** 183 + 25   **c** 326 + 42   **d** 543 + 57
    e 486 + 23   **f** 196 – 23   **g** 388 – 36   **h** 285 – 32
    i 293 – 57   **j** 597 – 79   **k** 894 + 27   **l** 588 + 97
    m 430 – 76   **n** 113 – 78   **o** 427 – 89

You may need to use jottings for some of these.

8 **a** 170 + 360   **b** 520 – 360   **c** 480 + 350   **d** 830 – 470
    e 740 + 470   **f** 840 – 390   **g** 427 + 103   **h** 925 – 402
    i 586 + 278   **j** 350 – 289   **k** 630 – 376   **l** 327 + 366
    m 1064 + 2387   **n** 7013 – 4875   **o** 5860 + 3429

9 **a** What is the sum of 83 and 58?
   b Find the difference between 93 and 76.
   c Increase 86 by 19.
   d Decrease 72 by 39.

*10 **a** Three consecutive numbers add to 81. What are they?
   b Find four consecutive odd numbers which add to 80.
      7, 9, 11, 13 are four consecutive odd numbers.
   c Is it possible to find four consecutive odd numbers which add to 511?
      Explain your answer.

Consecutive numbers come one after the other.

Review 1
a 5 + 7 + 12 + 3 + 9   **b** 8 + 6 + 15 + 7 + 11   **c** 8 + 12 − 3 − 5 + 2
d 9 − 3 + 16 − 4 − 2   **e** 40 + 70 + 60   **f** 90 − 50 + 20
g 56 + 20   **h** 86 + 34 + 14   **i** 92 − 49
j 289 + 43   **k** 572 − 79   **l** 596 − 412

Review 2   Use a copy of these arithmagons. Fill them in.

a 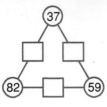   **b** 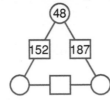   **c**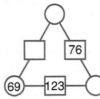

*### Review 3   Use all the numbers on these cards to make two 2-digit numbers.
   a How many different pairs can you make?
   b Which pairs give the greatest sum?
   c Which pair gives the greatest difference?
   d Which pair gives a difference closest to 50?

| 7 | 6 | 3 | 9 |

? Puzzle

1 Use all of the digits 3, 4, 5, 6, 7 to make the smallest answer possible.

☐ ☐ ☐ − ☐ ☐ = ?

2 Choose numbers from the green box to make these true.
   In each question use a number only **once**.
   a ☐ + ☐ − ☐ = 9
   b ☐ ☐ + ☐ ☐ = 100   **c** ☐ ☐ + ☐ ☐ = ☐ ☐
   Is there more than one way for each?

| 1 | 2 | 3 | 4 |
| 5 | 6 | 7 | 8 |

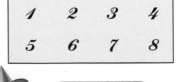
Do them mentally.

3 Use only the digits 3, 5, 7 and 9 to make each sum correct. You may use each digit as often as you like.
   a ☐ ☐ + ☐ ☐ = 114   **b** ☐ ☐ + ☐ ☐ = 90   **c** ☐ ☐ + ☐ ☐ = 172
   d ☐ ☐ + ☐ ☐ = 136   **e** ☐ ☐ + ☐ ☐ = 130

⊤

Investigation

Magic Squares

1 In a magic square the numbers in each **row**, each **column** and each **diagonal** add to the same total.

The numbers 1, 2, 3, 4, 5, 6, 7, 8 and 9 are placed as shown to make a magic square.

Investigate other ways of placing these numbers to make a magic square.

2	9	4
7	5	3
6	1	8

2 Can a magic square be made from the first 9 even numbers, 2, 4, 6, 8, 10, 12, 14, 16, 18? **Investigate**.

What if odd numbers were used instead of even numbers?

*3 Can you make a magic square with the numbers 16, 17, 18, 19, 20, 21, 22, 23, 24? **Investigate**.

Adding and subtracting decimals

Here are some ways to **add and subtract decimals mentally**.

Example   4·8 + 5·7

$4·8 + 5·7 = 4 + 0·8 + 5 + 0·7$ $= 4 + 5 + 0·8 + 0·7$ $= 9 + 1·5$ $= \mathbf{10·5}$ **Partitioning**	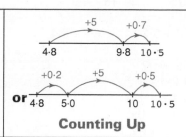 **Counting Up**	$4·8 + 5·7 = 5 - 0·2 + 6 - 0·3$ $= 5 + 6 - 0·2 - 0·3$ $= 11 - 0·2 - 0·3$ $= 10·8 - 0·3$ $= \mathbf{10·5}$ **Nearly numbers**

Example   12·3 − 5·5

$12·3 - 5·5 = 12·3 - 5 - 0·5$ $= 7·3 - 0·5$ $= 7·3 - 0·3 - 0·2$ $= 7 - 0·2$ $= \mathbf{6·8}$ 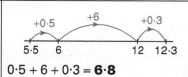 **Partitioning**	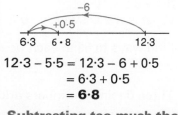 $0·5 + 6 + 0·3 = \mathbf{6·8}$ **Counting Up**	$12·3 - 5·5 = 12·3 - 6 + 0·5$ $= 6·3 + 0·5$ $= \mathbf{6·8}$ **Subtracting too much then adding back (compensation)**

Number

Look for **complements in 1** to help you add.

Examples $0.37 + 0.63 = 1$

$3.4 + 4.6 = 3 + 4 + 0.4 + 0.6$
$\qquad\quad = 3 + 4 + 1$
$\qquad\quad = 8$

$5.7 + 2.9 = 5.7 + 2.3 + 0.6$
$\qquad\quad = 8 + 0.6$
$\qquad\quad = 8.6$

Exercise 2 **This exercise is to be done mentally.**

1 a $0.7 + 0.8$ b $0.4 + 0.6$ c $0.9 + 0.8$ d $0.5 + 0.7$
   e $1.3 + 0.4$ f $4.7 + 0.6$ g $2.9 + 0.8$ h $0.2 + 0.8$
   i $0.7 + 1.3$ j $2.7 + 1.3$ k $4.3 + 6.9$ l $4.5 + 4.8$
   m $2.8 + 2.9$ n $4.8 + 7.2$ o $5.7 + 2.9$ p $0.59 + 0.41$
   q $0.41 + 0.57$ r $0.58 + 0.74$ s $0.97 + 0.68$

Remember to look for complements.

2 a $0.9 - 0.5$ b $0.8 - 0.3$ c $1.7 - 0.8$ d $1.3 - 0.7$
   e $6.7 - 2.4$ f $3.9 - 1.5$ g $8.2 - 1.9$ h $7.3 - 3.8$
   i $5.6 - 3.9$ j $3 - 0.7$ k $7 - 5.2$ l $0.78 - 0.39$
   m $0.65 - 0.36$ n $0.73 - 0.39$

3 a $0.6 + 0.7 + 0.4$ b $3 + 0.3 + 4 + 0.7$ c $7 + 0.7 + 5.3 + 0.3$
   d $9.6 + 7 + 0.4$ e $3.7 + 2 + 5.1 + 0.3$

4 a $18.6 + 7.2$ b $6.3 + 1.4$ c $15.3 + 4.2$ d $8.3 - 4.2$
   e $14.3 - 5.5$ f $17.8 + 3.6$ g $16.4 + 2.9$ h $15.3 - 5.7$
   i $16.4 + 3.8$ j $11.9 + 2.6$ k $12.8 - 4.9$ l $18.1 - 7.6$
   m $5.0 - 1.5$ n $16.0 - 2.7$ o $18 - 4.9$

You may need to use jottings for some of these.

5 a $8.6 + 8.4 + 0.8$ b $0.6 + 3.4 + 8.2$ c $4.7 + 5.4 - 0.2$

6 In which colour square is the answer?
   a $1.4 + 2.3 + 3.6$ b $4.6 + 2.7 + 1.2$
   c $3.1 + 4.2 + 1.5$ d $5.3 + 1.6 + 2.3$
   e $9.2 - 1.1 - 0.4$ f $9.6 - 1.3 - 0.4$
   g $5.2 + 1.3 - 0.5$ h $15.2 - 8.6 + 2.1$
   i $3.1 - 0.9 + 5.4$

7.6	9.2	6
7.9	7.3	8.8
7.7	8.7	8.5

7 $3.4 + 6.8 = 10.2$.
   What is the answer to $10.2 - 6.8$?

*8 Find ways to fill in each box with one digit to make these true.
   a $\square.\square + \square.\square = 14.6$ b $\square.\square - \square.\square = 1.3$

*9 Three decimal numbers add to 10. What might the three numbers be?

Review 1
a $0.5 + 0.7$ b $0.9 - 0.4$ c $8.7 + 0.9$ d $6.3 - 0.7$ e $4.2 + 3.9$
f $7.5 - 3.8$ g $14.6 - 3.7$ h $0.68 + 0.32$ i $15.6 - 7.9$ j $0.82 + 0.97$

Review 2   Find ways to fill in each box with one digit to make these true.
a $\square.\square + \square.\square = 9.2$ b $\square.\square - \square.\square = 2.7$

Investigation

Decimal Paths

Investigate paths from the red square to the blue square.

You could **investigate**
- the sums of paths that go through exactly 7 squares
- paths that have the same total
- the greatest difference between any two paths that go through the same number of squares

You may move down or sideways but not up.

You may go through each square just once on each path.

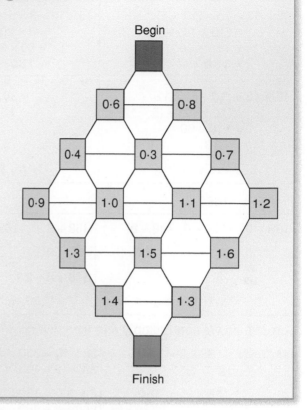

Begin

| 0·6 | | 0·8 |

| 0·4 | 0·3 | 0·7 |

0·9 | 1·0 | 1·1 | 1·2

| 1·3 | 1·5 | 1·6 |

| 1·4 | 1·3 |

Finish

Multiplying and dividing whole numbers

Discussion

Claire had 40 beads each 11 mm long. She wanted to know how long a necklace would be if she threaded them all onto a nylon thread.

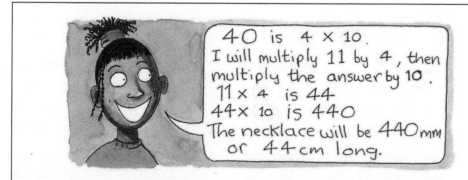

40 is 4 × 10.
I will multiply 11 by 4, then multiply the answer by 10.
11 × 4 is 44
44 × 10 is 440
The necklace will be 440 mm or 44 cm long.

Discuss Claire's method.

What other ways could you use to find the answer to 40 × 11 mentally?

Number

Here are some **ways to multiply and divide mentally**.

Example 15×12

We can multiply in any order.

| $15 \times 12 = 15 \times (10 + 2)$
 $= (15 \times 10) + (15 \times 2)$
 $= 150 + 30$
 $= \mathbf{180}$
 or
 $15 \times 12 = (10 + 5) \times 12$
 $= (10 \times 12) + (5 \times 12)$
 $= 120 + 60$
 $= \mathbf{180}$

 Partitioning | $15 \times 12 = 3 \times 5 \times 4 \times 3$
 $= 3 \times 20 \times 3$
 $= 60 \times 3$
 $= \mathbf{180}$
 or $15 \times 12 = 15 \times 2 \times 6$
 $= 30 \times 6$
 $= \mathbf{180}$

 Using Factors | $15 \times 12 = 30 \times 6$
 $= 180$
 Double 15 is 30
 Half of 12 is 6

 Doubling one number and halving the other |

Example $168 \div 8$

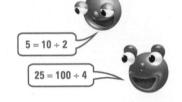
$8 = 2 \times 2 \times 2$

| $168 \div 8 = (160 \div 8) + (8 \div 8)$
 $= 20 + 1$
 $= \mathbf{21}$

 Partitioning | $168 \div 8$ $168 \div 2 = 84$
             $84 \div 2 = 42$
             $42 \div 2 = \mathbf{21}$
 $168 \div 8 = \mathbf{21}$

 Using Factors |

Sometimes we can multiply and then divide.

Examples $38 \times 5 = \mathbf{190}$ $38 \times 10 = 380$
                                   $380 \div 2 = 190$ $5 = 10 \div 2$

            $18 \times 25 = \mathbf{450}$ $18 \times 100 = 1800$
                                    $1800 \div 4 = 450$ $25 = 100 \div 4$

Exercise 3   **This exercise is to be done mentally.**

1   **a** $2 \times 3 \times 4$ **b** $8 \times 5 \times 3$ **c** $5 \times 4 \times 2$ **d** $12 \times 2 \times 2$
    e $8 \times 7 \times 2$ **f** $5 \times 3 \times 2 \times 4$ **g** $4 \times 6 \times 5$

2   What goes in the box?
    a $2 \times \square \times 4 = 40$ **b** $3 \times 6 \times \square = 36$ **c** $7 \times \square \times 3 = 63$
    d $5 \times \square \times 6 = 150$ **e** $\square \times 7 \times 2 = 28$

3   **a** 45×2 **b** 72×3 **c** 35×5 **d** 64×5 **e** 57×4
    f 62×8 **g** 63×7 **h** 49×5 **i** 21×9

4   **a** 540×2 **b** 860×2 **c** 370×2 **d** 590×2 **e** 830×2
    f 1500×2 **g** 1700×2 **h** 120×3 **i** 140×4 **j** 250×8

5   **a** 30×30 **b** 50×60 **c** 80×70 **d** 500×20 **e** 600×50
    f 24×50 **g** 150×20 **h** 52×40 **i** 32×30 **j** 120×60
    k 37×40 **l** 58×60 **m** 156×50 **n** 138×50

6   **a** 72×11 **b** 45×12 **c** 25×16 **d** 75×12
    e 23×11 **f** 23×21 **g** 42×25 **h** 24×31

7 **a** 84 ÷ 2　　　**b** 92 ÷ 2　　　**c** 120 ÷ 4　　　**d** 200 ÷ 5
　　　e 420 ÷ 5　　　**f** 208 ÷ 8　　　**g** 196 ÷ 4　　　**h** 900 ÷ 6
　　　i 156 ÷ 6　　　**j** 940 ÷ 2　　　**k** 340 ÷ 2　　　**l** 580 ÷ 2
　　　m 1700 ÷ 2　　**n** 5600 ÷ 2　　＊**o** 128 ÷ 16

8 64 × 17 = 1088. What is the answer to 1088 ÷ 17?

9 **a** Find the product of 8 and 4 and 3.
　　　b Find the product of 86 and 4.

> Remember:
> We find the *product* by multiplying
> the numbers together.

＊**10** Sophie divided 170 ÷ 12 like this.　　100 ÷ 12 = 8 R 4
　　　　　　　　　　　　　　　　　　　　　70 ÷ 12 = 5 R 10
　　　　　　　　　　　　　　　　　　　　　　　　 = 13 R 14
　　　　　　　　　　　　　　　　　　　　　　　　 = **14 R 2**
　　　Use Sophie's method to find the answers to these.
　　　a 150 ÷ 12　　**b** 160 ÷ 14　　**c** 260 ÷ 11　　**d** 430 ÷ 13
　　　e 470 ÷ 12　　**f** 190 ÷ 21　　**g** 370 ÷ 16

＊**11** Choose any three digits from 1 to 9.
　　　Make a 2-digit number and a 1-digit number from them.
　　　Multiply your numbers together.
　　　Use the same digits to make other pairs of 2-digit and 1-digit numbers.
　　　Which pair gives the biggest product?
　　　Which pair gives the smallest product?
　　　Try other sets of 3 digits.
　　　What is the biggest product you can make?

＊**12** Find ways of filling in the box and the circle. □ × ○ = 216.

Review 1
a 3 × 4 × 5　　　**b** 5 × 6 × 4　　　**c** 35 × 2　　　**d** 48 × 2　　　**e** 42 × 5
f 51 × 3　　　　**g** 64 × 6　　　　**h** 53 × 4　　　**i** 62 × 7　　　**j** 46 × 20
k 57 × 60　　　**l** 120 × 30　　　**m** 13 × 11　　**n** 21 × 12　　**o** 24 × 25

Review 2
a 68 ÷ 2　　　**b** 96 ÷ 2　　　**c** 124 ÷ 4　　　**d** 300 ÷ 5　　　**e** 540 ÷ 5
f 168 ÷ 8　　　**g** 144 ÷ 6　　　**h** 960 ÷ 2　　　**i** 1500 ÷ 2　　**j** 7200 ÷ 2

  Puzzle

1 Choose numbers from the box to put in the boxes below.
　　Use each number only once in each question.
　　Find as many different ways as you can.

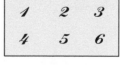

　　a □ × □ − □ = 2　　　　　　**b** (□ + □) × □ = 24
　　c (□ + □) ÷ (□ + □) = 1　　**d**　□ □
　　　　　　　　　　　　　　　　　　　× ___□
　　　　　　　　　　　　　　　　　　　 7 2

2 Put one of the numbers 1, 2, 4, 6, 8 and 12 in each circle.
　　Use each number only once. The three numbers along
　　each side of the triangle must multiply together to give 48.

*3 What digits do ▲, ■ and ★ stand for?

a
$$\begin{array}{r} ▲2 \\ \times\ \ 3 \\ \hline 18▲ \end{array}$$

b
$$\begin{array}{r} ■■ \\ \times\ \ 4 \\ \hline 396 \end{array}$$

c
$$\begin{array}{r} 2★ \\ \times\ ★ \\ \hline 224 \end{array}$$

*4 What values might A, B and C have?

a
$$\begin{array}{r} AB \\ \times\ B \\ \hline BC \end{array}$$

b
$$\begin{array}{r} AB \\ \times\ C \\ \hline CB \end{array}$$

*5 Make up some more sums in which the numbers are replaced by letters. Give them to someone else to solve.

Multiplying and dividing decimals

Here are some ways of **multiplying and dividing decimals mentally**.

Example $4·1 \times 50$

$50 = 100 \div 2$

Partitioning	Using Factors	Multiply then divide
$4·1 \times 50 = 4 \times 50 + 0·1 \times 50$ $= 200 + 5$ $= \textbf{205}$	$4·1 \times 50 = 4·1 \times 10 \times 5$ $4·1 \times 10 = 41$ $41 \times 5 = 40 \times 5 + 5$ $= 200 + 5$ $= 205$ $4·1 \times 50 = \textbf{205}$	$4·1 \times 50$ $4·1 \times 100 = 410$ $410 \div 2 = \textbf{205}$

Sometimes we can double one number and halve the other to make the calculation easier.

Example $3·5 \times 12 = 7 \times 6$
$= 42$

Double 3·5 is 7 and half of 12 is 6.

We can use **place value** to help us multiply.

Worked Example
a $2·1 \times 40$ b $\square \times 0·2 = 10$

Answer
a $2·1 \times 40 = 21 \times 4$
    $= \textbf{84}$

b We know $5 \times 2 = 10$.
    So $\textbf{50} \times 0·2 = 10$.

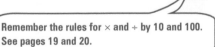

Remember the rules for $\times$ and $\div$ by 10 and 100.
See pages 19 and 20.

Worked Example
a $2·4 \div 6$ b $3·4 \div 2$

Answer
a We know $24 \div 6 = 4$
    So $2·4 \div 6 = \textbf{0·4}$.

b We know $34 \div 2 = 17$
    So $3·4 \div 2 = \textbf{1·7}$

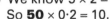

Exercise 4 **This exercise is to be done mentally.**

1 a 1·2 × 2   b 3·1 × 3   c 3·2 × 4   d 5·3 × 2
   e 2·6 × 2   f 4·3 × 3   g 2·5 × 3   h 4·5 × 4
   i 7·5 × 6   j 4·7 × 2   k 5·3 × 5   l 6·3 × 8
   m 4·9 × 7   n 2·3 × 9

2 a 0·2 × 8   b 0·3 × 4   c 6 × 0·5   d 3 × 0·2
   e 4 × 0.5   f 0·3 × 6   g 8 × 0·4   h 0·04 × 8
   i 0·02 × 6   j 0·05 × 3   k 6 × 0·07   l 9 × 0·02
   m 8 × 0.04   n 6 × 0·05

3 What goes in the box?
   a ☐ × 0·2 = 1   b ☐ × 0·2 = 10   c ☐ × 0·4 = 2   d ☐ × 0·3 = 1·8
   e 5 × ☐ = 0·5   f 3 × ☐ = 1·2   g 5 × ☐ = 4

4 a 1·2 × 20   b 3·1 × 20   c 5·3 × 40   d 4·2 × 30
   e 5·1 × 50   f 4·5 × 20   g 2·6 × 30   h 3·7 × 60
   i 4·5 × 60

5 a 4·1 × 11   b 3·2 × 12   c 8·4 × 11   d 3·2 × 13
   e 5·2 × 14   f 7·5 × 12   g 3·5 × 12

You may need to use jottings.

6 a 2·4 ÷ 4   b 3·6 ÷ 6   c 2·1 ÷ 7   d 1·8 ÷ 3
   e 2·5 ÷ 5   f 3·2 ÷ 8   g 5·6 ÷ 7   h 2·7 ÷ 9
   i 4·5 ÷ 5   j 4·8 ÷ 6   k 8·1 ÷ 9   l 3·6 ÷ 2
   m 8·4 ÷ 2   n 5·4 ÷ 2

7 3 × 2·6 = 7·8
   What does 7·8 ÷ 3 equal?

*8 ☐ × ◯ = 3·6
   Find ways of filling in the box and the circle.

T | **Review**

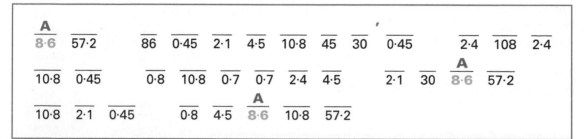

A												
8·6	57·2	86	0·45	2·1	4·5	10·8	45	30	0·45	2·4	108	2·4

								A	
10·8	0·45	0·8	10·8	0·7	0·7	2·4	4·5	2·1	30	8·6	57·2

				A		
10·8	2·1	0·45	0·8	4·5	8·6	10·8	57·2

Use a copy of this box. Write the letter beside each question above its answer in the box.

A 4·3 × 2 = 8·6   I 1·8 × 6   T 0·3 × 7   S 9 × 0·05   O 4·3 × 20
Y 3·6 × 30   N 5·2 × 11   H 2·5 × 12   C 0·9 × 50   E 4·8 ÷ 2
R 1·5 × 3   G 5·6 ÷ 8   B 7·2 ÷ 9

Solving problems mentally

You must decide whether to add, subtract, multiply or divide to solve these problems.

Look for clues in the wording.

Worked Example

Lucy's netball team raised £86 with a cake stall and £194 selling pizza. How much did they raise in total?

Answer

We must add £86 and £194.
One way to do this is $\quad$ £86 + £194 = £86 + £200 − £6
$$= £286 − £6$$
$$= £280$$

Lucy's team raised **£280** in total.

Exercise 5

This exercise is to be done mentally. You may use jottings.

1 Tia scored 68 in her first game on the computer and 182 in her second. What was her total score?

2 There are 104 pupils on a school trip. 57 are boys.
How many are girls?

3 Joel has saved £262. He buys a CD player for £89 and a CD for £8. How much does he have left?

4 At a bird sanctuary the keeper weighs injured birds before and after they feed to see how much they eat. Find the amount these birds ate at feeding time.
  a **Seamus** before 590 g after 647 g
  b **Pecker** before 389 g after 417 g
  c **Tramp** before 1864 g after 1958 g
  d **Rat** before 1479 g after 1601 g

5 The tallest ferris wheel in the world has 60 arms. Each arm seats 8 people.
How many people does the ferris wheel seat in total?

6 The Quarayak Glacier in Greenland moves about 24 metres per day.
How far does the glacier move in
  a 4 days $\quad$ b 1 week $\quad$ c 12 hours?

Link to Geography.

7 There are 52 cards in a pack.
How many cards will there be in
  a 9 packs $\quad$ b 50 packs?

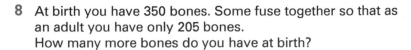

8 At birth you have 350 bones. Some fuse together so that as an adult you have only 205 bones.
How many more bones do you have at birth?

9 Brad ran 6·5 km one day and 3·8 km the next day.
How far did he run altogether?

10 Lucy had 4·6 kg of chicken pieces.
She used 2·8 kg for a party.
How many kilograms did she have left?

11 A pelican eats about 14·7 kg of fish each week.
   a About how much does it eat each day?
   b About how much fish would it eat in four days?

12 a Naim bought five rolls.
   How much did this cost?
   b Ralph bought two packets of crisps and a pie.
   How much did this cost?
   c Penny bought a pie and some crisps.
   How much did this cost?
   d How many drinks can you buy for £5.50?
   e Annie had £2 to spend on lunch.
   What could she buy?
   Give at least two different answers.

Roll	£0.90
Pie	£1.50
Drink	£0.55
Crisps	£0.45

13 Melissa had a piece of rope 3·5 m long. She cut it into five equal pieces to make the steps of a rope ladder.
How long was each piece?

14 Beth is making costumes for the school play.
   a A hat for a guard needs 52 cm of material.
   How many centimetres are needed for four guards?
   b A costume for a soldier needs 3·6 m of material.
   How many metres are needed for six soldiers?
   c Beth bought 325 cm of lace for five maid's dresses.
   How much can she use for each dress?
   d Beth also bought 3·6 m of material for four aprons for
   ladies in waiting.
   How much can she use for each apron?

Review 1 Sue had £215. She bought some speakers for £87 and a cabinet for £39.
How much has she got left?

Review 2 A bag was filled with 3·5 kg of apples and 2·7 kg of oranges.
How many kilograms of fruit were in the bag?

Review 3 A turtle can move 26 m each minute.
How far could the turtle move in
a 2 minutes   **b** 5 minutes   **c** 8 minutes   **d** 1 hour?

Review 4 Chris stacked 96 bricks into 4 equal piles. How many were in each pile?

Review 5
a How much would 5 kg of apples cost?
b How much would 3 kg of oranges cost?
c How many kg of oranges could you buy for £18·50?

Order of operations

$8 + 5 \times 2 \qquad 4 + 9 \times 2 - 3 \qquad 19 - 8 \div 2$

When there is more than one operation in a calculation we do the **multiplication and division before** the **addition and subtraction**.

Examples

$7 + 3 \times 2 = 7 + 6$
$\qquad\qquad = 13$

$4 + 6 \times 5 - 3 = 4 + 30 - 3$
$\qquad\qquad\qquad = 31$

× first

$17 - 6 \div 2 = 17 - 3$
$\qquad\qquad = 14$

$20 \div 4 \div 2 = 5 \div 2$
$\qquad\qquad = 2 \cdot 5$

÷ first

Work from left to right, doing × and ÷ first then work from left to right doing + and −.

Exercise 6

1
a $4 + 6 - 3$	**b** $4 + 2 \times 3$	**c** $2 + 3 \times 2$	**d** $5 - 2 + 4$
e $6 \times 3 - 1$	**f** $7 \times 2 - 4$	**g** $8 \times 3 + 3$	**h** $6 \times 4 - 7$
i $2 \times 3 + 5$	**j** $11 - 3 \times 2$	**k** $16 - 2 \times 3$	**l** $20 - 5 \times 3$
m $24 \div 6 \div 2$	**n** $4 \times 2 + 3 \times 4$	**o** $3 \times 5 + 2 \times 4$	**p** $8 \times 6 + 7 \times 2$
q $12 \div 3 \div 2$	**r** $12 + 8 \div 4$	**s** $11 + 16 \div 4$	**t** $5 \times 3 - 2 \times 4$
u $7 \times 9 - 3 \times 10$	**v** $16 \div 2 + 3$	**w** $20 \div 4 - 1$	**x** $5 + 2 \times 3 - 1$

Review

a $5 \times 2 - 3$	**b** $6 \times 7 - 5$	**c** $3 + 4 \times 2$	**d** $12 - 3 \times 3$
e $16 - 5 \times 2$	**f** $5 \times 3 + 2 \times 4$	**g** $16 \div 2 + 4$	**h** $20 - 8 \div 4$

If there are **brackets** in a calculation, work these out **first**.

A number immediately before a bracket means we multiply what is inside the bracket by this number.

Examples

$5(2 + 7) = 5 \times (2 + 7)$
$\qquad\qquad = 5 \times 9$
$\qquad\qquad = 45$

$15 \div (3 + 2) = 15 \div 5$
$\qquad\qquad\qquad = 3$

In $\frac{7+3}{2}$ the horizontal line acts as a bracket.

So $\quad \frac{7+3}{2} = \frac{(7+3)}{2}$
$\qquad\qquad = \frac{10}{2}$
$\qquad\qquad = 5$

Work out the numerator first, then divide.

Examples

$\frac{18-3}{5} = \frac{(18-3)}{5}$
$\qquad\quad = \frac{15}{5}$
$\qquad\quad = 3$

$\frac{16+4}{8-3} = \frac{(16+4)}{(8-3)}$
$\qquad\quad = \frac{20}{5}$
$\qquad\quad = 4$

Work out both the numerator and denominator, then divide.

Exercise 7

1
a $3(7 + 2)$
b $5(6 - 4)$
c $7(8 - 3)$
d $5(6 - 1)$
e $4(8 - 3)$
f $6(5 + 6)$
g $(15 - 3) \times 2$
h $(6 - 3) \times 4$
i $(14 - 6) \times 3$
j $5(20 - 8)$
k $(8 + 6) \times 2$
l $13(20 - 19)$
m $5 \times (20 + 5 - 16)$
n $(18 + 3 - 14) \times 2$
o $2(10 - 2 \times 3)$
p $3(8 - 2 \times 3)$
q $21 - 3(4 - 1)$
r $2(4 + 3) - 7$
s $18 - 2(3 \times 2)$
t $14 - 3(2 \times 2)$
u $(10 - 3) \times (12 - 4)$
v $(20 - 5) \div (8 - 3)$
w $(18 - 6) \div (9 - 6)$

2
a $\frac{16 + 4}{2}$
b $\frac{8 + 12}{4}$
c $\frac{5 + 9}{7}$
d $\frac{24 - 3}{7}$
e $\frac{15 - 9}{6}$
f $\frac{24}{3 \times 8}$
g $\frac{100}{2 \times 5}$
h $\frac{60}{5 \times 6}$
i $\frac{81}{3 \times 3}$
j $\frac{120}{2 \times 15}$
k $\frac{100}{5 \times 5}$
l $\frac{14}{6 - 4}$
m $\frac{7 + 9}{10 - 6}$
n $\frac{25 + 5}{7 - 2}$

3
a $3 \times 4 - 2(5 - 1)$
b $5 \times 4 - 3(6 - 3)$
c $9 \times 8 + 3(2 \times 2)$
d $5 \times 4 + 6(3 \times 0)$
e $6 \times 2 + 4(3 + 0)$
f $5 \times 8 - 3(8 - 2 \times 2)$

*4 $2 + 3 \times 6 = 20$ $2(6 - 3) = 6$
What other answers can you make using 2, 3 and 6.
You may use $+$, $-$, $\times$, $\div$ and brackets.
What is the biggest answer you can make?

*5 Choose any four different numbers from 1 to 9.
a What answers can you make using your four numbers together with $+$, $-$, $\times$, $\div$ or brackets?
b What is the biggest answer you can make?
c What is the second biggest?

T

Review 1

$\overline{12}$	$\overline{1}$	$\overline{12}$	$\overline{5}$	$\overline{05}$		$\overline{6}$	$\overline{4}$	$\overline{10}$	$\overline{5}$	$\overline{7}$	
$\overline{36}$	$\overline{1}$	$\overline{81}$		$\overline{42}$	$\overline{21}$	$\overline{35}$	$\overline{81}$	$\overline{32}$		$\overline{14}$	$\overline{81}$ $\overline{5}$ $\overline{28}$
$\overline{21}$	$\overline{10}$		$\overline{16}$	$\overline{3}$	$\overline{21}$	$\overline{10}$	$\overline{1}$				

Use a copy of this box.
Write the letter that is beside each question above its answer.

I $3(5 + 2) = 21$
W $4(5 + 4)$
F $6(9 - 2)$
U $7(8 - 6)$
D $(8 + 6) \times 2$
T $(6 - 2) \times 8$
S $9(12 - 4 + 1)$
R $5(1 + 2 \times 3)$
C $4(12 - 2 \times 4)$
H $19 - 2(4 \times 2)$
M $\frac{8 + 4}{2}$
Y $\frac{20 + 8}{4}$
O $\frac{32 - 4}{7}$
N $\frac{80}{2 \times 4}$
E $\frac{26 + 4}{2 \times 3}$
A $(20 - 16) \div (9 - 5)$
P $36 - 2(3 \times 4)$

*Review 2 Use the digits 2, 3 and 7 together with $+$, $-$, $\times$, $\div$ or brackets.
What is the second largest answer you can make?

71

We work out squares (indices) **before** multiplication and division but after brackets.

The order in which we do operations is

> To help you remember this you could use the word BIDMAS.

Brackets
Squares (**I**ndices)
Division and **M**ultiplication
Addition and **S**ubtraction.

> In $1^2, 2^2, 3^2, 4^2$ ... the little 2 is called an index. The plural of index is indices.

Example

$$3 \times 4^2 = 3 \times 16$$
$$= 3 \times (10 + 6)$$
$$= 3 \times 10 + 3 \times 6$$
$$= 30 + 18$$
$$= 48$$

$$\frac{3^2 - 1}{2^2} = \frac{(3^2 - 1)}{2^2}$$
$$= \frac{9 - 1}{4}$$
$$= \frac{8}{4}$$
$$= 2$$

$$(3^2 - 2^2)^2 = (9 - 4)^2$$
$$= 5^2$$
$$= 25$$

Exercise 8

1 a 2×3^2 b $5^2 - 2$ c 4×2^2 d $3(2^2 - 1)$ e $5(3^2 + 1)$

   f $4^2 \times 2 - 3 \times 4$ g $4^2 \div 4 + 2$ h $5^2 - 12 \div 3$ i $3(4^2 - 3 \times 2)$ j $12 + 3 \times 2^2$

   k $(2^2 + 1^2)^2$ l $\frac{5^2 - 1}{10 - 2}$ m $\frac{3 \times 2^2}{2^2 - 1}$ n $\frac{4^2 - 7}{2^2 - 1}$

Review

a 4×2^2 b $4(5^2 - 5)$ c $(1^2 + 3^2)^2$ d $\frac{3^2 + 5}{4 + 3}$ e $\frac{4^2 - 1}{2^2 + 1}$

Puzzle

Choose numbers from the ring to fill in the boxes.
In each question use a number only once.

1 ☐ − ☐ + ☐ = 4   2 ☐ + ☐ − ☐ = 6   3 ☐ × ☐ − ☐ = 7

4 (☐ + ☐) ÷ ☐ = 4   5 (☐ + ☐) ÷ (☐ + ☐) = 3   6 ☐² + ☐ − ☐ = 18

Calculation Full House – a game for a group

To prepare
On a piece of paper each player writes down the numbers 1 to 20.
Beside each number write a calculation that has this number as the answer.

Example

$1 = 3 + 4 + 5 - 9 + 1 - 3$ $3 = 21 - 18$

$2 = 8 \times 3 - 22$ $4 = 14 - 2 \times 5 \ldots$

> These will be used if you are the leader.

To Play

- Choose a leader.
- Everyone except the leader, writes down five numbers from 1 to 20.
- The leader calls out one of the calculations (but not the answer) he or she wrote when preparing for the game.
- If the answer to the calculation is one of the five numbers you have, cross it out.

Example   Pete had the numbers 2, 3, 8, 10 and 18.
The leader called the calculation $7 + 9 - 6$.
The answer is 10. Pete crossed out 10.

- The first student to cross out all five numbers is the winner.
This student is the leader for the next round.

Investigation

Number Chains

1 Begin with any number.

If the number is even, divide it by 2.
If the number is odd, multiply it by 3 and add 1.
Repeat with the new number until you get the digit 1.

Example   If we begin with 20 we get

$$20 \div 2 = 10$$
$$10 \div 2 = 5$$
$$5 \times 3 + 1 = 16$$
$$16 \div 2 = 8$$
$$8 \div 2 = 4$$
$$4 \div 2 = 2$$
$$2 \div 2 = 1$$

The number chain is 10, 5, 16, 8, 4, 2, 1.
There are seven numbers in this chain.

Find all the pairs of numbers from 2 to 20 which have number chains of the same length. For example 3 and 20 have number chains of length 7.

2 Begin with any number less than 40.
Multiply the units digit by 4.
Add the tens digit to this.
Repeat until you get back to the number you started with.

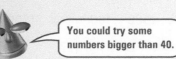

You could try some numbers bigger than 40.

Example $32 \rightarrow 11 \rightarrow 5 \rightarrow 20 \rightarrow 2 \rightarrow 8 \rightarrow 32$

What is the longest number chain you can make this way?

Estimating

Before we do a calculation we always **estimate** the answer.
An estimate is an **approximate answer**.

Example   504 – 98 is **approximately** 500 – 100 = 400.
An estimate of the answer to 504 – 98 is 400.

Worked Example
Which is the best approximation for 30.6 – 19.8?
A 306 – 198   B 3·0 – 1·9   C 31 – 19   D 31 – 20

Answer
30·6 is about 31.
19.8 is about 20.
So **31 – 20** is the best approximation.

Exercise 9   This exercise is to be done mentally.

Which is the best approximation for

1 4·1 – 2·83?
   A 4 – 2   B 41 – 28   C 4 – 3   D 4·0 – 2·8

2 8·94 + 6·07?
   A 9 + 6   B 8 + 6   C 9 + 7   D 8 + 7

3 32 × 68?
   A 30 × 60   B 40 × 70   C 40 × 60   D 30 × 70

4 7·38 × 4·92?
   A 8 × 4   B 7 × 4   C 7 × 5   D 8 × 5

5 1428 – 196?
   A 1500 – 100   B 1500 – 200   C 1400 – 200   D 1400 – 100

Review   8·72 × 3·12?
A 8 × 3   B 9 × 3   C 8 × 4   B 9 × 4

Sometimes there are different ways to find an estimate.

Example   An approximate answer for 512 – 146 can be
500 – 100 = 400 **or** 500 – 150 = 350.

Sometimes we only need a *rough* estimate and sometimes a more *exact* one.

Exercise 10 **This exercise is to be done mentally.**

1 Estimate the answers to these.
  a   804 − 396   b   285 + 716   c   3524 + 973   d   39 × 42
  e   81 × 77   f   396 × 32   g   804 ÷ 9   h   8·18 + 3·81
  i   50·8 − 39·7   j   596 ÷ 61   k   550 × 69

2 Write down two ways to find an approximate answer for these.
  a   815 − 254   b   862 + 304   c   44 × 19   d   9·23 + 8·5
  e   5·78 × 3·45   f   396 × 44   g   145 × 21

Review   Estimate the answers to these.
a   916 + 898   b   872 − 342   c   58 × 22   d   924 ÷ 31
e   4·68 + 3·02   f   17·47 − 5·62   g   4·71 × 21

You will use your estimating skills in lots of other chapters.

Summary of key points

 A   Some ways of **adding and subtracting mentally** are

partioning

counting up

adding too much then taking some away (compensation)

nearly doubles

nearly numbers

subtracting too much then adding back (compensation)

recognising complements

 B   Some ways of **multiplying and dividing mentally** are

partitioning

using factors

doubling one number and halving another

multiplying then dividing

using place value

 C   The order in which we do operations

Remember *BIDMAS* and this will help you.

Brackets

 ⇓

squares (Indices)

 ⇓

Division and **M**ultiplication

⇓

Addition and **S**ubtraction

Work from left to right doing brackets, then left to right doing squares, then left to right doing multiplication and division, then left to right doing addition and subtraction.

Number

Examples   **a** $7 + 20 \div 5 = 7 + 4$ **b** $3 \times 8 + 2 \times 5 = 24 + 10$
$$= 11 \qquad\qquad\qquad\qquad = 34$$

$$\frac{80}{4 \times 5} = \frac{80}{(4 \times 5)} \qquad\qquad \frac{4^2 + 8}{3^2 - 1} = \frac{(16 + 8)}{(4 \times 5)}$$

$$= \frac{80}{20} \qquad\qquad\qquad = \frac{24}{8}$$

$$= 4 \qquad\qquad\qquad\qquad = 3$$

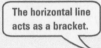
The horizontal line acts as a bracket.

D An **estimate** is an approximate answer.

Example $8 \cdot 16 \times 3 \cdot 98$ is approximately $8 \times 4 = 32$

E Sometimes there are **different ways to find an estimate**.

Example $348 + 197$ is approximately $300 + 200 = 500$ or $350 + 200 = 550$.

Test yourself   **Find the answers mentally. You may use jottings.**

1 **a** $15 + 3 - 11 + 16 - 2$ **b** $100 - 50 + 20$ **A**

2 **a** $96 - 50$ **b** $36 + 14$ **c** $76 - 52$ **d** $83 + 17$ **e** $59 + 58$ **A**
  f $186 + 39$ **g** $486 - 57$ **h** $411 - 379$ **i** $5200 - 384$

3 **a** $0 \cdot 9 + 0 \cdot 8$ **b** $2 \cdot 7 - 0 \cdot 8$ **c** $2 \cdot 9 + 6 \cdot 9$ **d** $5 \cdot 2 - 3 \cdot 5$ **e** $14 \cdot 7 + 8 \cdot 9$ **A**
  f $16 \cdot 3 - 4 \cdot 7$ **g** $1 \cdot 9 + 3 \cdot 2$ **h** $19 \cdot 0 - 5 \cdot 8$

4 This is a magic square.   **A**
  Copy it.
  Finish filling it in.

14		16
	17	
		20

5 **a** $3 \times 5 \times 4$ **b** 65×2 **c** 37×4 **d** 82×5 **e** 150×3 **B**
  f $78 \div 2$ **g** $128 \div 4$ **h** 40×20 **i** 120×30 **j** 82×11
  k 35×12 **l** $390 \div 3$ **m** 36×50 **n** $1850 \div 2$

6 **a** $1 \cdot 3 \times 2$ **b** $4 \cdot 3 \times 5$ **c** $5 \cdot 6 \times 7$ **d** $4 \cdot 8 \div 2$ **e** $6 \cdot 3 \div 9$ **B**
  f $0 \cdot 3 \times 8$ **g** $9 \times 0 \cdot 6$ **h** $5 \times 0 \cdot 04$ **i** $1 \cdot 8 \div 3$ **j** $4 \cdot 2 \div 6$
  k $5 \cdot 1 \times 11$ **l** $4 \cdot 3 \times 30$ **m** $5 \cdot 3 \times 50$ *n** $4 \cdot 7 \times 13$

7 **a** Find the sum of $8 \cdot 6$ and $2 \cdot 4$.   **A** **B**
  b Find the difference between 784 and 396.
  c Find the product of 5 and 97.

8 Find the answers to these mentally.

   a Melanie's mother bought a suit for £386.
      She paid a deposit of £155.
      How much more does she have to pay?

   b Dann bought six batteries. Each battery cost £2·60.
      How much did the batteries cost altogether?

   c Seven chocolate bars cost £2·80.
      How much does one chocolate bar cost?

9 **a** $5 \times 3 - 4$ **b** $6 + 2 \times 4$ **c** $15 - 3 \times 4$ **d** $15 \times 2 - 3 \times 4$

    e $5 + 12 \div 2$ **f** $6(8 + 2)$ **g** $\frac{9+6}{5}$ **h** $\frac{24}{3 \times 2}$

    i 2×5^2 **j** $3(4^2 - 8)$ **k** $\frac{3^2 - 1}{2}$

10 Which is the **best** approximation for

    a $60·8 - 39·6$?   **A** $60 - 39$ **B** $608 - 396$ **C** $61 - 40$ **D** $6·0 - 3·9$

    b $8·14 \times 4·91$?   **A** 9×4 **B** 8×5 **C** 9×5 **D** 8×4

11 Write down two ways to find an approximate answer for these.

    a $451 + 187$ **b** $3·47 \times 4·96$ **c** $962 - 359$

12 Find numbers from the orange box to make this true.
In each question use each number only **once**.

    a $\square + \square - \square = 10$

    b $\square\square \times \square = 96$

   * **c** $\square\square + \square\square = \square\square\square$

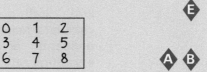

0	1	2
3	4	5
6	7	8

You need to know

✓ adding and subtracting 2 and 3-digit whole numbers   page 4
✓ adding and subtracting decimals with one decimal place   page 4
✓ multiplying 2 and 3-digit numbers by a single digit   page 4
✓ multiplying 2-digit by 2-digit numbers   page 4
✓ dividing 2 and 3-digit numbers by a single digit   page 4
✓ using the calculator   page 5

Key vocabulary

add, adding, approximately equal to, brackets, calculate, calculator: clear, display, enter, memory, change, difference, divide, dividend, division, divisor, equivalent calculation, estimate, exact, exchange rate, inverse, key, multiplication, multiply, operation, product, quotient, remainder, rough, subtract, subtracting, sum, total

Kaprekar's Number

Choose any 4-digit number such as 3259.

Step 1 Write the digits from largest to smallest   9532
          Write the digits from smallest to largest   −2359
                             Subtract   7173

Step 2 Take the answer you get and keep repeating Step 1.

7731	6543	8730	8532
−1377	−3456	−0378	−2358
6354	3087	8352	6174

6174 is **Kaprekar's number**.

What happens if you repeat Step 1 with Kaprekar's number?

Do you get Kaprekar's number if you begin with *any* 4-digit number?

Adding and subtracting whole numbers

Rosie loves stamps. She has collected 7654.
She bought her brother's collection of 278 stamps.
She wanted to know how many she had then.

We add and subtract whole numbers by lining up the units column.

```
  7654           5 14 1
+  278         7̶6̶5̶4̶
------         -  278
  7932         ------
   11            7376
```

Rosie has 7932 stamps now. She had 7376 more stamps than her brother before he sold her his collection.

Exercise 1

1 a 176 + 94   **b** 283 – 57   **c** 395 + 39   **d** 384 – 28   **e** 817 + 594
  f 588 – 197   **g** 982 + 97   **h** 342 – 169   **i** 1797 + 87   **j** 402 – 53
  k 7198 + 3265 **l** 8325 – 1214 **m** 8526 + 47   **n** 5107 – 56   **o** 8280 + 469
  p 6217 – 364 **q** 4386 – 59   **r** 4879 + 96   **s** 5003 – 586

2 a 84 + 37 + 586   **b** 47 + 392 + 487   **c** 369 + 52 + 1043 + 2
  d 4327 + 964 + 53 + 270 **e** 53 + 862 + 1957 + 2841

T

3 Use a copy of this.
The number in any square is found by adding the numbers in the two circles on either side of the square.

Find the missing numbers.

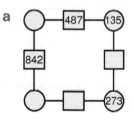

4 Between 1896 and 1996 Britain got 177 Gold, 233 Silver and 225 Bronze medals at the Olympics.
How many medals did they win altogether in that time?

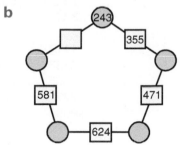
How many medals has Britain won now in total?

5 This table shows the number of trays of apples and pears packed at two orchards one weekend.

	Fruit Fields		Sunvalley	
	Apples	Pears	Apples	Pears
Sat	232	145	189	95
Sun	203	159	146	265

  a How many trays of apples did Fruit Fields pack on Sunday?
  b How many more trays of apples than pears did Sunvalley pack on Saturday?
  c On which day did Fruit Fields pack the most trays?
  d Which orchard packed more trays on Sunday and by how many?
  e A tray is 34 cm by 65 cm. What is the perimeter of a tray?

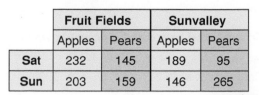
Remember the perimeter is the distance around the outside edge of the tray.

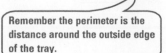

6 156 and 432 are 3-digit numbers that add to 588.
Use each of the digits 1, 2, 3, 4, 5, 6 once to make two other
3-digit numbers that add to 588.
How many ways are there of doing this?

$$\begin{array}{r} 156 \\ +432 \\ \hline 588 \end{array}$$

7 ☐☐☐ + ☐☐ = 213   ☐☐☐ – 123 = ☐☐☐
Find at least three ways to fill in these boxes. Put one digit in each box.

*8 Use all the digits 9, 8, 7, 6, 5, 4, 3, 2 to write two 4-digit numbers so that
   a The sum is as large as possible.   **b** The sum is as small as possible.
   c The difference between them is as small as possible.
   d The sum is exactly halfway between 16 585 and 16 605.
   Is there more than one answer to each of these?

> To find the sum, add the numbers.
> To find the difference, subtract.

Review 1
a 672 + 89   **b** 438 + 269   **c** 842 – 67   **d** 1857 + 396
e 2400 – 386   **f** 3592 + 89   **g** 6500 – 78

Review 2
Put the digits 1 to 8 in the boxes to make each true. Use each digit only once.
a ☐☐☐☐ + ☐☐☐☐ = 16 074
b ☐☐☐☐ – ☐☐☐☐ = 3367

? Puzzle

1
$$\begin{array}{r} \star2\star2 \\ -2\star2\star \\ \hline 3\star3\star \end{array}$$
Replace the ★ with the same digit to make this subtraction correct.

2
$$\begin{array}{r} 1111 \\ 3333 \\ 5555 \\ 7777 \\ +9999 \\ \hline \end{array}$$
Replace ten of the digits with 0 so that the sum comes to 1111.

Adding and subtracting decimals

To **add and subtract decimals** we line up the decimal points.

Worked Example
Calculate   **a** 862·3 + 9·87   **b** 283·6 – 27·57

Answer
a
$$\begin{array}{r} 862\cdot3 \\ +\ \ \ 9\cdot87 \\ \hline 872\cdot17 \\ \tiny{1\ 1} \end{array}$$

> The decimal points are lined up.

b
$$\begin{array}{r} {\scriptstyle 7\ 1\ \ 5\ 1} \\ 28\cancel{3}\cdot\cancel{6}0 \\ -\ 27\cdot57 \\ \hline 256\cdot03 \end{array}$$

> Fill the gaps with zeros.

Exercise 2

1   a   16·8 + 32·5   b   117·3 + 2·6   c   2·34 + 8·9   d   24·5 + 18·84
      e   4·37 + 18·5   f   0·86 + 2·3   g   52·7 + 0·79   h   343·6 + 52·87
      i   55 + 643·7   j   5371·8 + 98·45   k   72·03 + 1526·1

2   a   18·6 – 3·4   b   17·51 – 2·7   c   24·36 – 18·9   d   2·41 – 0·54
      e   471·7 – 50·26   f   304·6 – 5·79   g   24·9 – 6·83   h   521·3 – 0·69
      i   5027·3 – 83·65   j   2040·6 – 79·09   k   4086·5 – 126·57   l   185·3 – 0·64
      m   184 – 79·2   n   28 – 6·32   o   257 – 152·84

3   a   5·3 + 27·96 + 4·7   b   186·04 + 3·7 + 86·94   c   56·56 + 412·2 + 6·07
      d   83·75 + 1842·9 + 3·57   e   864·39 + 2·7 + 89·07

4   a   Find the sum of 83·3 and 5·27.
      b   Find the difference between 298·03 and 36·4
      c   Find the sum of 97·23 and 81·6.
      d   The difference between two numbers is 3·45.
         What might the numbers be?

5   Anthony wanted to fill in his car log book.
     He made four short journeys in his car. These were 6·8 km, 10·7 km, 9 km and 4·4 km.
     The reading on the speedometer at the end of these trips was 69 435·1 km.
     What was the reading before the trips?

6   Here is a table of record throws.

Frisbee	190·07 m	Boomerang	121 m
Wellington	52·73 m	**Cricket ball**	128·6 m
Slingshot	437·13 m	**Haggis**	55·11 m
Rolling pin	53·4 m	**Brick**	44·54 m

Find out some more records for throwing.

     i   Find the difference between the throws of
         a   a haggis and a brick   b   a frisbee and a cricket ball
         c   a rolling pin and a slingshot   d   a boomerang and a wellington.
     ii   Find the difference between the longest and shortest throws in the table.

*7   [3] [6] [7] [1] [2] [7] [9] [5] [.] [.]

     Use these number cards to make two 4-digit decimal numbers so that
     a   the sum of the numbers is as large as possible
     b   the difference between the two numbers is as small as possible.
     Each number must have at least one digit before and after the decimal point.

Review 1
a   13·4 + 16·9   b   14·5 – 3·9   c   18·7 + 3·94   d   5·6 – 5·27
e   12·04 – 6·32   f   5·7 + 3·94   g   8·6 – 2·54   h   15·01 – 3·8
i   24 – 3·59   j   56·42 + 118·3   k   3 – 1·62   l   178·32 + 56·7
m   179·2 – 99·04   n   875·87 – 68·3   o   68·56 + 412·3 + 6·09

Review 2
a I bought these goods. How much did they cost in total?
b Three months later I sold them both for a total of £25·50.
   How much did I lose?

 Puzzle

$36 + 12 = 372$ $136 - 4 = 96$ $281 + 12 - 145 = 31$

Where should the decimal points be placed to make these true?
Is there more than one answer?

  Eight Digits – a game for a group

You will need a copy of this diagram.

To play
● Choose a leader.
● The leader calls out eight digits one
   at a time.

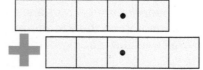

● As each digit is called, the other students write it in one of the eight squares.
● Once a digit is written down it cannot be changed.
● When all eight digits have been called, the students add their two numbers
   together.
● The student with the **highest** total is the leader for the next round.

Note: You could subtract the numbers instead.

Multiplying whole numbers

Discussion

● Sandy worked out 7×18 like this. $18 + 18 + 18 + 18 + 18 + 18 + 18 = 126$
   Paul worked out 7×18 like this.
$$\begin{array}{r} 18 \\ \times\ 7 \\ \hline 126 \end{array}$$

Which way is quicker? **Discuss**.
Who is less likely to make a mistake? **Discuss**.

● What is the answer to 3×1, 447×1, 8960×1, $52·4 \times 1$?
   Does this work for all numbers? **Discuss**.

● What is the answer to 7×0, 45×0, 874×0, $66·2 \times 0$?
   Does this work for all numbers? **Discuss**.

Always **estimate** the answer first.

Worked Example
Calculate 423 × 76.
423 × 76 is **approximately** equal to 400 × 80 = 32 000.

```
   423
 ×  76
 29 610  ← 423 × 70
  2 538  ← 423 × 6
 32 148
   11
```

or

76 is 70 + 6 so we multiply by 70 then by 6.

	400	20	3
70	28 000	1400	210
6	2400	120	18

Answer 28 000 + 1400 + 210 + 2400
+ 120 + 18 = **32 148**

This is probably correct because it is close to our estimate.

Exercise 3

Estimate the answers, then calculate.

1 a 89 × 6   **b** 324 × 4   **c** 652 × 7   **d** 708 × 9

2 a 29 × 82   **b** 97 × 33   **c** 986 × 32   **d** 562 × 28
   e 385 × 71   **f** 783 × 85   **g** 846 × 23   **h** 792 × 43
   i 205 × 64   **j** 309 × 54   **k** 630 × 19   **l** 840 × 56
   m 729 × 27   **n** 461 × 82   **o** 587 × 39

3 Use a copy of this crossnumber.
   Fill it in.

Across
   1. 341 × 38
   3. 235 × 21
   5. 133 × 18
   7. 769 × 89
   8. 15 × 11
   9. 11 × 23
   12. 33 × 19
   14. 99 × 27
   15. 266 × 34
   16. 43 × 22
   17. 103 × 14

Down
   1. 59 × 19
   2. 211 × 45
   3. 128 × 35
   4. 447 × 86
   6. 118 × 31
   10. 131 × 40
   11. 28 × 13
   12. 102 × 62
   13. 214 × 33
   15. 12 × 8

4 There are 687 earth days in a year on Mars.
   How many earth days are there in these Mars years
   a 28   **b** 16   **c** 34?

How old are you in Mars years?

5 Tim's class is raising money for a trip to a wildlife park.
   They sold 384 cups of soup at 35p per cup.
   How much did they raise selling soup?

6 A seacat holds 435 people.
Last week every trip it made was full.

Day	Number of Trips
Mon	12
Tue	16
Wed	18
Thur	24
Fri	29

  a How many people did it carry on each day?
  b How many people did it carry in total from Monday to Friday?
  c The wharf the seacat docks at is a rectangle, 118 m by 15 m. What is the area of this wharf?

*7 Make up word problems for these calculations.
  a $36 \times 41 = 1476$ **b** $862 \times 21 = 18\,102$ **c** $15 \times 21 \times 33 = 10\,395$

*8 Find two consecutive numbers with a product of
  a 552   **b** 1406.

To find the *product* we multiply numbers.

*9 Find three consecutive numbers with a product of 19 656.

T

Review 1

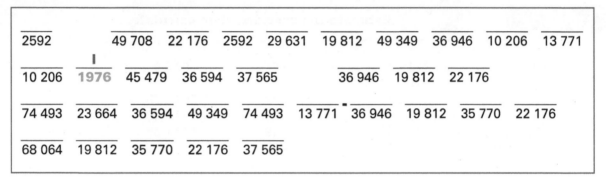

2592		49 708	22 176	2592	29 631	19 812	49 349	36 946	10 206	13 771

10 206	1976	45 479	36 594	37 565		36 946	19 812	22 176

74 493	23 664	36 594	49 349	74 493	13 771	36 946	19 812	35 770	22 176

68 064	19 812	35 770	22 176	37 565

Use a copy of the box above. Write the letter beside each question above its answer.

I $52 \times 38 = 1976$ **A** 96×27 **O** 381×52 **U** 730×49 **N** 809×61 **G** 357×83
E 642×57 **W** 986×24 **V** 511×89 **S** 683×55 **H** 709×96 **T** 837×89
Y 47×293 **L** 21×486 **R** 72×308 **D** 68×731 **F** 58×637

Review 2 In a computer game, Sophie had 382 points.
She got the 'Jackpot' which means her points are multiplied by 15.
How many points did she have then?

* Review 3 Find two consecutive numbers with a product of 756.

Puzzle

Put the digits 3, 4, 5, 6 and 7 in the boxes to make each true.
Use each digit only once.

1 ☐☐☐ × ☐☐ = 25 228   **2** ☐☐☐ × ☐☐ = 24 310

Investigation

Digit Reversals

24 × 63 = 1512   42 × 36 = 1512

same answer

42 is 24 with the digits reversed
36 is 63 with the digits reversed

This doesn't work for all multiplications.
What other pairs of two-digit numbers does it work for? **Investigate**.

Multiplying with decimals

Discussion

● Christine worked out 84 ÷ 7 like this, 84 − 7 − 7 − 7 − 7 − 7 − 7 − 7 − 7 − 7 − 7 − 7 − 7 = 0

84 ÷ 7 = 12

 There are 12 sevens.

Peter worked out 84 ÷ 7 like this.

```
     7)84
      −70    7 × 10
      ‾‾‾
       14              Answer 12
       14    7 × 2
      ‾‾‾
        0
```

Which way is quicker? Who is less likely to make a mistake? **Discuss**.

● Diana worked out 58 ÷ 0 on the calculator.
This is what she got on her screen.
Why? **Discuss**.

ERROR

Worked Example
3·74 × 8

Answer
3·74 × 8 is approximately 4 × 8 = 32.
Two ways of finding the exact answer are shown.

 Always estimate the answer first.

3·74 × 8			
×	**3**	**0·7**	**0·04**
8	24	5·6	0·32
Answer 24 + 5·6 + 0·32 = **29·92**			

3·74 × 8 is equivalent
to 374 × 8 ÷ 100.

```
   374
 ×   8
 ‾‾‾‾‾
  2992      2992 ÷ 100 = 29·92
   5 3
```

 Use the estimate to check that your answer is about the right size.

Number

Worked Example
562·7 × 4

Answer

562·7 × 4 is approximately 600 × 4 = 2400
Two ways of finding the answer are shown.

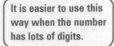

It is easier to use this way when the number has lots of digits.

562·7 × 4					562·7 × 4 is equivalent to 5627 × 4 ÷ 10

×	**500**	**60**	**2**	**0·7**
4	2000	240	8	2·8

Answer 2000 + 240 + 8 + 2·8 = **2250·8**

5627
×   4
———
22 508   22 508 ÷ 10 = **2250·8**
2 1 2

Exercise 4

Estimate the answers, then calculate.

1
a 3·42 × 3	b 4·61 × 2	c 8·17 × 3	d 3·76 × 4	e 5·93 × 6
f 12·7 × 3	g 16·4 × 4	h 21·3 × 5	i 2·46 × 7	j 25·62 × 8
k 19·34 × 5	l 17·21 × 6	m 121·8 × 3	n 115·6 × 4	o 135·2 × 6
p 83·61 × 4	q 152·7 × 9	r 309·5 × 4	s 443·5 × 8	

2 A pack of bath salts costs £2·36.
How much do eight packs cost?

3 Johnny bought 4 m² of wood at £3·59 per square metre.
How much did it cost?

4 Shabir measured one of her paces as 78·3 cm.
Her bedroom is six paces long and four paces wide.
What is the length and width of Shabir's bedroom?

5 How much juice is in five bottles each with 1.05 ℓ?

6 A table tennis ball travels 47·22 m in one second.
A squash ball travels 64·64 m in one second.
How much further would a squash ball travel in 5 seconds than a table tennis ball?

7 a A chair is 42·63 cm wide. Nine of these chairs are joined side by side. How long is the row of nine chairs?
b How long is a row of 27 chairs if they are put in three sets of 9 chairs with a space of 0·89 m between each set of 9 chairs?

*8 ☐·☐ × ☐ = ☐☐·☐
Find four different ways to fill in the boxes.

Review 1
a 5·61 × 4	b 14·6 × 5	c 1·64 × 7	d 24·69 × 9
e 11·35 × 8	f 187·3 × 4	g 196·7 × 3	h 352·4 × 9

Review 2
a Pat bought 6 ℓ of this juice.
How much did it cost?

b Peter bought 9 ℓ of this juice.
How much did it cost?

Fill your own Juice £1·84 per litre

Dividing with whole numbers

Worked Example
3269 ÷ 7

We round to the nearest number you can divide easily by 7.

Answer
3269 ÷ 7 is approximately 3500 ÷ 7 = 500

```
7 ) 3269
   -2800   7 × 400   because 7 × 4 = 28
     469
   - 420   7 × 60    because 7 × 6 = 42
      49
   -  49   7 × 7
       0
```

or

```
        467
        4 4
7 ) 3269
```

In 3269 ÷ 7 = 467
7 is the *divisor*.
3269 is the *dividend*.
467 is the *quotient*.

Answer **467**

Worked Example
Tim buys a 756 cm length of wood to make 21 shelf ends.
The shelf ends are to be as large as possible.
How much wood does Tim use for each?

Answer
To find the length of wood used for each shelf end we need to **divide** 756 by 21.

756 ÷ 21 is approximately 800 ÷ 20 = 40.

```
21 ) 756
    -630   21 × 30   because 21 × 3 = 63
     126
     126   21 × 6
       0
```

Answer **36**

Tim uses **36 cm** for each shelf end.

Exercise 5   **Estimate the answers, then calculate.**

1 a 312 ÷ 4   b 295 ÷ 5   c 234 ÷ 6   d 486 ÷ 9
   e 1715 ÷ 7   f 1707 ÷ 3   g 2008 ÷ 8   h 1832 ÷ 4
   i 3378 ÷ 6   j 5094 ÷ 9   k 3724 ÷ 7

2 Find the quotient.
   a 156 ÷ 12   b 325 ÷ 25   c 672 ÷ 21   d 624 ÷ 26
   e 418 ÷ 22   f 722 ÷ 38   g 648 ÷ 36   h 832 ÷ 52
   i 432 ÷ 18   j 483 ÷ 21   k 966 ÷ 23   l 756 ÷ 36
   m 720 ÷ 24   n 780 ÷ 39   o 494 ÷ 26

3 The apples from a large box are packed into 35 bags. The same number is packed into
   each bag.
   If there were 595 apples in the box, how many are in each bag?

87

4 Sixteen friends went on holiday. They paid a total of £3392 for their travel.
How much did each pay?

5 The 912 students at Yew House School are seated in 38 equal rows in the school hall.
How many are in each row?

*6 Find the quotient for these.
   a dividend 486 divisor 9
   b dividend 742 divisor 53
   c dividend 966 divisor 42

Link to mean on page 376 and Geography.

*7 304 people live in an area of 16 square miles.
What is the mean number of people per square mile?

*8 The area of a rectangular field of experimental wheat is 26 400 m². It is 96 m wide.
How long is it?

*9 Lyn's car exhaust was tested.
On a 96 km journey it emitted 10 752 g of carbon monoxide.
What is the mean rate of emission per kilometre?

Review 1 Calculate
a 522 ÷ 9   b 2208 ÷ 6   c 3392 ÷ 8   d 3626 ÷ 7.

Review 2 Calculate
a 851 ÷ 23   b 675 ÷ 25   c 513 ÷ 19   d 544 ÷ 32
e 882 ÷ 98   f 756 ÷ 36   g 504 ÷ 42   h 777 ÷ 37
i 490 ÷ 14   j 374 ÷ 17   k 507 ÷ 13.

Review 3 Yolande was knitting a plain scarf.
One night she knitted nine rows and a total of 612 stitches.
How many stitches were in each row?

Review 4 The Rawlins family had 432 CDs.
These were kept in CD racks which each held 36.
How many racks did they have?

 Puzzle

Copy this diagram.

☐☐☐ ÷ ☐☐ = 58

Put the numbers 1, 2, 6, 8 and 9 in your boxes to make this true.

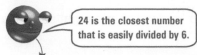
Dividing a decimal

Worked Example
26·28 ÷ 6

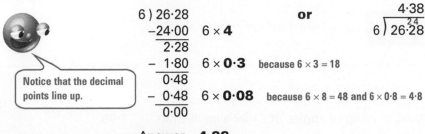

24 is the closest number that is easily divided by 6.

Answer
26·28 ÷ 6 is approximately 24 ÷ 6 = 4

```
6 ) 26·28
  −24·00    6 × 4
    2·28
  − 1·80    6 × 0·3    because 6 × 3 = 18
    0·48
  − 0·48    6 × 0·08   because 6 × 8 = 48 and 6 × 0·8 = 4·8
    0·00
```

or

```
        4·38
        24
6 ) 26·28
```

Notice that the decimal points line up.

Answer   **4·38**

Exercise 6   **Estimate the answers then calculate.**

1 a 5·5 ÷ 5   b 6·78 ÷ 2   c 4·84 ÷ 4   d 45·5 ÷ 5   e 12·72 ÷ 6
   f 9·38 ÷ 7   g 34·5 ÷ 5   h 74·88 ÷ 6   i 69·02 ÷ 7   j 740·4 ÷ 6
   k 393·4 ÷ 7   l 510·3 ÷ 9   m 438·9 ÷ 3   n 826·36 ÷ 4   o 762·88 ÷ 8
   p 499·44 ÷ 6   q 457·24 ÷ 7   r 589·95 ÷ 9   s 496·85 ÷ 5

2 Eight equal sized books were stacked on top of one another. The stack was 49·6 cm high.
   How thick was each book?

3 Nine people shared the cost of a meal equally.
   The meal cost £239·22.
   How much did each pay?

There is more about the average (mean) on page 376.

*4 Five people stood on a bridge. Their total mass was 384·25 kg.
    a What was the average mass of each person?
    b The bridge mass limit is 700 kg.
       Can nine people of the average mass you found in **a** safely stand on the bridge?

T **Review 1**

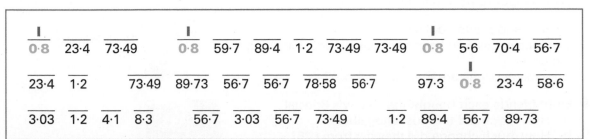

Use a copy of this box.
Put the letter that is beside the division above its answer. The first one is done.

I	6·4 ÷ 8 = 0·8	**O**	10·8 ÷ 9	**B**	39·2 ÷ 7	**R**	49·8 ÷ 6
U	32·8 ÷ 8	**Y**	21·21 ÷ 7	**T**	210·6 ÷ 9	**H**	351·6 ÷ 6
W	486·5 ÷ 5	**M**	238·8 ÷ 4	**L**	563·2 ÷ 8	**P**	268·2 ÷ 3
E	226·8 ÷ 4	**N**	448·65 ÷ 5	**S**	587·92 ÷ 8	**Z**	314·32 ÷ 4

Review 2   A block of apricot cheese is to be divided into eight equal pieces.
The block weighs 98·4 g.
How much will each piece weigh?

Mixed calculations

Discussion

● Roy worked out the cost of 26 kg of apples at £1·99 a kg like this.

$$\begin{array}{r} 1{\cdot}99 \\ \times\ 26 \\ \hline 39\ 80 \\ 11\ 94 \\ \hline 51{\cdot}74 \end{array}$$

Sarah worked out the cost like this. $26 \times £2 - 26 \times 1p = £52 - 26p$
                                                    $= £51{\cdot}74$

Which way is easier? **Discuss**.
Who is less likely to make a mistake? **Discuss**.

In the next exercise you will need to decide whether to **add**, **subtract**, **multiply** or **divide**. In most questions you will need to do more than one of these.

Exercise 7

1   What goes in the box, +, −, × or ÷ ?
   a 990 ☐ 120 = 1110   **b** 990 ☐ 120 = 118 800   **c** 990 ☐ 120 = 870
   d 990 ☐ 120 = 8·25

2   Alex cut 12 pieces of wool, each 28 cm long, from a 400 cm length.
   What length of wool was left?

3   Mr Bascand prints books.
   Claire pays £6·05 to have her book printed, including
   the cover.
   How many pages are there in her book?

Book Printing

Each page	4p
Cover	85p

4   Eight friends each bought the 'Today's Special'.
   a How much did this cost altogether?
   b How much change did they get from £75?
   c Maddison bought a pudding for £3·75 and a
        Today's Special.
        How much change did she get from £20?

TODAY'S
SPECIALS
£8.65

5 A drink and a chocolate bar cost 80p.
Two drinks and a chocolate bar cost £1·25.
How much does the chocolate bar cost?

6 Anna bought six fancy soaps for £14·10.
  a How much does one cost?
  b How much do nine cost?

7 Pembrook the snail crawls 41 cm in a hour.
How far would Pembrook crawl in a week?

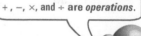

8 Chris bought 4 kg of beef for £8.64 per kg.
  a How much did it cost?
  b How much change did he get from £40?

9 A mini-bus can carry eight people. How many trips will it need to take 184 people to a soccer game?

* **10** Choco Treats sells chocolates.
Saskia pays £11·75 for a gift wrapped box of chocolates.
How many chocolates are in Saskia's box?

> Chocolates 85p each
> Gift wrapped in a box
> an extra £1.55 per box

* **11** Make up some word problems for these calculations.
  a 186 × 7 = 1302   **b** 19·4 × 3 = 58·2   **c** 183·4 ÷ 4 = 45·85   **d** 9·6 + 3·7 − 2·4 = 10·9

Review 1 Which operation, +, −, ×, or ÷, does * stand for?
a 306 * 85 = 26 010   **b** 306 * 85 = 221   **c** 306 * 85 = 3·6
d 306 * 85 = 391

+, −, ×, and ÷ are *operations*.

Review 2
a Along St Catherine's church wall there are eight small windows side by side. Each window is 1·14 m wide.
How long is the row of eight windows?
b The church wall is 15·5 m long.
What length of wall has no window?

Review 3 Grace bought
6 soaps at £0·85 each
9 brushes at £1·84 each
7 face-cloths at £1·40 each.
a How much did these cost altogether?
b What change would she get from £50?

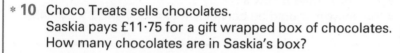

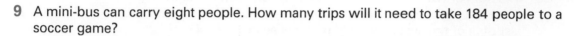

Checking if answers are sensible

Always check that your answer to a calculation is **sensible**.
Ask yourself 'Is the answer sensible and about what I would expect?'

Worked Example
Gareth calculated the distance between his classroom and the school gate.
He got an answer of 2·6 km.
Is this a sensible answer?

Answer
No it is not sensible because 2·6 km is much too far.

Worked Example
Jason worked out how many 24p pens he could buy for £5.
His answer was 20.
Do you think Jason was right or wrong? Why?

Answer
We check if the answer is sensible by estimating.
He can buy four pens for just under £1.
So he can buy 4 × 5 = 20 for under £5.
His answer is about right.

You can check the answer is sensible in other ways.

Examples   The sum of two odd numbers should be an even number.
If the last digits of two numbers you have multiplied together are 6 and 7 then the
last digit of the answer should be **2** because 6 × 7 = **42**.
If you multiply by 3, the answer should be divisible by 3.
If you multiply 1·59 × 24 the answer should be between 1 × 24 = 24 and 2 × 24 = 48.

Exercise 8   **Answer these questions *without* doing the calculation.**

1   Adam worked out the cost of two chocolate bars at 48p each. His answer was £8·16. Is he
correct or incorrect? Say why.

2   Annabel added up these amounts to find the total length of wood needed for a garden
border.

    3·7 m   4·8 m   2·1 m   3·4 m

    She got an answer of 57·2 m.
Is this answer about what you would expect? Say why or why not.

3   Rebecca wrote down how much she had earned doing jobs for her parents.

    £4·50   £5·25   £8·70   £9·85

    She added these up and got £82·50.
Is she correct or incorrect? Say why.

4   David worked out that he could buy 25 of these easter eggs for £5.
Could this answer be right? Explain your answer.

5   Mrs Street worked out how many litres of paint were needed to paint the school pool.
She got 0·57 ℓ.
Is this answer sensible? Say why or why not.

6   Hitesh worked out £3·52 + 86p on the calculator. He got 89·52 as his answer.
How can you tell this answer is wrong?

7   Mikey can run 100 m in 15 seconds.
He worked out it would take him 16 minutes 15 seconds to run 650 m.
Could this answer be right? Explain your answer.

8 Maha bought a pie for £2·68, a muffin for £1·20 and a drink for 84p.
She gave £10 and got £2·26 change. Is this correct or incorrect? Say why.

9 Pete multiplied 864 × 529 and got 447 534. How can you tell this answer is wrong?

10 Nick multiplied 783 × 467 and got 36 810. How can you tell this answer is wrong?

Review 1 Morgan worked out the cost of four model aeroplanes.
He got £42·16.
Is he correct or incorrect? Say why.

Special Model Aeroplanes £8.79

Review 2 Josh added up these times he had spent watching TV.

8 min   27 min   53 min   35 min   15 min   42 min

He got 6 hours and 32 minutes.
Is this answer about what you would expect? Say why or why not.

Review 3 Raj worked out that he needed 864 m of wood to make a picture frame.
Is this answer sensible? Say why or why not.

Review 4
a Pam gave a shop assistant a £10 note and got £3·86 change.
She bought items costing 89p, £1·08, £2·03, 26p, 45p, 28p.
Did she get the correct or incorrect change? Say why.
b Pam bought 6 stamps for 19p each. She worked out this would cost £1·15.
How can you tell she is wrong?

Checking answers using inverse operations

We can check the answer to a calculation using **inverse operations**.

Addition and subtraction are inverse operations.
Multiplication and division are inverse operations.

> Remember inverse operations undo each other.

Examples   check 2·4 × 6 = 14·4   using 14·4 ÷ 6 or 14·4 ÷ 2·4
              check 42·4 × 28·5 = 1208·4   using 1208·4 ÷ 42·4   or   1208·4 ÷ 28·5
              check 6 ÷ 7 = 0·857 14 28 ... using 7 × 0·857 142 8
              check 96·3 + 5·89 = 102·19   using 102·19 − 5·89   or   102·19 − 96·3

Sometimes we check the answer to a calculation using **estimating** *and* **inverse operations**.

Examples

6783 ÷ 19 = 357 appears to be about right because 350 × 20 = 7000.

9357 − 4216 = 5141 appears to be about right because 5000 + 4000 = 9,000.

Exercise 9

1 Find the missing numbers.
  a Check 3864 − 258 = 3606   using 3606 + ☐ = 3864.
  b Check 9432 + 4176 = 13 608   using 13 608 − ☐ = 4176.
  c Check 563 × 14 = 7882   using 7882 ÷ ☐ = 563.
  d Check 3705 ÷ 65 = 57   using ☐ × 65 = 3705.
  e Check 152·73 + 81·06 = 233·79   using 233·79 − ☐ = 81·06.
  f Check 8·75 × 25·3 = 221·375   using ☐ ÷ 25·3 = ☐.
  g Check 3 ÷ 7 = 0·428571428 ...   using ☐ × 7 = ☐.
  h Check 96·483 + 2·79 = 99·273   using ☐ − ☐ = 96·483.

2 i Find the answer to these calculations.
  ii Write down a calculation using inverse operations that you could do to check your answer.
  a 9613 + 9423   b 5873 − 1659   c 321 × 89   d 2080 ÷ 65   e 20·7 ÷ 9
  f 834·62 + 93·1   g 8·5 ÷ 2·5   h 97 − 3·842   i 92·3 × 4·1   j 8 ÷ 14

3 What calculation could you do to check these? Use estimation *and* inverse operations.
  a 80 231 − 4975 = 75 256   b 5421 ÷ 65 = 83·4   c 89·6 − 30·5 = 59·1
  d 242·73 ÷ 2·9 = 83·7   e 581 × 62 = 36 022   f 9842 + 3754 = 13 596
  g 78·72 ÷ 3·2 = 24·6

4 Neroli wanted to know how much 18 bags of crisps would cost. Each bag costs £1·49.
  a What calculation does she need to do?
  b How much would 18 bags of crisps cost?
  c What calculation could she do to check her answer?

5 Rosalie had 86·4 m² of grass matting. She wanted to cut it into 16 equal pieces.
  a What calculation would she do to find out what area each piece would be?
  b What area would each piece be?
  c What calculation could she do to check the answer?

Review 1   Find the missing numbers.
a Check 96 423 + 82 471 = 178 894   using 178 894 − ☐ = 96 423.
b Check 83 × 96 = 7968   using ☐ ÷ ☐ = 83.
c Check 11 ÷ 7 = 1·571428571 ...   using ☐ × ☐ = 11.
d Check 89·06 − 5·79 = 83·27   using ☐ + ☐ = 89·06.

Review 2   Using inverse operations check which of these answers are wrong.
a 5·6 × 3·7 = 20·72   b 8·64 ÷ 5 = 1·86   c 187·9 − 3·02 = 184·88

Review 3   What calculation could you do to check these?
Use estimation *and* inverse operations.
a 50 964 − 3827 = 47 137   b 237·38 ÷ 8·3 = 28·6

Puzzle

Using the numbers in the squares write

four divisions such as $16 \div 2 = 8$
four multiplications such as $8 \times 3 = 24$
four additions such as $16 + 2 = 18$
and four subtractions such as $12 - 2 = 10$.

Each square must be used only once.
The 16 calculations will use *all* of the numbers.

16	8	2	2	8	6
5	6	18	16	4	40
6	18	2	6	10	32
32	2	24	3	3	4
8	4	6	8	24	8
18	32	5	8	2	3
12	18	40	6	10	24
12	24	2	9	6	12

Checking answers using an equivalent calculation

We can check the answer to a calculation by doing an **equivalent calculation**.

Example $494 \times 5 = 2470$ can be checked by doing one of these calculations.

$(500 - 6) \times 5 = 500 \times 5 - 6 \times 5$

or $494 \times 10 \div 2$

or $400 \times 5 + 90 \times 5 + 4 \times 5$

Example $20{\cdot}8 \div 8 = 2{\cdot}6$ can be checked by doing one of these calculations.

$20{\cdot}8 \div 4 \div 2$

or $20 \div 8 + 0{\cdot}8 \div 8$

Exercise 10

1   Check the answers to these by doing an equivalent calculation. If the answers are wrong
give the correct answer. Write down the calculation you did.
  a $304 \times 5 = 1520$ **b** $893 \times 6 = 5358$ **c** $52 \times 40 = 2080$ **d** $705 \times 50 = 35\,250$
  e $388 \div 4 = 48{\cdot}5$ **f** $495 \div 5 = 99$ **g** $86 \times 20 = 1270$ **h** $45 \times 98 = 4410$

Review   Check the answers to these by doing an equivalent calculation.
Write down the calculation you did.

a $594 \times 9 = 5346$ **b** $38 \times 24 = 912$ **c** $729 \div 9 = 81$ **d** $54 \times 80 = 4320$

Using the calculator

Jan did a calculation on the calculator. The answer was displayed as ⟨ **45.2** ⟩.

How she writes down the answer depends on what the question was.

Question	**Answer**
Pip bought five pictures for £226. How much did each cost?	**£45·20**

Madhu took 226 minutes to run five laps. How long did she take to run each lap?

45·2 minutes is 45 minutes and 0·2 of a minute.
Subtract 45 to leave 0·2 displayed.
There are 60 seconds in a minute.
Multiply by 60 to get 12 displayed.
The answer is **45 minutes and 12 seconds**.

Peter bought five lengths of fencing. Each length was 9·04 m long. What is the total length he bought?

45·2 m is 45 m and 0·2 of a metre.
There are 100 centimetres in a metre.
0·2 × 100 = 20 cm
The answer is **45 m and 20 cm**.

6 m and 74 cm = 6·74 m
We enter 6 m 74 cm into the calculator as ⟨ **6.74** ⟩

3 hours 15 minutes = $3 + \frac{15}{60}$ hours
$= 3\frac{1}{4}$ hours
$= 3·25$ hours

We enter 3 hours 15 minutes into the calculator as ⟨ **3.25** ⟩

Exercise 11 Always check your answer.

1 Robert paid £164 for five tickets to a show.
   How much did each ticket cost?

2 Chloe bought 164 cm of ribbon. She cut it into five equal pieces. How long, in cm and mm, was each piece?

3 Freda took 164 minutes to walk round the block five times.
   How long did she take to walk round once? Give the answer in minutes and seconds.

4 Helen did a calculation and got the answer 16·4.
   What should she write down if the question asked for
   a £   b cm and mm   c minutes and seconds
   d hours and minutes   e m and cm?

5 Rosemary took 2 hours and 15 minutes to do an experiment. How long would she take to do the same experiment five times? Assume it takes the same time for each repeat.

6 Matthew took 5 minutes and 24 seconds to run round the track once. How long would he take to run round four times?

7 Pritesh is 7 years 3 months old. His mother is four times as old. How old is his mother?

8 Tiffany measured one of her pet stick insects. It was 3 cm and 4 mm long. If eight of these stick insects lined up one behind the other, how long would the line be?

Review 1 Samuel did a calculation and got the answer 42·6. What should he write down if the question asked for
a £   **b** cm and mm   **c** minutes and seconds   **d** hours and minutes?

Review 2 Libby spends 3 hours and 45 minutes at work each day. How long does she spend at work each week (5 days)?

Discussion

51 ÷ 8 = 6 **remainder** 3 or 6R3
The calculator gives the answer to 51 ÷ 8 as 6·375.
Sally said she could get the remainder 3 on the calculator.
She subtracted 6 from 6·375 to get 0·375 on the screen.
Then she multiplied by the divisor, 8.
Does this give the remainder, 3? **Discuss**.

`0.375`

What do you need to key to give the remainder in these? $\frac{34}{8}$ $\frac{71}{4}$ $\frac{102}{5}$
Discuss.

Worked Example
Find the remainder. **a** $\frac{128}{5}$ **b** $\frac{60}{7}$ **c** $\frac{123}{9}$

Answer
a Key 1 2 8 ÷ 5 = – 2 5 = × 5 = to get remainder **3**.
b Key 6 0 ÷ 7 = – 8 = × 7 = to get remainder **4**.
c Key 1 2 3 ÷ 9 = – 1 3 = × 9 = to get remainder **6**.

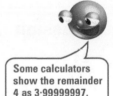

> Some calculators show the remainder 4 as 3·99999997.

* *Worked Example*
Write 259 hours in days and hours.

Answer
Key 2 5 9 ÷ 2 4 = to get 10·79166667 days. There are 10 whole days.
Now – 1 0 = × 2 4 = to get remainder 19. There are 19 hours left over.
259 hours = **10 days 19 hours**.

---

* Discussion

Discuss how to write   259 minutes in hours and minutes
259 weeks in years and weeks
259 months in years and months.

Exercise 12

1 Find the remainder.

a $\frac{154}{5}$ b $\frac{261}{8}$ c $\frac{342}{4}$ d $\frac{411}{6}$ e $\frac{279}{16}$

f $\frac{326}{3}$ g $\frac{249}{7}$ h $\frac{178}{19}$ i $\frac{806}{35}$ j $\frac{903}{41}$

*2 Convert
   a 227 hours to days and hours   b 1862 hours to days and hours
   c 527 minutes to hours and minutes d 613 minutes to hours and minutes
   e 426 weeks to years and weeks   f 379 weeks to years and weeks
   g 195 months to years and months.

*3 Simone took 569 minutes to play the same CD 6 times. In minutes and seconds, how long does the CD take to play?

*4 Wasim spent the same amount of time at the gym each week.
    At the end of the year he worked out he had spent a total of 4108 minutes there.
    a How many hours and minutes is this?
    b How many hours and minutes did Wasim spend at the gym each week?

Review 1 Find the remainder. a $\frac{261}{4}$ b $\frac{307}{8}$ c $\frac{247}{6}$ d $\frac{834}{17}$

Review 2 Write
a 417 minutes in hours and minutes
b 375 hours in days and hours
c 572 months in years and months.

Brackets on the calculator

Discussion

- $\frac{24}{6+2}$

  Emily and Tom did this on a calculator.
  Emily keyed 2 4 ÷ 6 + 2 = and got the answer 6.
  Tom keyed 2 4 ÷ (6 + 2) = and got the answer 3.
  Who is right? **Discuss**.

- $\frac{12+6}{3-1}$

  Janita, Asad, Mel and Peter did this on a calculator.
  Janita got 5.
  Asad got 9.
  Mel got 13.
  Peter got 15.
  Asad got the right answer. He keyed

  ((1 2 + 6) ÷ (3 − 1)) =

  What did the others key? **Discuss**.

Sometimes to find the answer to a calculation, we need to use the **brackets** on a calculator.

Worked Example
Calculate
a $(46 + 15) \times (26 - 7)$
b $3\cdot7 - (2\cdot09 - 1\cdot6)$
c $\frac{8\cdot2}{3\cdot2 - 1\cdot8}$
d $\frac{9\cdot6 - 4\cdot7}{3\cdot8 + 2\cdot1}$

brackets

Answer
a Key
   to get **1159**

b Key
   to get **3·21**

Remember the horizontal line acts as a bracket.

c Key
   to get 5·857142857 or **5·9 (1 d.p.)**

d Key ((9 · 6 – 4 · 7) ÷ (3 · 8 + 2 · 1)) =
   to get 0·830508474 or **0·8 (1 d.p.)**

Both the numerator and denominator must have brackets.

Exercise 13

1 a $7 + 3(14 + 8)$ b $10 + 4(27 + 9)$ c $48 - 2(3 + 11)$ d $124 - 5(14 + 9)$
   e $47 \times (396 - 72)$ f $52 \times (851 - 36)$ g $(7 + 14) \times (23 + 28)$ h $(83 + 27) \times (94 - 39)$
   i $(52 + 81) \times (37 - 19)$ j $314 - (827 - 619)$ k $816 - (327 - 214)$ l $\frac{512}{63 + 65}$
   m $\frac{475}{82 - 57}$ n $\frac{16 + 92}{112 - 94}$ o $\frac{183 + 293}{89 - 61}$

2 Calculate
   a $5\cdot2 \times (9\cdot6 - 3\cdot85)$ b $7\cdot2 \times (15\cdot73 - 8\cdot6)$ c $(5\cdot2 + 1\cdot89) \times (12 - 3\cdot8)$
   d $(8\cdot1 - 2\cdot97) \times (53\cdot4 - 48\cdot63)$ e $364\cdot7 - (27\cdot92 - 8\cdot3)$ f $8\cdot2 + 5(6\cdot4 + 1\cdot04)$
   g $15\cdot2 - 2(3\cdot6 - 1\cdot8)$ h $57\cdot3 - 4(19\cdot7 - 14\cdot32)$ i $4\cdot2 \times 3 - (5\cdot6 - 2\cdot39)$
   j $3\cdot7 \times 8\cdot6 + (8\cdot3 - 0\cdot75)$ k $4\cdot2 \times 6\cdot8 - (3\cdot2 - 1\cdot6)$

3 Calculate and give the answers to 1 d.p.
   a $\frac{86}{24 + 17}$ b $\frac{133 + 89}{416 - 329}$ c $\frac{12\cdot7}{5\cdot2 + 1\cdot6}$ d $\frac{7\cdot2 + 8\cdot6}{5\cdot2 - 3\cdot4}$

Review Calculate
a $53 \times (584 - 79)$ b $(46 + 81) \times (187 - 152)$ c $4\cdot2 \times (3\cdot8 - 1\cdot26)$
d $\frac{368}{90 - 67}$ e $\frac{8\cdot4 + 2\cdot6}{9\cdot6 - 7\cdot4}$ f $4\cdot2 \times 5 - (6\cdot4 + 2\cdot3)$
g $18\cdot7 - 3(1\cdot89 - 1\cdot5)$ h $7\cdot2 \times 3\cdot6 - (5\cdot2 - 4\cdot85)$

Summary of key points

 We **add and subtract** whole numbers by first lining up the units (ones) column.
We add and subtract decimals by lining up the decimal points.

Examples

```
  85247            8·73          14·060
+  3952         +16·059         − 8·327
 ─────          ───────         ──────
  89199          24·789          5·733
      1               1
```

 Multiplying whole numbers

Example 541 × 72

541 × 72 is approximately 500 × 70 = 35 000.

 Always estimate first.

```
    541
  ×  72
  ─────
  37870  ← 541 × 70
   1082  ← 541 × 2
  ─────
  38952  ← add
```

 Multiplying with decimals

Example 4·79 × 6

4·79 × 6 is approximately 5 × 6 = 30 and is equivalent to 479 × 6 ÷ 100.

×	4	0·7	0·09
6	24	4·2	0·54

Answer 24 + 4·2 + 0·54 = **28·74**

```
   479
  ×  6
  ────
  2874
   4 5
```

2874 ÷ 100 = **28·74**

 Dividing with whole numbers

Example 646 ÷ 19

646 ÷ 19 is approximately 600 ÷ 20 = 30.

```
19 ) 646
    −570   19 × 30
    ────
      76
     −76   19 × 4
     ───
       0
```

Answer **34**

E **Dividing a decimal**

Example 194·88 ÷ 6

194·88 ÷ 6 is approximately

180 ÷ 6 = 30

 180 is the closest number we recognise easily that will divide by 6.

```
6 ) 194·88
   −180·00   6 × 30
   ───────
    14·88
   −12·00   6 × 2
   ──────
     2·88
    −2·40   6 × 0·4
    ─────
     0·48
    −0·48   6 × 0·08
    ─────
     0·00
```

Answer **32·48**

Always check that your **answer is sensible**.

Example　Beth buys items for 89p, £1·24, £1·06, 35p and 92p.

She gives £10 and gets £2·58 change.

This is not reasonable because the items would add to less than £5.

 We can check the answer to a calculation using **inverse operations**.

　　　　addition and **subtraction** are inverse operations

　　　　multiplication and **division** are inverse operations

Examples　1184 − 397 = 787 so 787 + 397 = 1184

　　　　197 × 31 = 6107 so 6107 ÷ 31 = 197 and 6107 ÷ 197 = 31

 We can check the answer to a calculation by doing an **equivalent** calculation.

Example　56·4 × 5 = 282 can be checked by doing one of these calculations.

　　　　56·4 × 10 ÷ 2 = 564 ÷ 2　　　50 × 5 + 6 × 5 + 0·4 × 5 = 250 + 30 + 2

　　　　　　　　= 282　　　　　　　　　　　　　　= 282

 Sometimes we need to **interpret** the answer given by a calculator.

Example　　| 5.6 |　might be £5·60 or 5 hours 36 minutes or 5 metres

60 centimetres and so on.

 To find the **remainder** when dividing, subtract the whole number part of the

answer and then multiply by the divisor.

Example　387 ÷ 63

Key　[3] [8] [7] [÷] [6] [3] [=]　to get 6·142857143.

Key　[−] [6] [=]　to get 0·142857143.

Key　[×] [6] [3] [=]　to give the remainder 9.

 Sometimes we use **brackets** on the calculator.

Example　4·2 − (3·6 − 1·2)

Key　[4] [·] [2] [−] [(] [3] [·] [6] [−] [1] [·] [2] [)] [=]　to get 1·8.

In a fraction, the horizontal line acts as a bracket.

Example　$\frac{8·6 + 3·2}{4·5 − 3·8}$

Key　[(] [8] [·] [6] [+] [3] [·] [2] [)] [÷] [(] [4] [·] [5] [−] [3] [·] [8] [)] [=]

to get 16·9 to 1 d.p.

Test yourself **except questions 18, 19 and 20.**

1 Calculate these.
  a 1723 + 589   b 76 423 + 8659   c 7·96 + 8·3   d 15·01 + 9·63   **A**
  e 36·8 + 5·709   f 9806 − 394   g 4509 − 3689   h 20·6 − 19·24
  i 113·5 − 24·68   j 314·7 − 19·64   k 800·5 − 3·69

2 Ben Nevis is the highest mountain in Britain. It is 1343 m high.
Mount Everest is 8843 m high.
How much higher is Mount Everest than Ben Nevis?   **A**

3 An empty lorry weighs 4·26 tonne.
When this lorry is loaded with wheat its weight is 11·4 tonne.
How much wheat is on this lorry?   **A**

4 Calculate these.   **B**
  a 56 × 34   b 56 × 78   c 345 × 45   d 678 × 34
  e 569 × 89   f 789 × 56   g 457 × 38   h 987 × 34
  i 289 × 67   j 678 × 73   k 569 × 34

5 For each of the 52 weeks in a year, a supermarket is open for 86 hours.
What is the total number of hours it is open?   **B**

6 Find the answers to these. Estimate first.   **C**
Write down your estimate and your answer.
  a 4·6 × 7   b 18·3 × 5   c 1·68 × 9   d 4·83 × 7
  e 16·42 × 5   f 18·24 × 6   g 142·7 × 8

7 Thomas bought these take aways for his friends.   **A** **C**
    3 scoops of chips   £2·70
    5 sausages   £5·40
    2 beefburgers   £1·60
How much did this cost altogether?

8 Bev's garden needed 128·78 m of edging.   **A**
On Saturday she edged 25·6 m and on Sunday she edged 36·83 m.
How many more metres did she need to edge?

9 Estimate the answers, then calculate.   **D**
  a 224 ÷ 4   b 712 ÷ 8   c 1410 ÷ 6   d 672 ÷ 12
  e 885 ÷ 15   f 984 ÷ 24   g 992 ÷ 32

10 There are 45 chocolates in a box.   **D**
How many boxes can be filled if there are 810 chocolates?

11 Estimate the answers, then calculate.   **E**
  a 5·6 ÷ 7   b 13·8 ÷ 6   c 58·4 ÷ 4   d 170·4 ÷ 8   e 212·04 ÷ 9

12 Shirley bought   **A** **C**
  3 toothbrushes   4 bars of soap
  1 deodorant   1 sunscreen
  a How much did this cost?
  b How much change did she get from £25?

13 a The ⑧ key on your calculator doesn't work.   **A**
     How could you make your calculator display 858?
  b The ②, ④, ⑥ and ⑧ keys on your calculator don't work.
     How could you make your calculator display 846?

14 Answer these questions **without** doing the calculation.
   a Megan added up these amounts to find the total amount of money made on a sponsored walk.
     £8·69   £12·70   £9·50   £13·64   £6·35   £8·24
     She got an answer of £167·12. Is she correct or incorrect? Say why.
   b Anna worked out she would need to order 39 coaches to take 204 people to a concert. Each coach holds 52 people. Is she correct or incorrect? Explain your answer.

15 Using inverse operations, write down a calculation you could do to check the answer to these.
   a $836 \times 42 = 35\ 112$ **b** $364·8 \div 1·9 = 192$ **c** $8·36 + 19·72 = 28·08$

16 Write down an equivalent calculation you could do to check these.
   a $603 \times 27 = 16\ 281$ **b** $42 \times 59 = 2478$ **c** $632 \div 4 = 158$

17 185   56   589   423   156   86
   Put one of the numbers above in each box to make a true statement
   a $\square - \square + \square = 323$ **b** $\square + \square - \square = 490$

18 Find the remainder when 396 is divided by 27.

19 Calculate. Give the answer to 1 d.p. if you need to round.
   a $(53 + 21) \times 75$ **b** $4·2 + (6·8 - 0·87)$ **c** $\frac{17·6}{4·2 + 3·7}$ **d** $\frac{8·3 + 4·2}{8·3 - 5·7}$

20 Convert
   a 186 hours to days and hours   **b** 373 seconds to minutes and seconds

5 Fractions

You need to know

✓ Fractions   — numerator and denominator   page 4

                       — know that $\frac{3}{5}$ means 3 out of 5

                       — equivalent fractions

                       — fraction and division

                       — fraction of

Key vocabulary

cancel, convert, denominator, diagram, equivalent fraction, lowest terms, mixed number, numerator, proper/improper fraction, simplest form

Bits and pieces

About $\frac{1}{4}$ of all meat eaten in the world is meat from pigs.

Four out of five teenagers worry about exams.

Ninety-seven hundredths of water on earth is in the ocean.

Find some more real-life fraction facts.
Make a poster or collage of your facts.
You could show your fractions on a number line or diagram.

Fractions of shapes

One out of nine **parts** of this diagram is shaded.
One ninth or $\frac{1}{9}$ is shaded.

Three out of eight pieces of cake have been iced.
$\frac{3}{8}$ have been iced.

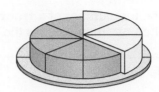

Exercise 1

What fraction of each of these is coloured?

a   **b**   **c** 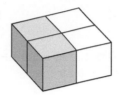   ***d**

Review   What fraction of each of these is coloured?

a 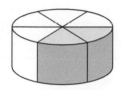   **b** 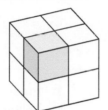   ***c**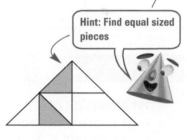

Hint: Find equal sized pieces

⭐ Practical

You will need copies of the five shapes.

1 Tim coloured half of this rectangle as shown.
Find as many ways as you can to colour half of the rectangle.
Find as many ways as you can to colour three-quarters of it.

2 Colour half of these shapes.

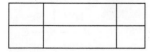   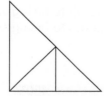   

You may draw extra lines if you want to.

*3 Ben divided this square into quarters as shown.

Find as many ways as you can to divide the square into quarters.

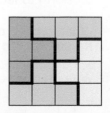

One number as a fraction of another

Remember
The top number of a **fraction** is the **numerator**.
The bottom number is the **denominator**.

$\dfrac{3}{8}$ ← numerator
← denominator

Jason had a piece of wood 1 m long.
He cut off 43 cm.

1 metre = 100 centimetres

43 cm as a fraction of a metre = $\frac{43}{100}$ ←

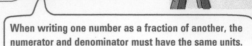
When writing one number as a fraction of another, the numerator and denominator must have the same units.

Worked Example
What fraction of a turn does the minute hand turn through between 6:25 p.m. and 6:40 p.m.?

Answer
Between 6:25 p.m. and 6:40 p.m. the hand turns through 15 minutes. This is $\frac{1}{4}$ of a turn.

Exercise 2

1 What fraction of
   a 1 metre is 51 centimetres
   b 1 metre is 89 centimetres
   c 1 kilogram is 37 grams
   d 1 kilogram is 59 grams
   e 1 hour is 13 minutes
   f 1 hour is 47 minutes
   g 1 metre is 17 centimetres
   h 1 minute is 17 seconds
   i £2 is 87p
   j £3 is 61p
   k 1 yard is a foot
   l 1 kilometre is 253 metres
   m 1 litre is 183 millilitres?

2 What fraction of a turn does the minute hand turn through between
   a 8 p.m. and 8:25 p.m.
   b 2:20 a.m. and 2:40 a.m.
   c 11:10 a.m. and 11:38 a.m.
   d 7:35 p.m. and 8:20 p.m.
   *__e__ 5:35 a.m. and 7:25 a.m.?

3 What fraction of a turn takes you from facing
   a north to facing south
   b south to facing west
   c north to facing north-west
   d south to facing south-west
   e north to facing south-west?

Turn the way of the red arrow.

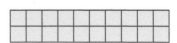

4 What fraction of a turn is
   a 90°   **b** 180°   **c** 120°   **d** 36°   *__e__ 450°?

5 What fraction of the large shape is the small one?

   a   **b**   *__c__

Review 1 What fraction of
a 1 metre is 27 centimetres   **b** 1 hour is 49 minutes?

Review 2 What fraction of a turn does the minute hand turn through between
a 2:15 p.m. and 2:35 p.m.   **b** 12:45 p.m. and 1:20 p.m.?

Review 3 What fraction of the large shape is the small one?

a   **b**

Equivalent fractions

 Practical

This diagram shows that
$\frac{1}{2}, \frac{2}{4}, \frac{4}{8}, \frac{8}{16}$ are equal.

Draw diagrams to show that these sets of
fractions are equal.

1 $\frac{1}{2}, \quad \frac{3}{6}, \quad \frac{5}{10}, \quad \frac{6}{12}$ **2** $\frac{1}{4}, \quad \frac{2}{8}, \quad \frac{3}{12}, \quad \frac{4}{16}$

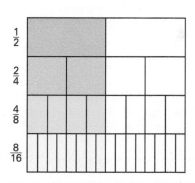

Equal fractions are called **equivalent fractions**.

Examples
$\frac{1}{2}, \frac{2}{4}, \frac{4}{8}, \frac{16}{32}, \frac{50}{100}$ are equivalent fractions. They all equal $\frac{1}{2}$.
$\frac{1}{1}, \frac{4}{4}, \frac{8}{8}, \frac{100}{100}$ are equivalent fractions. They all equal 1.

We make equivalent fractions by multiplying or dividing
both the numerator and denominator by the same number.

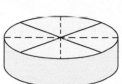

This diagram shows that
$\frac{1}{1}, \frac{4}{4}$ and $\frac{8}{8}$ are all equal.

Example

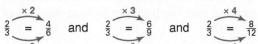

$\frac{2}{3} = \frac{4}{6}$ and $\frac{2}{3} = \frac{6}{9}$ and $\frac{2}{3} = \frac{8}{12}$

$\frac{2}{3} = \frac{4}{6} = \frac{6}{9} = \frac{8}{12}$ $\frac{2}{3}, \frac{4}{6}, \frac{6}{9}, \frac{8}{12}$ are equivalent fractions.

Example

$\frac{24}{32} = \frac{12}{16}$ and $\frac{24}{32} = \frac{6}{8}$ and $\frac{24}{32} = \frac{3}{4}$

$\frac{24}{32} = \frac{12}{16} = \frac{6}{8} = \frac{3}{4}$ $\frac{24}{32}, \frac{12}{16}, \frac{6}{8}, \frac{3}{4}$ are equivalent fractions.

Number

Exercise 3   **This exercise is to be done mentally.**

1 What number goes in the box to make an equivalent fraction?

a $\dfrac{1}{8} \overset{\times\,2}{\underset{\times\,2}{=}} \dfrac{\square}{16}$

b $\dfrac{1}{3} \overset{\times\,3}{\underset{\times\,3}{=}} \dfrac{3}{\square}$

c $\dfrac{4}{18} \overset{\div\,2}{\underset{\div\,2}{=}} \dfrac{\square}{9}$

d $\dfrac{10}{100} \overset{\div\,10}{\underset{\div\,10}{=}} \dfrac{1}{\square}$

e $\dfrac{4}{10} = \dfrac{\square}{100}$

f $\dfrac{6}{15} = \dfrac{2}{\square}$

g $\dfrac{4}{3} = \dfrac{16}{\square}$

h $\dfrac{15}{9} = \dfrac{\square}{18}$

i $\dfrac{24}{5} = \dfrac{\square}{20}$

j $\dfrac{45}{18} = \dfrac{\square}{2}$

k $\dfrac{24}{20} = \dfrac{\square}{5}$

2 What is the missing number?

a $\dfrac{3}{10} = \dfrac{?}{30}$

b $\dfrac{2}{3} = \dfrac{?}{12}$

c $\dfrac{4}{5} = \dfrac{?}{15}$

d $\dfrac{3}{4} = \dfrac{6}{?}$

e $\dfrac{6}{15} = \dfrac{36}{?}$

f $\dfrac{1}{7} = \dfrac{4}{?}$

g $\dfrac{3}{4} = \dfrac{?}{100}$

h $\dfrac{16}{20} = \dfrac{4}{?}$

i $\dfrac{3}{8} = \dfrac{15}{?}$

j $\dfrac{12}{8} = \dfrac{?}{4}$

k $\dfrac{25}{100} = \dfrac{1}{?}$

l $\dfrac{1}{1} = \dfrac{5}{?}$

3 Copy and complete.

a $\dfrac{2}{5} = \dfrac{4}{...} = \dfrac{10}{...} = \dfrac{16}{...}$

b $\dfrac{3}{8} = \dfrac{...}{16} = \dfrac{...}{24} = \dfrac{...}{40} = \dfrac{...}{80}$

c $\dfrac{2}{3} = \dfrac{...}{9} = \dfrac{12}{...} = \dfrac{...}{12}$

d $\dfrac{4}{5} = \dfrac{...}{20} = \dfrac{40}{...} = \dfrac{...}{40}$

e $\dfrac{3}{4} = \dfrac{6}{...} = \dfrac{...}{12} = \dfrac{12}{...}$

f $1 = \dfrac{...}{5} = \dfrac{6}{...} = \dfrac{...}{10} = \dfrac{100}{...}$

4 Choose the two fractions that are equivalent.

a $\dfrac{2}{4}, \dfrac{3}{5}, \dfrac{1}{2}$

b $\dfrac{2}{5}, \dfrac{5}{10}, \dfrac{1}{2}$

c $\dfrac{1}{3}, \dfrac{1}{5}, \dfrac{5}{15}$

d $\dfrac{1}{7}, \dfrac{2}{15}, \dfrac{3}{21}$

e $\dfrac{3}{5}, \dfrac{3}{4}, \dfrac{6}{10}$

f $\dfrac{2}{3}, \dfrac{6}{9}, \dfrac{5}{8}$

g $\dfrac{12}{20}, \dfrac{4}{5}, \dfrac{12}{15}$

h $\dfrac{2}{9}, \dfrac{6}{24}, \dfrac{12}{48}$

i $\dfrac{14}{5}, \dfrac{28}{10}, \dfrac{3}{4}$

j $\dfrac{40}{100}, \dfrac{4}{10}, \dfrac{3}{5}$

k $\dfrac{17}{10}, \dfrac{17}{100}, \dfrac{170}{100}$

5 Write two equivalent fractions for each of these.

a $\dfrac{1}{3}$

b $\dfrac{1}{4}$

c $\dfrac{3}{8}$

d $\dfrac{3}{4}$

e $\dfrac{5}{6}$

f $\dfrac{8}{10}$

g $\dfrac{20}{25}$

h $\dfrac{32}{48}$

i 1

You can use jottings for question 5 if you need to.

*6 Write the equivalent fraction which has a denominator of 24.

a $\dfrac{1}{2}$

b $\dfrac{2}{3}$

c $\dfrac{3}{4}$

d $\dfrac{5}{6}$

e $\dfrac{3}{8}$

*7 Caroline is putting a border on a poster.
First she divided it into five equal rectangles and made three of them red. But then she decided to divide the border into 15 rectangles and keep an equivalent fraction of them red. How many out of the 15 should she make red?

Review 1   Find the missing numbers.

a $\dfrac{3}{5} = \dfrac{12}{\square}$

b $\dfrac{2}{3} = \dfrac{\square}{12}$

c $\dfrac{24}{30} = \dfrac{\square}{5}$

d $\dfrac{36}{45} = \dfrac{12}{\square}$

e $1 = \dfrac{\square}{18}$

f $\dfrac{45}{50} = \dfrac{9}{\square}$

g $\dfrac{2}{3} = \dfrac{\square}{300}$

h $\dfrac{8}{7} = \dfrac{64}{\square}$

i $\dfrac{72}{54} = \dfrac{\square}{6}$

Review 2   Copy and complete.

a $\dfrac{3}{5} = \dfrac{6}{...} = \dfrac{12}{...} = \dfrac{...}{25} = \dfrac{...}{100}$

b $\dfrac{5}{9} = \dfrac{...}{18} = \dfrac{15}{...} = \dfrac{25}{...} = \dfrac{...}{90}$

Review 3

$$\underset{24}{15} \quad \underset{50}{35} \quad \underset{36}{24} \quad \overset{A}{\underset{4}{2}} \quad \underset{45}{12} \qquad \underset{64}{24} \quad \underset{27}{12} \quad \overset{A}{\underset{4}{2}} \quad \underset{45}{12} \quad \underset{100}{60} \qquad \overset{A}{\underset{4}{2}} \quad \underset{45}{12} \quad \underset{27}{12}$$

$$\underset{36}{24} \quad \underset{27}{12} \quad \underset{9}{5} \quad \underset{48}{12} \quad \underset{4}{3} \quad - \overset{A}{\underset{4}{2}} \quad \underset{28}{28} \quad \underset{5}{4} \quad \underset{27}{12} \quad \underset{5}{4}$$

Use a copy of this box.
Write the letter beside each fraction, above an equivalent fraction in the box.

A $\frac{1}{2} = \frac{2}{4}$ **S** $\frac{3}{5}$ **L** $\frac{2}{3}$ **D** $\frac{12}{15}$ **N** 1 **T** $\frac{25}{100}$ **E** $\frac{4}{9}$

P $\frac{5}{8}$ **O** $\frac{7}{10}$ **R** $\frac{4}{15}$ **B** $\frac{6}{16}$ **F** $\frac{45}{81}$ **H** $\frac{48}{64}$

? Puzzle

1 I am equivalent to $\frac{1}{2}$.
The sum of my numerator and denominator is 15.
What fraction am I?

2 I am equivalent to $\frac{2}{5}$.
The product of my numerator and denominator is 40.
What fraction am I?

3 I am equivalent to $\frac{2}{3}$.
My denominator is 10 more than my numerator.
What fraction am I?

4 I am equivalent to $\frac{80}{100}$.
My denominator is a prime number.
What fraction am I?

Investigation

Patterns in Equivalent Fractions

$\frac{1}{2} = \frac{2}{4} = \frac{3}{6} = \frac{4}{8} = \frac{5}{10} = \frac{6}{12} = \ldots$
The numbers in the numerator make the pattern 1, 2, 3, 4, 5, 6, ...
The numbers in the denominator make the pattern 2, 4, 6, 8, 10, 12, ...
(even numbers).

$\frac{2}{3} = \frac{4}{6} = \frac{6}{9} = \frac{8}{12} = \frac{10}{15} = \ldots$
What patterns are made by these equivalent fractions?

Investigate the patterns made by other equivalent fractions.
As part of your investigation, you could **investigate** the patterns made when the numerator of each equivalent fraction is subtracted from the denominator.

Cancelling a fraction to its lowest terms

A fraction is in its **lowest terms** if the numerator and denominator have no common factors other than 1.

We **cancel** a fraction to its lowest terms by looking for the highest common factor (HCF).

This is called putting a fraction in its simplest form.

Worked Example

Use cancelling to write these in their lowest terms.

a $\frac{15}{20}$ **b** $\frac{6}{42}$

There is more about HCF on page 43.

Answer

a $\frac{15^3}{20^4} = \frac{3}{4}$ We divide 15 and 20 by 5.
5 is the highest common factor of 15 and 20.

b $\frac{6^1}{42^7} = \frac{1}{7}$ We divide 6 and 42 by 6.
6 is the highest common factor of 6 and 42.

If you find it hard to find the HCF, you could cancel in steps.

Example $\frac{28^4}{42^6} = \frac{4}{6}$ 7 is a common factor of 28 and 42. We can cancel again.
$\frac{4^2}{6^3} = \frac{2}{3}$ 2 is the highest common factor of 4 and 6.

Exercise 4

1 Cancel these to their simplest form.

a $\frac{3}{6}$ **b** $\frac{3}{9}$ **c** $\frac{5}{15}$ **d** $\frac{3}{12}$ **e** $\frac{4}{20}$ **f** $\frac{6}{18}$

g $\frac{4}{6}$ **h** $\frac{10}{15}$ **i** $\frac{9}{12}$ **j** $\frac{15}{45}$ **k** $\frac{16}{20}$ **l** $\frac{24}{40}$

m $\frac{20}{25}$ **n** $\frac{12}{18}$ **o** $\frac{24}{36}$ **p** $\frac{35}{42}$ **q** $\frac{56}{84}$ **r** $\frac{48}{64}$

s $\frac{56}{72}$ ***t** $\frac{36}{96}$ ***u** $\frac{48}{108}$ ***v** $\frac{80}{144}$

2 Give the answers as fractions in their lowest terms.
Hollydale School is organising a fair.

  a 4 of the 28 outside stalls are to be covered. What fraction will be covered?

  b 10 of the 150 volunteers are working full-time organising the fair. What fraction are not working full-time?

  c 15 of the 21 inside stalls need two volunteers on the day. What fraction is this?

  d The school secretary made 27 phone calls yesterday. 21 of them were about the fair. What fraction were not about the fair?

  e In a survey about providing help on the day of the fair, 6 out of 45 families said they could not help. What fraction is this?

Review 1   Use cancelling to write these fractions in their lowest terms.

a $\frac{4}{8}$ **b** $\frac{7}{21}$ **c** $\frac{6}{8}$ **d** $\frac{16}{20}$ **e** $\frac{10}{15}$ **f** $\frac{80}{100}$ **g** $\frac{18}{24}$

h $\frac{20}{24}$ **i** $\frac{5}{45}$ **j** $\frac{15}{40}$ **k** $\frac{42}{56}$ **l** $\frac{24}{40}$ **m** $\frac{27}{45}$ **n** $\frac{42}{63}$

Review 2   Give the answers as fractions in their simplest form.

a Tim had £18. He spent £6 at the fair. What fraction did he spend at the fair?

b There are 28 pupils in Lorna's class. 21 of them gave something for the fair raffle. What fraction did not give something?

Mixed numbers and improper fractions

Remember $\frac{7}{8}$ means the same as $7 \div 8$.

$\frac{7}{8}$ is called a **proper fraction**. The numerator (top) is smaller than the denominator (bottom).

$\frac{23}{4}$ is called an **improper fraction**. The numerator is larger than the denominator.

$3\frac{2}{5}$ is called a **mixed number**. A mixed number has a whole number and a fraction.

Worked Example

Write these as mixed numbers. **a** $\frac{25}{6}$ **b** $\frac{47}{7}$

Answer

a $25 \div 6 = 4$ with remainder 1
$\frac{25}{6} = \mathbf{4\frac{1}{6}}$

b $47 \div 7 = 6$ with remainder 5
$\frac{47}{7} = \mathbf{6\frac{5}{7}}$

Worked Example

Write these as improper fractions. **a** $4\frac{2}{3}$ **b** $6\frac{3}{5}$

Answer

a $4\frac{2}{3}$ is 4 wholes plus 2 thirds.
There are 12 thirds in 4 wholes.

$4\frac{2}{3}$ is 14 thirds.
$4\frac{2}{3} = \frac{12}{3} + \frac{2}{3} = \mathbf{\frac{14}{3}}$

b $6\frac{3}{5}$ is 6 wholes plus 3 fifths.
There are 30 fifths in 6 wholes.

$6\frac{3}{5}$ is 33 fifths.
$6\frac{3}{5} = \frac{30}{5} + \frac{3}{5} = \mathbf{\frac{33}{5}}$

Discussion

Michael changed mixed numbers to improper fractions like this.

$$5\frac{2}{3} = \frac{5 \times 3 + 2}{3} = \frac{17}{3} \qquad 4\frac{5}{7} = \frac{4 \times 7 + 5}{7} = \frac{33}{7}$$

Does this method always work? **Discuss**.

Exercise 5

1 | $\frac{7}{5}$ | $\frac{5}{7}$ | $\frac{2}{5}$ | $\frac{3}{4}$ | $\frac{4}{3}$ | $\frac{17}{4}$ | $\frac{8}{9}$ | $\frac{5}{6}$ | $\frac{6}{5}$ | $\frac{9}{8}$ | $\frac{10}{3}$ | $\frac{3}{10}$ |

  a Which of these are proper fractions?   **b** Which are improper fractions?

2 Write these as mixed numbers.

a $\frac{13}{5}$ **b** $\frac{17}{4}$ **c** $\frac{5}{3}$ **d** $\frac{9}{2}$ **e** $\frac{11}{5}$

f $\frac{19}{6}$ **g** $\frac{25}{2}$ **h** $\frac{25}{3}$ **i** $\frac{36}{7}$ **j** $\frac{14}{9}$

k $\frac{27}{4}$ **l** $\frac{36}{5}$ **m** $\frac{74}{9}$ **n** $\frac{107}{10}$ **o** $\frac{77}{8}$

p $\frac{63}{5}$ **q** $\frac{59}{7}$ **r** $\frac{55}{6}$ **s** $\frac{61}{8}$

3 How many fifths are in these?

a 8   **b** 12   **c** 7   **d** $6\frac{4}{5}$ **e** $7\frac{1}{5}$ **f** $8\frac{3}{5}$

4 How many quarters are in these?

   a 3   **b** 7   **c** 8   **d** $1\frac{1}{4}$ **e** $2\frac{3}{4}$ **f** 16

   g $3\frac{1}{4}$ **h** $4\frac{3}{4}$ **i** $6\frac{2}{4}$ **j** $5\frac{3}{4}$ ***k** $8\frac{1}{2}$ ***l** $5\frac{1}{2}$

5 a Maria cut $3\frac{5}{8}$ pizzas into eighths.
How many pieces did she get?

   b Maria poured 8 litres of juice into $\frac{1}{4}$ ℓ jugs.
How many jugs did she fill?

6 Write these as improper fractions.

   a $2\frac{3}{4}$ **b** $3\frac{1}{2}$ **c** $2\frac{2}{5}$ **d** $1\frac{5}{8}$ **e** $5\frac{1}{4}$ **f** $2\frac{5}{6}$

   g $5\frac{3}{8}$ **h** $7\frac{1}{6}$ **i** $3\frac{4}{9}$ **j** $5\frac{3}{10}$ **k** $3\frac{1}{4}$ **l** $4\frac{3}{4}$

   m $5\frac{1}{8}$ **n** $1\frac{2}{7}$ **o** $3\frac{5}{6}$ **p** $4\frac{5}{8}$ **q** $3\frac{2}{9}$ **r** $5\frac{3}{4}$

7 Write the answer to these as a mixed number.

   a $17 \div 3$ **b** $51 \div 8$ **c** $51 \div 4$ **d** $73 \div 3$

   e $83 \div 8$ **f** $94 \div 5$ **g** $100 \div 9$ **h** $201 \div 5$

***8** Which fraction is bigger?

   a $\frac{25}{6}$ or $4\frac{5}{6}$ **b** $\frac{13}{4}$ or $2\frac{3}{4}$ **c** $5\frac{3}{5}$ or $\frac{29}{5}$ **d** $3\frac{5}{8}$ or $\frac{27}{8}$ **e** $\frac{24}{3}$ or $7\frac{1}{3}$ **f** $6\frac{4}{9}$ or $\frac{57}{9}$ **g** $8\frac{1}{3}$ or $\frac{26}{3}$

Review 1   Which of these are **a** proper fractions **b** improper fractions?

$\frac{17}{4}$ $\frac{8}{9}$ $\frac{11}{10}$ $\frac{3}{5}$ $\frac{25}{29}$ $\frac{31}{7}$ $\frac{19}{6}$ $\frac{21}{30}$

Review 2   How many thirds are in these?

   a 4   **b** 10   **c** $2\frac{1}{3}$ **d** $1\frac{2}{3}$ **e** $4\frac{2}{3}$

Review 3   Peter took $3\frac{1}{4}$ oranges to share with his football team.
How many quarters will these oranges make?

***Review 4**   Which fraction is bigger?

   a $3\frac{1}{4}$ or $\frac{15}{4}$ **b** $\frac{23}{5}$ or $4\frac{2}{5}$ **c** $6\frac{2}{3}$ or $\frac{19}{3}$ **d** $\frac{50}{8}$ or $6\frac{3}{8}$

T

Review 5

| $9\frac{1}{3}$ | $\frac{20}{3}$ | $8\frac{1}{7}$ | $\frac{52}{7}$ | $\frac{71}{9}$ | $\overset{\textbf{E}}{5\frac{3}{5}}$ | | $8\frac{5}{6}$ | $8\frac{1}{7}$ | $8\frac{4}{9}$ | | $6\frac{7}{8}$ | $8\frac{5}{6}$ | $\overset{\textbf{E}}{5\frac{3}{5}}$ |

| $\frac{9}{2}$ | $\frac{20}{3}$ | $\overset{\textbf{E}}{5\frac{3}{5}}$ | $8\frac{1}{7}$ | $6\frac{7}{8}$ | $\overset{\textbf{E}}{5\frac{3}{5}}$ | $8\frac{4}{9}$ | $6\frac{7}{8}$ | | $10\frac{3}{10}$ | $\overset{\textbf{E}}{5\frac{3}{5}}$ | $\frac{52}{7}$ | $\frac{9}{2}$ | $6\frac{7}{8}$ | $8\frac{5}{6}$ |

| $9\frac{5}{6}$ | $9\frac{1}{3}$ | | $\frac{37}{3}$ | $8\frac{1}{7}$ | $\frac{49}{9}$ | $\overset{\textbf{E}}{5\frac{3}{5}}$ | $\frac{23}{6}$ | | $\frac{20}{3}$ | $9\frac{5}{6}$ | $8\frac{1}{7}$ | $\frac{23}{6}$ | $8\frac{4}{9}$ |

Use a copy of this box. Write the letter that is beside the question above its answer.

Write these as mixed numbers.

E $\frac{28}{5} = 5\frac{3}{5}$ **A** $\frac{57}{7}$ **T** $\frac{55}{8}$ **H** $\frac{53}{6}$ **S** $\frac{76}{9}$ **L** $\frac{103}{10}$ **O** $\frac{59}{6}$ **F** $\frac{84}{9}$

Write these as improper fractions.

G $4\frac{1}{2}$ **R** $6\frac{2}{3}$ **D** $3\frac{5}{6}$ **N** $7\frac{3}{7}$ **V** $5\frac{4}{9}$ **C** $7\frac{8}{9}$ **P** $12\frac{1}{3}$

Practical

You will need 4 dice (2 **red** dice and 2 **green** dice).
a Toss the two **red** dice.
   Add the numbers.
   This gives you the numerator.
Example   The dice shown give the numerator 7.
b Toss the two **green** dice.
   Add the numbers.
   This gives you the denominator.
Example   The dice shown give the denominator 3.

c Make the fraction, $\dfrac{\text{\textbf{red} dice total}}{\text{\textbf{green} dice total}}$.

Example   The dice shown give the fraction $\frac{7}{3}$.

Find the following.
i   The smallest fraction it is possible to toss.
ii The largest improper fraction it is possible to toss.
iii The largest proper fraction it is possible to toss.
iv The fractions that can be written as whole numbers.
v   The number of different fractions that can be tossed.

Note: Count $\frac{2}{2}$ as a fraction.
       Treat $\frac{2}{4}$ as the same fraction as $\frac{3}{6}$.

Investigation

Three-Digit Fractions

- $\frac{1}{2} = \frac{8}{16}$ $\frac{8}{16}$ has 3 digits, 8, 1 and 6.
  Which other 3-digit fractions are equal to $\frac{1}{2}$?
  Have you found them all? How do you know?
  $\frac{1}{3} = \frac{4}{12}$
  Find all the 3-digit fractions equal to $\frac{1}{3}$.
  Have you found them all? How do you know?
  What if you began with $\frac{1}{4}$?   **What if** you began with $\frac{1}{5}$?

- $2\frac{1}{2} = \frac{5}{2} = \frac{10}{4}$ $\frac{10}{4}$ is a 3-digit fraction equal to $2\frac{1}{2}$.
  Find all the 3-digit fractions equal to $2\frac{1}{2}$.
  What if you began with $1\frac{1}{2}$?   **What if** you began with $3\frac{1}{2}$?

  What if you began with $4\frac{1}{2}$?   **What if** ...

*● $1\frac{1}{3} = \frac{4}{3} = \frac{12}{9}$
  $\frac{12}{9}$ is the only 3-digit fraction equal to $1\frac{1}{3}$.

  What if you began with $2\frac{1}{3}$, $3\frac{1}{3}$, ... ?

Number

Fractions and decimals

Remember

$$0 \cdot 1 = \tfrac{1}{10} \qquad\qquad 0 \cdot 01 = \tfrac{1}{100}$$
$$0 \cdot 2 = \tfrac{2}{10} \qquad\qquad 0 \cdot 02 = \tfrac{2}{100}$$
$$0 \cdot 3 = \tfrac{3}{10} \qquad\qquad 0 \cdot 03 = \tfrac{3}{100}$$
$$1 \cdot 7 = 1\tfrac{7}{10} \qquad\qquad 2 \cdot 33 = 2\tfrac{33}{100}$$
$$3 \cdot 05 = 3\tfrac{5}{100}$$

To **convert** a **decimal to a fraction** make the denominator 10 or 100.

Cancel, if possible, to give the fraction in its simplest form.

Examples $\qquad 0 \cdot 6 = \dfrac{\cancel{6}^{3}}{\cancel{10}^{5}} \qquad\qquad 0 \cdot 71 = \dfrac{71}{100} \qquad\qquad 6 \cdot 25 = 6\dfrac{\cancel{25}^{1}}{\cancel{100}^{4}}$

$\qquad\qquad\qquad\quad = \tfrac{3}{5} \qquad\qquad\qquad\qquad\qquad\qquad\qquad\qquad = 6\tfrac{1}{4}$

Exercise 6

Try to do these mentally.

1 Write these as fractions in their simplest form.

a	0·3	**b**	0·7	**c**	0·9	**d**	0·8	**e**	0·5	**f**	0·6
g	0·41	**h**	0·77	**i**	0·83	**j**	0·25	**k**	0·14	**l**	0·18
m	0·64	**n**	0·75	**o**	0·45	**p**	0·72	**q**	0·85	**r**	0·94
s	1·7	**t**	1·4	**u**	2·43	**v**	2·78	**w**	3·65		

2 Use a copy of these diagrams. Beside each is a decimal. Show the decimal number by shading the diagram. **a** is done for you.

a

0·3

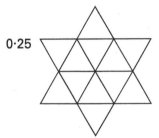

b

0·8

c

0·25

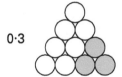

d

0·75

3 Which of the following does *not* show 0·6?

A

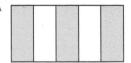

B

C

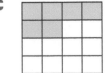

D

114

T **Review 1**

$$\frac{87}{100} \qquad 3\frac{1}{4} \quad 8\frac{7}{10} \quad 1\frac{2}{5} \quad 1\frac{3}{5} \quad \frac{11}{50} \qquad 3\frac{13}{20} \quad 8\frac{7}{10} \quad \frac{87}{100} \quad 1\frac{2}{5} \quad 1\frac{3}{5}$$

$$\overset{\textbf{E}}{} \qquad \overset{\textbf{E}}{}$$

$$3\frac{1}{4} \quad \frac{11}{20} \quad \frac{1}{5} \quad \frac{11}{50} \quad \frac{18}{25} \quad \frac{3}{20} \qquad \frac{1}{5} \quad 1\frac{2}{5} \quad \frac{17}{25} \quad \frac{11}{20}$$

Use a copy of this box.
Write these as fractions in their lowest terms. Write the letter beside each above its answer.

E $0.2 = \frac{1}{5}$ **A** 0·87   **S** 0·22   **T** 0·15   **H** 0·68   **O** 0·55

N 0·72   **K** 1·6   **U** 8·7   **C** 1·4   **D** 3·25   **Q** 3·65

T **Review 2**   Use a copy of these diagrams. Shade the diagram to show the decimal.

a   0·7

b   0·75

Remember

$\frac{1}{2} = 0·5 \qquad \frac{1}{4} = 0·25 \qquad \frac{3}{4} = 0·75 \qquad \frac{1}{5} = 0·2$

To **convert** a **fraction to a decimal** we can

1 use one of the facts above

$$\frac{2}{5} = 2 \times \frac{1}{5} \qquad \frac{1}{8} = \frac{1}{4} \div 2 \qquad \frac{9}{12} = \frac{3}{4}$$
$$\phantom{\frac{2}{5}} = 2 \times 0·2 \qquad \phantom{\frac{1}{8}} = 0·25 \div 2 \qquad \phantom{\frac{9}{12}} = 0·75$$
$$\phantom{\frac{2}{5}} = 0·4 \qquad \phantom{\frac{1}{8}} = 0·125$$

Divide top and bottom by 3.

2 make an equivalent fraction with denominator 10 or 100.

Multiply top and bottom by 5.

$$\frac{7}{20} = \frac{35}{100} \qquad \frac{12}{40} = \frac{3}{10}$$
$$\phantom{\frac{7}{20}} = 0·35 \qquad \phantom{\frac{12}{40}} = 0·3$$

Divide top and bottom by 4.

Exercise 7

1 Write these as decimals.

a $\frac{7}{10}$ **b** $\frac{9}{10}$ **c** $\frac{31}{100}$ **d** $\frac{77}{100}$ **e** $\frac{8}{100}$ **f** $\frac{4}{100}$ **g** $1\frac{3}{10}$ **h** $1\frac{7}{10}$

i $2\frac{42}{100}$ **j** $3\frac{67}{100}$ **k** $7\frac{9}{100}$ **l** $\frac{25}{10}$ **m** $\frac{35}{10}$ **n** $\frac{216}{100}$ **o** $\frac{315}{100}$

2 Convert these to decimals. Find the answers mentally.

a $\frac{1}{2}$ **b** $\frac{1}{4}$ **c** $\frac{3}{4}$ **d** $\frac{3}{5}$ **e** $\frac{4}{8}$ **f** $\frac{3}{12}$ **g** $1\frac{1}{2}$ **h** $\frac{9}{12}$

i $\frac{1}{8}$ **j** $2\frac{1}{4}$ **k** $2\frac{1}{5}$ **l** $\frac{12}{16}$ **m** $3\frac{3}{6}$ **n** $5\frac{9}{12}$

3 Write these as decimals by first writing them with a denominator of 10 or 100.

a $\frac{4}{20}$ **b** $\frac{4}{5}$ **c** $\frac{4}{50}$ **d** $\frac{2}{5}$ **e** $\frac{14}{20}$ **f** $\frac{3}{20}$ **g** $\frac{7}{50}$ **h** $\frac{40}{200}$

i $\frac{9}{20}$ **j** $\frac{8}{40}$ **k** $\frac{36}{60}$ **l** $\frac{36}{200}$ **m** $\frac{9}{30}$ **n** $\frac{24}{80}$ **o** $\frac{16}{40}$

4 What fraction of these shapes is shaded?
Write your answer as a decimal.

a   **b** 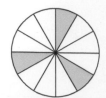 **c**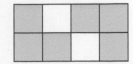

Review Write these as decimals.

a $\frac{1}{5}$ **b** $\frac{6}{8}$ **c** $\frac{79}{100}$ **d** $\frac{4}{100}$ **e** $\frac{24}{10}$ **f** $\frac{227}{100}$

g $3\frac{1}{2}$ **h** $\frac{14}{20}$ **i** $\frac{6}{25}$ **j** $\frac{13}{50}$ **k** $1\frac{3}{4}$

Comparing fractions

A diagram is a good way to **compare fractions**.

We can see from this diagram that

$\frac{1}{2} > \frac{1}{3} > \frac{1}{4} > \frac{1}{5} > \frac{1}{6} > \frac{1}{7} > \frac{1}{8}$.

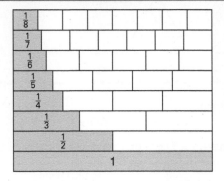

We could also use lines divided into parts.

Worked Example
Write these fractions in order, smallest to largest.

$\frac{3}{4}, \quad \frac{4}{5}, \quad \frac{2}{3}$

Answer
We can see from the lines that the
fractions in order from smallest to
largest are $\frac{2}{3}, \frac{3}{4}, \frac{4}{5}$.

A line of 60 mm is
easy to divide up.

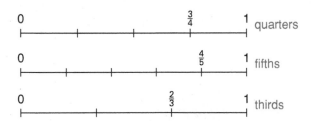

Exercise 8

1 Which of < or > goes in the box?

a $\frac{1}{2}\square\frac{2}{3}$ **b** $\frac{4}{5}\square\frac{1}{2}$ **c** $\frac{2}{7}\square\frac{1}{2}$ **d** $\frac{3}{8}\square\frac{1}{2}$ **e** $\frac{2}{3}\square\frac{4}{5}$ **f** $\frac{5}{6}\square\frac{7}{8}$ **g** $\frac{6}{7}\square\frac{3}{4}$

2 Write these fractions in ascending order.

a $\frac{1}{2}, \frac{1}{4}, \frac{3}{8}$ **b** $\frac{5}{8}, \frac{2}{3}, \frac{3}{4}$ **c** $\frac{5}{6}, \frac{6}{7}, \frac{2}{3}$,   **d** $\frac{7}{8}, \frac{3}{4}, \frac{4}{5}$ **e** $\frac{2}{3}, \frac{5}{6}, \frac{3}{5}$ **f** $\frac{1}{4}, \frac{2}{5}, \frac{3}{8}$

3 Write these fractions in order, smallest first.

  a $1\frac{1}{4}$, $1\frac{1}{2}$, $1\frac{2}{3}$ **b** $3\frac{1}{4}$, $3\frac{3}{8}$, $3\frac{2}{5}$, $3\frac{1}{5}$ **c** $2\frac{3}{10}$, $1\frac{7}{8}$, $2\frac{3}{5}$, $1\frac{3}{5}$, $1\frac{5}{6}$

  d $1\frac{1}{2}$, $1\frac{2}{3}$, $1\frac{5}{6}$, $1\frac{3}{4}$, $1\frac{4}{5}$ **e** $4\frac{1}{4}$, $4\frac{1}{5}$, $4\frac{2}{7}$, $4\frac{3}{8}$, $4\frac{1}{3}$

* 4 Which of these fractions is closer to 1? Explain how you can tell.

  a $\frac{3}{4}$ or $\frac{4}{3}$ **b** $\frac{7}{8}$ or $\frac{8}{7}$

Review   Write these fractions from largest to smallest.

a $\frac{3}{4}$, $\frac{2}{3}$, $\frac{5}{6}$ **b** $\frac{1}{2}$, $\frac{7}{8}$, $\frac{1}{5}$, $\frac{1}{6}$ **c** $4\frac{1}{10}$, $4\frac{1}{2}$, $3\frac{3}{5}$, $3\frac{3}{4}$, $3\frac{1}{2}$

Adding and subtracting fractions

We can **add and subtract simple fractions mentally**.

$\frac{1}{4} + \frac{1}{2} = \frac{3}{4}$

$\frac{1}{4}$ $\frac{1}{2}$ $\frac{3}{4}$

$\frac{3}{4} + \frac{3}{4} = 1\frac{1}{2}$

$\frac{3}{4}$ $\frac{3}{4}$ $1\frac{1}{2}$

$\frac{1}{8} + \frac{1}{8} = \frac{1}{4}$

$\frac{1}{8}$ $\frac{1}{8}$ $\frac{1}{4}$

It is easy to add and subtract fractions when the **denominators are the same**.

Examples

$\frac{2}{5} + \frac{1}{5} = \frac{2+1}{5}$
$= \frac{3}{5}$

$\frac{5}{8} - \frac{3}{8} = \frac{5-3}{8}$
$= \frac{2}{8}^{\,1}_{\,4}$
$= \frac{1}{4}$

Cancel the fraction to its lowest terms.

$\frac{3}{10} + \frac{2}{10} + \frac{4}{10} + \frac{3}{10} = \frac{3+2+4+3}{10}$
$= \frac{12}{10}$
$= 1\frac{2}{10}^{\,1}_{\,5}$
$= 1\frac{1}{5}$

Change improper fractions to mixed numbers.

Exercise 9

1 Find the answers to these mentally.

  a $\frac{1}{2} + \frac{1}{4}$ **b** $\frac{3}{4} + \frac{1}{2}$ **c** $\frac{1}{2} - \frac{3}{8}$ **d** $\frac{3}{4} - \frac{1}{2}$ **e** $\frac{1}{8} + \frac{1}{8} + \frac{1}{8}$

  f $\frac{1}{2} - \frac{1}{8}$ **g** $\frac{1}{8} + \frac{1}{8} + \frac{1}{4}$ **h** $\frac{3}{4} + \frac{1}{8}$ **i** $\frac{5}{8} - \frac{1}{4}$

Remember: $\frac{3}{3} = 1$

2 a $\frac{1}{3} + \frac{1}{3}$ **b** $\frac{1}{4} + \frac{1}{4}$ **c** $\frac{1}{8} + \frac{2}{8}$ **d** $\frac{2}{5} + \frac{2}{5}$ **e** $\frac{6}{7} - \frac{4}{7}$

  f $\frac{3}{10} + \frac{2}{10}$ **g** $\frac{5}{8} + \frac{3}{8}$ **h** $\frac{9}{12} - \frac{3}{12}$ **i** $\frac{11}{20} + \frac{3}{20}$ **j** $\frac{7}{12} - \frac{5}{12}$

  k $\frac{13}{15} - \frac{8}{15}$ **l** $\frac{3}{10} + \frac{5}{10}$ **m** $\frac{7}{12} - \frac{1}{12}$ **n** $\frac{17}{20} - \frac{13}{20}$

3 a $\frac{1}{3} + \frac{1}{3} + \frac{2}{3}$ **b** $\frac{4}{5} + \frac{1}{5} + \frac{2}{5}$ **c** $\frac{7}{8} + \frac{5}{8} + \frac{3}{8}$ **d** $\frac{5}{6} + \frac{5}{6} + \frac{1}{6} + \frac{1}{6}$

  e $\frac{7}{10} + \frac{3}{10} + \frac{1}{10} + \frac{4}{10}$ **f** $\frac{3}{12} + \frac{5}{12} + \frac{7}{12} + \frac{4}{12}$ **g** $\frac{5}{20} + \frac{17}{20} + \frac{8}{20} + \frac{6}{20}$

4 a $\frac{7}{10} + \frac{2}{10} - \frac{4}{10}$ **b** $\frac{4}{20} + \frac{13}{20} - \frac{8}{20}$ **c** $\frac{3}{12} + \frac{5}{12} + \frac{4}{12} - \frac{3}{12}$ **d** $\frac{5}{9} + \frac{7}{9} - \frac{3}{9}$

  e $\frac{6}{15} + \frac{7}{15} - \frac{8}{15}$ **f** $\frac{11}{12} + \frac{11}{12} + \frac{5}{12} - \frac{1}{12}$ **g** $\frac{19}{20} + \frac{3}{20} - \frac{7}{20}$ **h** $\frac{8}{10} + \frac{3}{10} - \frac{5}{10}$

5 $\frac{1}{8}$ of the pupils at Jake's school walked to school. $\frac{3}{8}$ came by bus. What fraction either walked or came by bus?

*6 Debs had $\frac{7}{12}$ of her bus card left. She used another $\frac{3}{12}$ of it. What fraction did she have left then?

*7 $\frac{1}{5}$ of a stall at a fair is used for sweets and $\frac{2}{5}$ for cakes. The rest is to be used for biscuits. What fraction of the stall is left for biscuits?

Review 1 Find the answers to these mentally.

a $\frac{1}{2}+\frac{1}{8}$ **b** $\frac{1}{2}+\frac{3}{8}$ **c** $\frac{1}{2}-\frac{1}{4}$ **d** $\frac{1}{4}+\frac{1}{8}$ **e** $\frac{3}{4}-\frac{1}{8}$ **f** $\frac{1}{4}+\frac{1}{4}+\frac{1}{8}$

Review 2   **a** $\frac{3}{8}+\frac{4}{8}$ **b** $\frac{9}{12}-\frac{5}{12}$ **c** $\frac{5}{6}+\frac{1}{6}$ **d** $\frac{3}{5}+\frac{3}{5}$

e $\frac{5}{8}+\frac{3}{8}+\frac{1}{8}$ **f** $\frac{4}{10}+\frac{7}{10}+\frac{5}{10}+\frac{1}{10}$ **g** $\frac{15}{20}-\frac{5}{20}+\frac{7}{20}$

Review 3 $\frac{5}{9}$ of a class went on a trip in the morning. $\frac{2}{9}$ went in the afternoon. What fraction of the class went on the trip?

We **add and subtract fractions with different denominators** using equivalent fractions.

Examples $\frac{1}{3}+\frac{5}{6}=\frac{2}{6}+\frac{5}{6}$

$=\frac{2+5}{6}$

$=\frac{7}{6}$

$=1\frac{1}{6}$

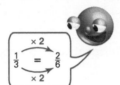

$\frac{7}{8}-\frac{1}{4}=\frac{7}{8}-\frac{2}{8}$

$=\frac{7-2}{8}$

$=\frac{5}{8}$

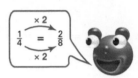

Exercise 10

1 a $\frac{1}{2}+\frac{3}{10}$ **b** $\frac{1}{3}+\frac{1}{6}$ **c** $\frac{1}{3}+\frac{4}{9}$ **d** $\frac{3}{8}+\frac{1}{4}$ **e** $\frac{3}{7}+\frac{5}{14}$

   f $\frac{7}{10}-\frac{1}{5}$ **g** $\frac{7}{8}-\frac{1}{4}$ **h** $\frac{9}{10}-\frac{1}{2}$ **i** $\frac{5}{12}-\frac{1}{6}$

2 a $\frac{1}{2}+\frac{1}{4}+\frac{3}{8}$ **b** $\frac{3}{10}+\frac{2}{5}+\frac{1}{10}$ **c** $\frac{3}{4}+\frac{1}{8}-\frac{1}{4}$ **d** $\frac{5}{9}+\frac{1}{3}-\frac{2}{9}$

   e $\frac{3}{4}+\frac{2}{3}+\frac{5}{12}$ **f** $\frac{7}{20}-\frac{1}{5}+\frac{1}{4}$ **g** $\frac{13}{15}-\frac{1}{5}+\frac{1}{3}$

3 Fran bought $\frac{2}{5}$ m of ribbon and then another $\frac{3}{10}$ m. What fraction of a metre did she buy altogether?

4 $\frac{1}{4}$ of a park is grass and $\frac{3}{8}$ is a playground. What fraction of the park is this altogether?

Review 1   **a** $\frac{1}{4}+\frac{3}{8}$ **b** $\frac{3}{4}+\frac{1}{12}$ **c** $\frac{2}{3}-\frac{2}{9}$ **d** $\frac{1}{8}+\frac{3}{16}$ **e** $\frac{7}{20}-\frac{1}{4}$

Review 2   **a** $\frac{1}{4}+\frac{7}{8}+\frac{1}{2}$ **b** $\frac{9}{10}-\frac{3}{5}+\frac{1}{2}$ **c** $\frac{11}{12}-\frac{3}{4}+\frac{5}{6}$

Fraction of

Remember

To find $\frac{1}{5}$ we divide by 5. $\frac{1}{5}$ of 30 = 30 ÷ 5 = 6

To find $\frac{1}{8}$ we divide by 8. $\frac{1}{8}$ of 56 = 56 ÷ 8 = 7

Worked Example

A test was out of 60.
Lorraine got $\frac{2}{3}$ of the possible marks.
How many marks did she get?

Answer

First find $\frac{1}{3}$ of 60 and then multiply this
answer by 2.

$\frac{1}{3}$ of 60 = 60 ÷ 3
              = 20

$\frac{2}{3}$ of 60 = **2** × 20
              = **40**

You can do ones like this mentally.

Worked Example

Find $\frac{1}{8}$ of 28.

Answer

Find $\frac{1}{4}$ of 28, then halve it.

$\frac{1}{4}$ of 28 = 7

$\frac{1}{2}$ of 7 = **3·5**

Worked Example

Find   **a** 0·75 of 48   **b** 1·25 of 3·6

Answer

a

$0·75 = \frac{3}{4}$

0·75 of 48 = $\frac{3}{4}$ of 48

$\frac{1}{4}$ of 48 = 48 ÷ 4
              = 12

$\frac{3}{4}$ of 48 = **3** × 12
              = **36**

b

$1·25 = 1\frac{1}{4}$

1·25 of 3·6 = $1\frac{1}{4}$ of 3·6

$\frac{1}{4}$ of 3·6 = 3·6 ÷ 4
              = 0·9

$1\frac{1}{4}$ of 3·6 = 3·6 + 0·9
              = **4·5**

Exercise 11

Questions 1, 2, 3, 4 and Review 1 are to be done mentally. You may use jottings.

1 a $\frac{1}{8}$ of £24   **b** $\frac{1}{6}$ of 30 cm   **c** $\frac{1}{7}$ of 56 g   **d** $\frac{2}{3}$ of 12 mm

  e $\frac{3}{4}$ of 20 kg   **f** $\frac{2}{5}$ of £25   **g** $\frac{3}{8}$ of £16   **h** $\frac{5}{6}$ of 42 p

  i $\frac{4}{5}$ of 35 m   **j** $\frac{3}{4}$ of 60 kg   **k** $\frac{2}{5}$ of 150 g   **l** $\frac{3}{4}$ of 200 p

  m $\frac{3}{8}$ of 160 mm   **n** $\frac{2}{3}$ of £120   **o** $\frac{1}{8}$ of 20 kg   **p** $\frac{1}{4}$ of £10

  q $\frac{9}{10}$ of 1 metre   **r** $\frac{3}{10}$ of 2 litres

2 a $1\frac{1}{2}$ of 20 kg   **b** $1\frac{1}{4}$ of 40 cm   **c** $1\frac{1}{4}$ of 36 g   **d** $1\frac{2}{4}$ of 20 ℓ

  e $2\frac{1}{2}$ of 10 p   **f** $4\frac{1}{2}$ of 8 m   **g** $2\frac{1}{4}$ of 8 ℓ   **h** $5\frac{1}{4}$ of 16 g

3 a 0·5 of 18   **b** 0·5 of 28   **c** 0·5 of 60   **d** 0·25 of 24

  e 0·25 of 48   **f** 0·75 of 32   **g** 0·75 of 60   **h** 1·5 of 12

  i 1·5 of 50   **j** 1·25 of 80   **k** 1·25 of 120

4 Find the answers to these mentally.
   a Only about $\frac{1}{6}$ of suitable organ donors actually donate their organs. Out of 48 suitable donors, how many will donate their organs?
   b About $\frac{3}{4}$ of Amy's drama group went to Pizza Hut at the end of term. If there were 20 people in the drama group, how many went to Pizza Hut?
   c Alison got $\frac{9}{10}$ of the multiple-choice questions correct. If there were 50 multiple-choice questions, how many did she get correct?
   d In the story Peter Pan, the children spent $\frac{1}{8}$ of a day in Never-Never Land. How many hours was this?
   e A Jupiter day is about $\frac{3}{8}$ of an Earth day. About how many hours are there in a Jupiter day?
   f On the moon, a person's weight is about $\frac{1}{6}$ of their weight on earth. About how many kilograms would a person weigh on the moon if their weight on earth is 54 kg?

5 a $\frac{3}{10}$ of £120   **b** $\frac{7}{10}$ of 210 km   **c** $\frac{3}{5}$ of 350 m   **d** $\frac{5}{9}$ of 450 cm
   e $\frac{3}{7}$ of 280 mℓ **f** $\frac{3}{2}$ of 12 kg   **g** $\frac{6}{5}$ of 25 ℓ **h** $\frac{5}{4}$ of 40 cm

> You may need to write some things down to do these.

6 a Write $\frac{3}{10}$ of 1 cm in millimetres.   **b** Write $\frac{7}{1000}$ of 1 metre in millimetres.
   c Write $\frac{163}{1000}$ of 1 ℓ in millilitres.   **d** Write $\frac{23}{1000}$ of 4 kg in grams.

7 Sasha bought 35 m of material. She used $\frac{4}{5}$ of it to make curtains and the rest to make two bedspreads.
   a How much did she use for curtains?
   b How much did she use for each bedspread?

*8 a $\frac{2}{5}$ of 4·5 m   **b** $\frac{3}{4}$ of 2·4 cm   **c** $\frac{2}{3}$ of £1·80   **d** $\frac{3}{8}$ of 1·6 kg
   e $\frac{5}{6}$ of 4·8 mm   **f** $1\frac{1}{2}$ of 6·2 g   **g** $1\frac{1}{4}$ of 1·2 m   **h** $1\frac{3}{4}$ of 8·4 m

*9 A basketball should bounce $\frac{2}{3}$ of the height from which it is dropped.
   a How high should a basketball bounce when it is dropped from a height of 1·8 metres?
   b A basketball is dropped from 6·3 m and it bounced to a height of 4·2 m. Is it bouncing as it should?

⊤ **Review 1**

												A			
100	45	50	50	150	21	60	18	40	150	180	50	16	180	50	150

27	40	50	90	50	90	150	40	21	60	150	150	50

Use a copy of this box.
Find the answer to these mentally. Write the letter that is beside each question above its answer.

A $\frac{2}{3}$ of 24 = 16   **W** $\frac{3}{4}$ of 36   **I** $\frac{2}{5}$ of 100   **R** $\frac{3}{8}$ of 56   **U** $\frac{5}{6}$ of 54
H $\frac{3}{7}$ of 210   **E** $\frac{5}{9}$ of 270   **L** $1\frac{1}{2}$ of 12   **F** 0·5 of 120   **T** 1·25 of 40
B $\frac{5}{7}$ of 140   **S** $\frac{3}{8}$ of 480

Review 2   Kelly made 45 small cakes for the school fair. $\frac{4}{9}$ of them were iced.
a   How many were iced?
b   How many were not iced?

* **Review 3**   Geoff bought 3·6 ℓ of paint to paint his room.
He used $\frac{5}{6}$ of it. How much paint did he have left?

Puzzle

1	2		3	4		5	6	7		8	9		10		11	12		13		14	15
WO																					

Use a copy of the chart.
Use the clues below to fill it in. The first one is done.

1 First two fifths of *women*.
3 First quarter of *blankets*.
5 Second half of *lane*.
7 Last third of *really*.
9 Last three quarters of *nice*.
11 First two fifths of *munch*.
13 First two thirds of *ask*.
15 Last fifth of *chairwomen*.

2 First three quarters of *mend*.
4 Last three fifths of *think*.
6 Last two thirds of *car*.
8 First two thirds of *two*.
10 Middle third of *Alaska*.
12 Last third of *crunch*.
14 Last third of *Sam*.

Multiplying fractions and integers

$\frac{1}{4}$ of 20 is the same as $\frac{1}{4} \times 20$ or $20 \times \frac{1}{4}$.
$\frac{1}{4}$ of 20 is also the same as $20 \div 4$.
$\frac{1}{4}$ of 20, $\frac{1}{4} \times 20$, $20 \times \frac{1}{4}$, $20 \div 4$ are all equivalent.

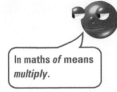

In maths *of* means *multiply*.

Worked Example
a $\frac{1}{8} \times 24$ b $\frac{5}{8} \times 24$ c 0.75×16

Answer
a $\frac{1}{8} \times 24 = \frac{1}{8}$ of 24
      $= 3$

b $\frac{5}{8} \times 24 = 5 \times \frac{1}{8} \times 24$
      $= 5 \times 3$
      $= 15$

c $0.75 \times 16 = \frac{3}{4} \times 16$
      $= 3 \times \frac{1}{4} \times 16$
      $= 3 \times 4$
      $= 12$

Worked Example
Find   a $\frac{1}{4} \times 3$ b $\frac{3}{5} \times 4$

Answer
a $\frac{1}{4} \times 3 = \frac{3}{4}$

b $\frac{3}{5} \times 4 = 3 \times \frac{1}{5} \times 4$
      $= 3 \times \frac{4}{5}$
      $= \frac{12}{5}$

We could write $\frac{12}{5}$ as $2\frac{2}{5}$.

Exercise 12

1 Write true or false for these.

a $\frac{1}{4}$ of $16 = \frac{1}{4} \times 16$ **b** $\frac{1}{3}$ of $18 = 18 \times 3$ **c** $\frac{1}{4} \times 24 = 24 \div 4$

d $1\frac{1}{2} \times 18 = 1\frac{1}{2}$ of 18 **e** $\frac{3}{4} \times 24 = 3 \times \frac{1}{4} \times 24$ **f** $8 \times 0{\cdot}2 = \frac{2}{5} \times 8$

2 Calculate

a $\frac{3}{5} \times 20$ **b** $\frac{3}{8} \times 16$ **c** $\frac{7}{10} \times 40$ **d** $\frac{5}{6} \times 24$ **e** $\frac{2}{3} \times 60$

f $\frac{3}{4} \times 48$ **g** $\frac{7}{8} \times 56$ **h** $\frac{3}{7} \times 210$ **i** $45 \times \frac{2}{5}$ **j** $32 \times \frac{5}{8}$

k $48 \times \frac{5}{6}$ **l** $100 \times \frac{3}{4}$ **m** $120 \times \frac{2}{3}$ **n** $1\frac{1}{2} \times 4$ **o** $1\frac{1}{2} \times 10$

p $1\frac{1}{4} \times 20$ **q** $2\frac{1}{2} \times 6$ **r** $2\frac{1}{4} \times 16$ **s** $1\frac{3}{4} \times 24$

3 Calculate

a $\frac{1}{3} \times 2$ **b** $\frac{1}{5} \times 3$ **c** $\frac{1}{6} \times 5$ **d** $\frac{1}{8} \times 5$

e $\frac{1}{4} \times 3$ **f** $\frac{2}{5} \times 3$ **g** $\frac{3}{8} \times 5$ **h** $\frac{5}{6} \times 5$

i $\frac{3}{5} \times 4$ **j** $1\frac{1}{2} \times 3$ **k** $1\frac{1}{4} \times 10$

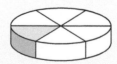

> Change the decimals to fractions first.

4 Calculate

a $0{\cdot}25 \times 12$ **b** $0{\cdot}75 \times 16$ **c** $0{\cdot}2 \times 40$ **d** $0{\cdot}4 \times 50$

e $0{\cdot}8 \times 25$ **f** $0{\cdot}1 \times 70$ **g** $0{\cdot}3 \times 120$ ***h** $1{\cdot}1 \times 60$

***i** $1{\cdot}3 \times 40$ ***j** $2{\cdot}7 \times 20$ ***k** $3{\cdot}3 \times 30$

Review 1   Calculate

a $\frac{2}{5} \times 15$ **b** $28 \times \frac{3}{7}$ **c** $1\frac{1}{2} \times 16$ **d** $1\frac{1}{4} \times 12$ **e** $\frac{1}{5} \times 2$

f $\frac{1}{8} \times 3$ **g** $\frac{2}{3} \times 4$ **h** $\frac{3}{4} \times 9$ **i** $1\frac{1}{2} \times 5$ **j** $2\frac{1}{2} \times 3$

Review 2   Calculate

a $0{\cdot}25 \times 28$ **b** $0{\cdot}75 \times 32$ **c** $0{\cdot}2 \times 120$ ***d** $2{\cdot}1 \times 70$

Summary of key points

A   One out of six is written as $\frac{1}{6}$.

$\frac{1}{6}$ of this shape is shaded.

B   When writing **one number as a fraction of another**, make sure the numerator and denominator have the same units.

Example   To write 37 minutes as a fraction of an hour we change one hour to 60 minutes.

37 minutes as a fraction of an hour $= \frac{37}{60}$.

$\frac{2}{3}, \frac{4}{6}, \frac{6}{9}, \frac{8}{12}, \frac{10}{15}, \ldots$ are **equivalent fractions**.

We find **equivalent fractions** by multiplying or dividing both the numerator and denominator by the same number.

Example

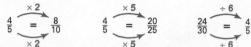

$\frac{4}{5}, \frac{8}{10}, \frac{20}{25}, \frac{24}{30}$ are equivalent fractions.

 A fraction is in its **lowest terms** or **simplest form** if the numerator and denominator have no common factors except 1.

We simplify a fraction to its lowest terms by **cancelling**.

Example $\frac{24^4}{30^5} = \frac{4}{5}$ Both 24 and 30 have been divided by 6. 6 is the highest common factor of 24 and 30.

 A **proper fraction** has a smaller numerator than denominator. *Example* $\frac{3}{4}$

An **improper fraction** has a larger numerator than denominator. *Example* $\frac{13}{6}$

A **mixed number** has a whole number and a fraction. *Example* $2\frac{4}{5}$

 To convert a **decimal to a fraction**, write it as tenths or hundredths and then cancel

$0.8 = \frac{8^4}{10^5}$ $0.93 = \frac{93}{100}$ $1.65 = 1\frac{65}{100}$

$= \frac{4}{5}$ $= 1\frac{13}{20}$

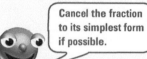

Cancel the fraction to its simplest form if possible.

 We can write **a fraction as a decimal** using facts we already know or by making an equivalent fraction with denominator 10 or 100.

Examples $\frac{3}{5} = 3 \times \frac{1}{5}$ $\frac{8}{16} = \frac{1}{2}$ $\frac{3}{20} = \frac{15}{100}$

  $= 3 \times 0.2$ $= 0.5$ $= 0.15$

  $= 0.6$

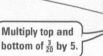

Multiply top and bottom of $\frac{3}{20}$ by 5.

 We can use this diagram to **compare fractions**.

We can also divide lines of the same length into parts.

Example

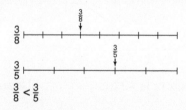

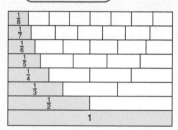

$\frac{3}{8} < \frac{3}{5}$

 We can **add (or subtract) fractions** which have the **same denominator** as follows.

Example $\frac{3}{5} + \frac{4}{5} = \frac{3+4}{5}$ ← keep the denominator

    $= \frac{7}{5}$

    $= 1\frac{2}{5}$

J To **add (or subtract) fractions** which have **different denominators** change to equivalent fractions with the same denominator.

Example $\frac{1}{2} + \frac{3}{10} = \frac{5}{10} + \frac{3}{10}$

$= \frac{5+3}{10}$

$= \frac{8}{10}$

$= \frac{4}{5}$

$\frac{1}{2} = \frac{5}{10}$

 K **Fraction of a quantity**

Example   To find $\frac{3}{5}$ of £45. $\frac{1}{5}$ of £45 $= \frac{1}{5} \times$ £45

$=$ £45 $\div$ 5

$=$ £9

$\frac{3}{5}$ of £45 $= 3 \times$ £9

$=$ £27

Do these mentally whenever you can.

 L $\frac{1}{3}$ of 12, $\frac{1}{3} \times 12$, $12 \times \frac{1}{3}$, $12 \div 3$ are all equivalent.

Examples $\frac{2}{3} \times 6 = 2 \times \frac{1}{3} \times 6$ $\frac{2}{7} \times 5 = \frac{10}{7}$

$= 2 \times 2$

$= 4$

Test yourself

1 Which of these has $\frac{1}{4}$ shaded?   **Ⓐ**

A 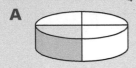   **B** 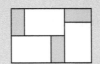   **C**   **D**

2 What fraction of   **Ⓑ**
   a 1 metre is 51 centimetres   **b** 1 kilogram is 49 grams   **c** 1 hour is 43 minutes?

3 What fraction of a complete turn does the minute hand make between 8:15 p.m. and   **Ⓑ**
   8:25 p.m.?

4 Copy and complete.   **Ⓒ**
   a $\frac{3}{5} = \frac{6}{} = \frac{}{15} = \frac{12}{}$ **b** $\frac{5}{8} = \frac{}{16} = \frac{}{24} = \frac{20}{} = \frac{25}{}$ **c** $\frac{4}{9} = \frac{8}{} = \frac{12}{} = \frac{}{45} = \frac{24}{}$ **d** $\frac{2}{7} = \frac{4}{} = \frac{}{21} = \frac{}{28} = \frac{}{35}$

5 Write four equivalent fractions for each of these.   **Ⓒ**
   a $\frac{2}{3}$ **b** $\frac{3}{4}$ **c** $\frac{1}{8}$ **d** $\frac{24}{32}$ **e** $\frac{20}{36}$

6 Use cancelling to write these as fractions in their simplest form.

   a $\frac{6}{12}$ **b** $\frac{6}{24}$ **c** $\frac{4}{12}$ **d** $\frac{7}{21}$ **e** $\frac{10}{15}$

   f $\frac{9}{15}$ **g** $\frac{20}{25}$ **h** $\frac{24}{30}$ **i** $\frac{28}{42}$ **j** $\frac{27}{45}$

7 Give the answers to these fractions in their lowest terms.

   a In a quiz, Michael got 80 out of 120 of the questions correct. What fraction did he get wrong?

   b On Monday 48 animals were seen by a vet. 36 were dogs. What fraction is this?

8 Write these as mixed numbers.

   a $\frac{24}{7}$ **b** $\frac{29}{4}$ **c** $\frac{37}{5}$ **d** $\frac{43}{8}$ **e** $\frac{51}{6}$

9 How many fifths are in these?

   a $4\frac{1}{5}$ **b** $3\frac{3}{5}$ **c** $2\frac{4}{5}$

10 Write these as improper fractions.

   a $1\frac{1}{4}$ **b** $2\frac{3}{5}$ **c** $4\frac{7}{8}$ **d** $3\frac{5}{9}$ **e** $6\frac{5}{7}$

11 Write these as fractions.

   a 0·5   **b** 0·1   **c** 0·6   **d** 0·43   **e** 0·73   **f** 0·24

   g 0·45   **h** 0·86   **i** 0·94   **j** 0·28   **k** 0·36

12 Write these as decimals.

   a $\frac{2}{10}$ **b** $\frac{57}{100}$ **c** $\frac{7}{20}$ **d** $1\frac{4}{5}$

13 Write these fractions in order, biggest first.

   a $\frac{1}{4}, \frac{2}{5}, \frac{3}{4}, \frac{3}{8}, \frac{1}{6}$ **b** $1\frac{4}{5}, 1\frac{3}{4}, 1\frac{7}{8}, 1\frac{5}{8}$

14 a $\frac{3}{7} + \frac{3}{7}$ **b** $\frac{1}{8} + \frac{1}{8} + \frac{1}{8}$ **c** $\frac{3}{5} + \frac{4}{5}$ **d** $\frac{3}{4} - \frac{1}{2}$

   e $\frac{7}{8} - \frac{3}{8}$ **f** $\frac{5}{12} + \frac{7}{12}$ **g** $\frac{12}{13} - \frac{6}{13}$

15 a $\frac{1}{3} + \frac{1}{6}$ **b** $\frac{7}{8} - \frac{1}{2}$ **c** $\frac{2}{3} + \frac{5}{6}$ **d** $\frac{7}{8} - \frac{3}{4}$

16 a $\frac{1}{10}$ of £420   **b** $\frac{1}{5}$ of 45 m   **c** $\frac{1}{7}$ of 140 ℓ **d** $\frac{1}{4}$ of 60 km

   e $\frac{2}{3}$ of £24   **f** $\frac{5}{8}$ of 56 cm   **g** $\frac{3}{7}$ of 21 m   **h** $\frac{7}{8}$ of 160 g

   i $\frac{1}{4}$ of £18   **j** $1\frac{1}{2}$ of 16 m   **k** 0·2 × 25 m

 Do as many as possible mentally.

17 Find the answers to these mentally.

   a How many minutes is $\frac{5}{6}$ of an hour?

   b Jack had 18 biscuits. He ate $\frac{2}{3}$ of them and gave the rest away. How many did he give away?

18 a $\frac{1}{3} \times 24$ **b** $16 \times \frac{1}{4}$ **c** $\frac{3}{8} \times 32$ **d** $70 \times \frac{3}{10}$

   e $1\frac{1}{4} \times 24$ **f** $\frac{1}{5} \times 2$ **g** $5 \times \frac{2}{3}$ **h** $\frac{2}{3} \times 10$

   i $1\frac{1}{2} \times 7$ **j** $0·4 \times 50$ **k** $2·1 \times 30$.

 Do as many as possible mentally.

6 Percentages, Fractions, Decimals

You need to know

⋯ **Key vocabulary** ⋯⋯⋯⋯⋯⋯⋯⋯⋯⋯⋯⋯⋯⋯

discount,   percentage (%),   sale price

 A Bit Off

- The advertisement in the centre is for the internet.
  What might the other advertisements be for?

  Find some other real-life percentages.
  Make a poster or collage of them.

- Labels on clothes often have the percentage of wool,
  polyester, cotton, ...

  Find as many labels as you can.

Percentages to fractions and decimals

Remember
Lucy got 87% in her French test.
'per cent' means out of 100.
87% is 87 out of 100 or $\frac{87}{100}$.

Sometimes we want to **write a percentage as a fraction or decimal**.
We do this by writing the percentages as the number of parts per hundred.

Example $23\% = \frac{23}{100} = 0{\cdot}23$

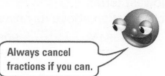

23% as   23% as
a fraction   a decimal

Worked Example
65% of Lucy's class passed the French test.
What fraction is this?

Answer
$65\% = \frac{^{13}\cancel{65}}{_{20}\cancel{100}} = \frac{13}{20}$

Always cancel
fractions if you can.

Example $142\% = \frac{^{71}\cancel{142}}{_{50}\cancel{100}} = \frac{71}{50} = 1\frac{21}{50}$

Exercise 1

1 Write these as fractions in their lowest terms.
a	1%	b	25%	c	50%	d	30%	e	70%	f	75%
g	90%	h	40%	i	60%	j	20%	k	5%	l	45%
m	62%	n	120%	o	154%	p	174%	q	185%		

2 Write the percentages in **question 1** as decimals.

3 In a survey it was found that 39% of people think we will find
   intelligent life on another planet by 2010.
   What fraction of people surveyed thought this?

4 Belinda got 95% in her Science test.
   Write this as a decimal.

5 Akbar paid a 20% deposit.
   Write this as a decimal.

6 78% of people in Britain listen to tapes, records or CDs.
   What fraction is this?

7 There are 100 pupils at Tara's school.
   76 of them eat lunch at school.
   a What percentage eat lunch at school?
   b Write your answer to a as a decimal.
   c What percentage do not eat lunch at school?
   d Write your answer to c as a decimal.
   e What do you notice about your answers to a and c?
   f What do you notice about your answers to b and d?

*8 Write as a decimal.
  a 8·5%   b 15·5%   c 24·5%

T Review 1

                      A

$\frac{17}{100}$ $\frac{11}{20}$ $\frac{11}{20}$ $\frac{67}{100}$ $\frac{3}{25}$ 0·15 0·36 $\frac{11}{20}$ $\frac{16}{25}$ 0·15

A   A

$\frac{67}{100}$ $\frac{27}{50}$ 0·15 $\frac{6}{25}$ $\frac{67}{100}$ $\frac{9}{25}$ 0·15   0·2 0·15 $\frac{6}{25}$ 0·36 $\frac{23}{25}$ $\frac{9}{20}$

0·36 0·15 $\frac{27}{50}$ 0·15 $\frac{9}{20}$ $\frac{22}{25}$ $\frac{17}{100}$ $\frac{9}{20}$ $\frac{4}{5}$ $\frac{11}{20}$ 0·15 0·36 $\frac{11}{20}$ $\frac{23}{25}$

       A   A

0·35 $\frac{67}{100}$ 0·78 0·78 $\frac{67}{100}$ 0·36 0·78 0·15 0·15 0·2

Use a copy of this box.
Write these as fractions in their lowest terms.
Write the letter that is beside each question above its answer.

A 67% = $\frac{67}{100}$ I 17%   H 64%   U 80%   O 92%   N 45%
M 88%   K 12%   T 55%   R 24%   G 36%   V 54%

Write these as decimals.
P 20%   F 35%   E 15%   L 78%   S 36%

Review 2   In a survey it was found that 95% of Americans had been to McDonald's in the last year.
Write 95% as a a decimal   b fraction.

Fractions and decimals to percentages

Discussion

    1 = 100%

How could you use this to show that these are true?
Discuss.

    10 = 1000%
    $\frac{1}{10}$ = 0·1 = 10%
    $\frac{1}{100}$ = 0·01 = 1%
    $\frac{1}{8}$ = 0·125 = 12$\frac{1}{2}$%
    $\frac{1}{4}$ = 0·25 = 25%
    $\frac{1}{3}$ = 0·333 ... = 33$\frac{1}{3}$%

I ate 100% of the pie. That's a whole pie.

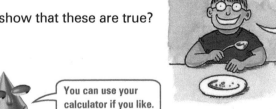

You can use your calculator if you like.

We can **write fractions and decimals as percentages** by writing them as fractions with denominators of 100.

Remember per cent means out of 100.

Examples

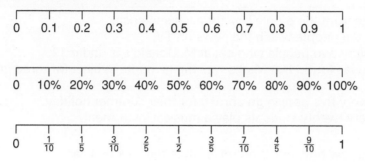

$$\frac{4}{5} \overset{\times 20}{=} \frac{80}{100} = 80\% \qquad \frac{9}{20} \overset{\times 5}{=} \frac{45}{100} = 45\%$$

$$0{\cdot}69 = \frac{69}{100} = 69\% \qquad 0{\cdot}7 = \frac{7}{10} = \frac{70}{100} = 70\%$$

You need to know these

$$\frac{1}{3} = 33\frac{1}{3}\% \qquad\qquad \frac{2}{3} = 66\frac{2}{3}\%$$

We could use these numbers lines to help.

Worked Example

Nicola entered a boomerang throwing competition.
She had to throw the boomerang and catch it again.
Nicola did this on four out of her ten throws.
a What fraction of the throws did the boomerang come back?
b Write the answer to **a** as a decimal.
c What percentage success did Nicola have?

Answer
a 4 out of 10 is $\frac{4}{10} = \frac{2}{5}$
b $\frac{2}{5} = \mathbf{0{\cdot}4}$
c $\frac{2}{5} = \mathbf{40\%}$

Exercise 2

1 Write these as percentages.
   a $\frac{1}{2}$ b $\frac{1}{4}$ c $\frac{3}{4}$ d $\frac{1}{10}$ e $\frac{7}{10}$ f $\frac{1}{5}$
   g $\frac{3}{10}$ h $\frac{1}{3}$ i $\frac{7}{20}$ j $\frac{3}{20}$ k 1   l $1\frac{1}{4}$
   m $2\frac{3}{4}$ n 2   o $2\frac{3}{10}$ p $3\frac{1}{5}$ q $1\frac{1}{3}$

2 Write these as percentages.
   a $0{\cdot}18$ b $0{\cdot}27$ c $0{\cdot}94$ d $0{\cdot}39$ e $0{\cdot}56$ f $0{\cdot}14$
   g $0{\cdot}43$ h $0{\cdot}8$ i $0{\cdot}6$ j $0{\cdot}08$ k $0{\cdot}02$

3 What percentage of these diagrams is blue?

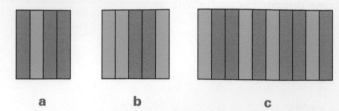

a   b   c

4 0.65 of a chemical compound is zinc. What percentage is this?

5

| 40% | 75% | 30% | 20% | 50% | 60% | 25% | 80% | 70% |

Find a percentage from the purple box for each sentence.
a   Seven out of ten people watch the news on TV.
b   One out of every two people who eat at McDonald's is under 12.
c   In a survey it was found that Maths was the favourite subject for three quarters of students.
d   Two out of every five people go abroad for their summer holiday.
e   Six out of every twenty students play a musical instrument.

6 These diagrams show how much petrol is left in the tank of a car.

E is empty. F is full.   What percentage is left in each case?

7 a   5p in 100p. What percentage is this?   b   25p in £1. What percentage is this?

8 a   What fraction of this shape is red?
   b   What percentage is red?
   c   What percentage is *not* red?

9 a   What fraction of the circles are yellow?
   b   Write this as a percentage.
   c   Write this as a decimal.

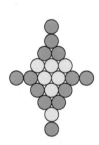

10 41% of a cardigan is cotton.
   a   What fraction is cotton?
   b   Write the answer to a as a decimal.

11 Rachel saved £2 out of her £5 pocket money.
   a   What percentage of her pocket money did she save?
   b   Write the answer to a as a decimal.

T

12 Use a copy of this.
   a   Colour $\frac{1}{5}$ of it blue.
   b   Colour 30% of it red.
   c   Colour 0·2 of it green.
   d   What percentage is *not* coloured?

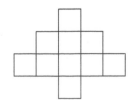

T

13 Use a copy of this table.
Fill it in. The first one is done for you.

Decimal	Percentage	Fraction	Decimal	Percentage	Fraction
0·39	39%	$\frac{39}{100}$		40%	
		$\frac{7}{10}$	0·45		
	75%				$\frac{8}{50}$
		$\frac{1}{5}$		35%	

14 Estimate the percentage that is **red**.

a

b

c

*15 This pie chart shows what sort of take-aways
Year 7 students like best.
  a   What percentage like fish and chips?
  b   What fraction like Burger King?
  c   What percentage do *not* like KFC best?

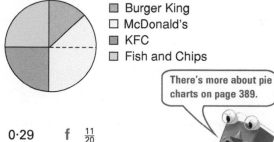

☐ Burger King
☐ McDonald's
☐ KFC
☐ Fish and Chips

There's more about pie charts on page 389.

Review 1   Write these as percentages.
a $\frac{2}{5}$ b 0·71   c 4   d $\frac{9}{10}$ e 0·29   f $\frac{11}{20}$
g 0·9   h 0·04   i $1\frac{3}{5}$

Review 2   Amanda's computer has 20 megabytes of memory. 15 megabytes are full.
a What percentage of the memory is full?
b Write the answer to a as a decimal.

Review 3
a What fraction of this shape is purple?
b What percentage is purple?
c What percentage is *not* purple?

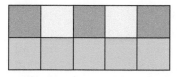

Review 4   A dam is at 0.45 of its maximum water level.
a Give the water level of the dam as a percentage.
b What fraction of the dam is *not* full?

Review 5   Estimate the percentage of the lace that is yellow.

a

b

Practical

Ask 20 pupils some questions like

What colour are your eyes?
How old are you?
What size shoes do you take?
What month were you born?

Work out some percentages.

> You could show your results on a pie chart.

Example
You could work out what percentage of the 20 pupils

have brown eyes
are boys
are older than you
are born in the first half of the year.

Make a poster or booklet showing your results.

Percentage of (mentally)

We sometimes use facts like those above to find percentages mentally.

Examples
10% of £50 = £5   dividing by 10
10% of 86 m = 8·6 m dividing by 10
5% of 8 ℓ = 0·4 ℓ finding 10% of 8 ℓ and halving
15% of 60 g = 9 g finding 10%, then 5% and adding
25% of 160 cm = 40 cm dividing by 4

Example Broomfield School raised £5600. 11% of this went to the RSPCA.
11% of £5600 = **£616** 10% of £5600 = £560
          1% of £5600 = £56
        11% of £5600 = £560 + £56
              = £616

Example 30% of 180 m = **54 m** 10% of 180 = 18
             30% = 3 × 18
               = 54

3 × 18 = 3 × 10 + 3 × 8
   = 30 + 24
   = 54

Exercise 3 **This exercise is to be done mentally.**

T

1 Use a copy of this.
Shade the answers to each of these.
Your shading will give you a decimal number.
What is it?

3	9	12	421	241	13	64	2	36
150	15	0.5	85	102	16	10	89	26
24	100	75	0.8		8.3	8	4	400
52	300	4.5	51	63	5.7	201	90	142
30	19.5	7.5	5.4	1	19	7.4	6	20

Find 10% of   **a** 30   **b** 80   **c** 74
Find 30% of   **a** 100   **b** 120   **c** 250
Find 25% of   **a** 80   **b** 400   **c** 30
Find 20% of   **a** 60   **b** 20   **c** 710
Find 50% of   **a** 18   **b** 104   **c** 9
Find 5% of   **a** 200   **b** 40   **c** 10
Find 100% of   **a** 150   **b** 6   **c** 19·5
Find 80% of   **a** 500   **b** 80   **c** 30

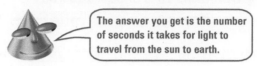

The answer you get is the number of seconds it takes for light to travel from the sun to earth.

2 Find
   a 10% of 50   **b** 20% of 30   **c** 25% of 24   **d** 5% of 300
   e 5% of 120   **f** 40% of 160   **g** 30% of 180   **h** 60% of 120
   i 80% of 230   **j** 25% of 48   **k** 75% of 16   **l** 20% of 54
   m 40% of 32   **n** 70% of 25   **o** 30% of 41   **p** 15% of 60
   q 15% of 120   **r** 11% of 7200   **s** 11% of 8500   *****t** 200% of 10
  *****u** 200% of 50   *****v** 300% of 100   *****w** 400% of 15.

3 a Jamie borrowed £160 from his aunt.
She charged him 10% interest.
How much interest did Jamie pay?
   b There are 180 pupils in a school.
30% went on a camp.
How many pupils went on camp?
   c Sudi sells computers.
She gets 5% of the value of each one she sells.
What does she get if she sells a computer for £840?

4 David's family went out for dinner.
The cost of the meal was £50.
They gave a 15% tip. How much did they tip?

*****5** The femur (thighbone) is a long bone.
Its length is about 25% of a person's height.
Ann is 1·6 m tall. About how long is her femur?

*****6** A jeweller took 10% off all gold rings for one week.
At the end of the week she added 10% of the sale price back.
These new prices were not the same as the original prices.
Why not?

Review 1
a 10% of 86 kg   **b** 30% of 20 g   **c** 25% of 36 ℓ   **d** 70% of £60
e 40% of £360   **f** 5% of 164 cm   **g** 15% of £240   **h** 11% of 3200 m
*****i** 200% of 25 m   *****j** 300% of 8 ℓ   *****k** 400% of £12

Review 2   Mr Hussein bought a flat for £50 000.
He sold it a year later and made a 15% profit.
How much profit did he make?

Percentage of — using the calculator

We can use fractions or decimals to find 23% of 55.

Using Fractions 23% of 55 = $\frac{23}{100} \times 55$

Key ②③÷①⓪⓪×⑤⑤= to get 12·65.

Using Decimals 23% of 55 = 0·23 × 55

Key ⓪·②③×⑤⑤= to get 12·65.

> Which way do you think is the quickest?

Key 14·5% as ①④·⑤÷①⓪⓪ if using fractions or
⓪·①④⑤ if using decimals.

Discussion

Danielle found 87% of £120 using the calculator.
Her answer was given on the screen as shown.
How should she write her answer? **Discuss**.

$$104.4$$

---

1 Use fractions to find these.
  a 26% of 8300 g   b 34% of 6500 m   c 18% of 6350 cm
  d 87% of £32   e 44% of 3250 mm   f 66% of 255 kg
  g 14% of 5660 kg   h 28% of £3650

2 Use decimals to find these.
  a 16% of £26·50   b 38% of £136   c 55% of £72·40
  d 84% of £425·50   e 98% of £24·50   f 46% of £199
  g 9% of £44   h 8% of £474·50   *i 14·5% of £52
 *j 17·5% of £82   *k 17·5% of £38

3 A TV survey was given to the 550 pupils at Chatfield College.
  a 78% of pupils watched TV every night. How many was this?
  b 46% watched for more than 1 hour every day. How many was this?

4 4550 people went to a play at the Sunset Theatre.
  Of these, 14% had seen the play before.
  How many had not seen the play before?

5 An adult's brain weighs about 2% of a person's total weight. What is the approximate
  weight of Sandy's brain if Sandy weighs 56 kg?

6 Mary bought a house for £135 000.
   She sold it 5 years later and made a 45% profit.
   a How much profit did Mary make?
   b How much did she sell the house for?

7 Daniel paid a 30% deposit on a £45·50 coat.
   a How much deposit was this?
   b How much more did he have to pay?

8 Sinita bought a bike for £75.
   When she sold it, she made a 35% loss.
   How much did Sinita lose when she sold the bike?

9 16% of a 125 g pot of 'Fruit Delight' is fruit.
   How many grams is *not* fruit?

10
   Was £25   Was £39·50   Was £89·40

SALE
30%
OFF

   Mark bought these three things in the sale.
   a How much did he save altogether?
   b How much did he pay altogether?

11 Which of these is true?
   A 85% of £9 < 90% of £8·50
   B 85% of £9 > 90% of £8·50
   C 85% of £9 = 90% of £8·50

12
Blood Type	% of population
O	42
A	44
B	10
AB	4

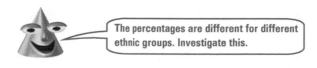
The percentages are different for different ethnic groups. Investigate this.

   a Out of 850 students at a school, how many would you expect to have type **O** blood?
   *b Out of 26 students in a class, how many would you expect to have type **A** blood?
   *c In Matthew's class of 28 pupils, three had **AB** blood type.
      Is this more or less than expected?

* 13 44% of the pupils on a technology visit are boys.
     25% of the boys and 50% of the girls are from Essex.
     What percentage of all the pupils on the visit are from Essex.

* 14 The value of a £60 000 painting increased by 5% last year.
     Its new value increased by a further 10% this year.
     What is the painting worth now?

Review 1
a 28% of 3600 kg
b 34% of 8150 m
c 16% of 8420 mm
d 82% of £3140
e 18% of 670 g
f 27% of 920 kg
g 48% of 7580 g
h 16% of 104 m

Review 2 84% of the profit made at a gala day is to go to charity. How much goes to charity if the total profit is £213·50?

∗ Practical

1 Design a floor plan of a bungalow or house of total floor area 100 m². Have the bedrooms take up 40% of the space, the lounge and dining room another 40% and the kitchen, bathroom, stairs and hallway 20%.

2 Investigate tipping. You could include some of these: where it began, the countries where you tip, when you tip, the wages of people who get tips, how much you should tip.

Summary of key points

 To write a **percentage as a fraction or decimal**, write the percentage as the number of parts per hundred.

Examples $57\% = \frac{57}{100}$ $85\% = \frac{85}{100}\,\frac{17}{20}$

$= 0·57$ $= \frac{17}{20}$

Cancel if you can.

 To write a **fraction or decimal as a percentage** write it with denominator of 100.

Examples $0·83 = \frac{83}{100} = 83\%$

$$\frac{7}{20} \xrightarrow{\times 5} = \frac{35}{100} = 35\%$$

C $10\% = \frac{1}{10}$ 5% is half of 10% $25\% = \frac{1}{4}$
We sometimes use these facts **to find the percentage of a quantity mentally**.

Examples   20% of 300 m = 60 m   finding 10% and doubling it
15% of 80 g = 12 g   finding 10%, then 5% and adding

 We sometimes need to **use the calculator** to find the **percentage of a quantity**.
Using Fractions   47% of 8250 $= \frac{47}{100} \times 8250$
     Key ④ ⑦ ÷ ① ⓪ ⓪ × ⑧ ② ⑤ ⓪ ⩵ to get 3877·5
Using Decimals   47% of 8250 = 0·47 × 8250
     Key ⓪ · ④ ⑦ × ⑧ ② ⑤ ⓪ ⩵ to get 3877·5

Test yourself **except for questions 12, 13 and 14.**

1 Write these as fractions in their lowest terms.
   a 70%   b 75%   c 35%   d 62%   e 175%

2 Write the percentages in **question 1** as decimals.

3 40% of the world's population is under 15.
   a What fraction is this?
   b Write your answer to **a** as a decimal.

4 Write these as percentages.
   a $\frac{1}{4}$ b $\frac{9}{10}$ c $\frac{4}{5}$ d $\frac{13}{20}$ e 3   f $1\frac{4}{5}$

5 Write these as percentages.
   a 0·63   b 0·84   c 0·05   d 0·11

6 a What fraction of this shape is red?
   b What percentage is red?
   c Write the answer to **b** as a decimal.

7 Find a percentage from the green box for each sentence.
   a Two in every five fast food stores sell burgers.
   b Three out of every ten students wear glasses.
   c Nine out of every twenty students have brown hair.

| 25% |
| 30% |
| 35% |
| 40% |
| 45% |

8 63% of the pupils at a band practice wore school uniform.
   a What fraction did *not* wear school uniform?
   b Write your answer to **a** as a decimal.

9 Estimate how full these jars are.
   Give your answer as a percentage.
   a   b   c

10 a 10% of 60   b 20% of 50   c 60% of 90   d 25% of 24
    e 50% of 3·6   f 25% of 2   g 15% of 40   h 11% of 8200

11 There are 80 pupils on a computer camp. 30% are under 14.
    How many pupils are under 14?

12 a 19% of £35   b 24% of 34 ℓ c 36% of 95 m
    d 83% of £1672   *e 6·5% of £8   *f 17·5% of £25

13 Which of these is true?   A 40% of £35 < 35% of £40
                               B 40% of £35 > 35% of £40
                               C 40% of £35 = 35% of £40

14 a Sam is 160 cm tall. Kim is 85% as tall as this.
       How tall is Kim?
    b Paul is 180 cm tall. Jessica is 94% as tall as Paul.
       How tall is Jessica?

15 Joshua got 94%. Mary got 19 out of 20.
    Who got the highest mark?

7 Ratio and Proportion

You need to know

✓ ratio and proportion page 5

Key vocabulary

proportion,   ratio,   equivalent ratio

▶▶ All Mixed Up

The chef in this picture is about to mix up some orange drink from concentrated juice. She has to mix 3 parts water to 1 part juice. What does this mean? What other things do we mix in parts?

Proportion

Worked Example
£1 is worth US$1·62.
How many US$ would Bethany get for £20?

Answer
£20 is worth 20 times as much as £1.
£20 is worth 20 × 1·62 = **US$32·40**.

Exercise 1   **except for questions 4 and 6**

1   Three packets of crisps cost £1·20.
    How much will these cost?
    a 6 packets   **b** 12 packets   **c** 18 packets   **d** 30 packets

2   Four chocolate bars cost £1·40.
    How much will these cost?
    a 8 bars   **b** 12 bars   **c** 20 bars   **d** 36 bars

3   1 litre of punch contains 300 mℓ of lemonade.
    How much lemonade is there in
    a 2 ℓ punch   **b** 3 ℓ punch   **c** $1\frac{1}{2}$ ℓ punch   **d** $7\frac{1}{2}$ ℓ punch?

4   £1 is worth 175 Japanese Yen.
    How many Yen will James get for
    a £30   **b** £50   **c** £150   **d** £225?

The Yen is called the currency of Japan

5   Raphael uses 12 tomatoes to make $\frac{1}{2}$ ℓ of sauce.
    a   How much sauce can he make with 36 tomatoes?
    b   How many tomatoes does he need to make 2 litres of sauce?

6   £1 = NZ$3·34
    How much are these amounts in NZ$?
    a £2   **b** £5   **c** £1·50   **d** £50   **e** £185

*7   Julian uses this recipe to make pizzas for four people.
    He wants to make a pizza for ten.
    What quantity of these does he need?
    a tomatoes   **b** bacon   **c** cheese

Pizza

2 pizza bases
2 onions
250 g cheese
4 tomatoes
120 g bacon

139

Review 1   5 kg of apples cost £6.
How much will these cost?
a 10 kg   **b** 15 kg   **c** 25 kg   **d** 40 kg

Curry
250g meat
1 onion
1 cup curry sauce
2 tbsp sultanas

* **Review 2**   Here is a recipe for curry for two people.
How many grams of meat would be needed to make
curry for three people?

Writing ratios

2 : 3   4 : 7   1 : 6   5 : 9

These are **ratios**. We read 2 : 3 as '2 to 3'.

Example   The ratio of flour to butter in pastry is 1 : 2.
This means for every one part of flour there are two parts of butter.

The **order** in a ratio is important. 1 : 2 is different from 2 : 1.

Worked Example
On a rock climbing course there are 17 adults and five children.

Find the ratio of   **a** adults to children
                       b children to adults.

Answer
a In our ratio we must have the number of adults first, then the number of children.
The ratio of adults to children is **17 : 5**.
b In our ratio we must have the number of children first, then the number of adults. The ratio
of children to adults is **5 : 17**.

Worked Example
In water (H_2O) there are twice as many hydrogen atoms as oxygen atoms. Find the ratio of the
number of hydrogen to oxygen atoms.

Answer
For every two hydrogen atoms there is one oxygen atom.
The ratio is **2 : 1**.

Exercise 2

1   A class has 18 boys and 11 girls.
Write the ratio of
a boys to girls   **b** girls to boys.

2   Orange paint has 13 parts yellow and 9 parts red.
Write the ratio of
a   yellow paint to red paint
b   red paint to yellow paint.

3 Mr Steven took his daughter Pru cross country skiing.
Pru is 103 cm tall and it was recommended she used
skis 120 cm long. What is the ratio of Pru's height to the length
of her skis?

4 The average rainfall in Jakarta is twice that in Manchester. What
is the ratio of the rainfall in Jakarta to the rainfall in Manchester?

5 Darren spent four times as long on his project as Charlotte.
Find the ratio of the time Charlotte spent on her project to the
time Darren spent.

6 a Julian pours one bottle of lemonade and three
bottles of orange into a large bowl.
What is the ratio of orange to lemonade?

  ***b** Olivia pours one bottle of lemonade and one
bottle of orange into a big jug.
She wants only half as much orange as lemonade
in her jug.
What should Olivia pour into the jug now?

Review 1   A drink is made by mixing one part of cordial with three parts of water.
Find the ratio of cordial to water in this drink.

Review 2   A blend of tea consists of two parts China tea to three parts Ceylon tea.
What is the ratio of Ceylon tea to China tea in this blend?

Review 3   Tracey got three times as many marks in a test as Judy.
What is the ratio of Tracey's marks to Judy's marks?

Equivalent ratios

Discussion

Discuss how to show the ratio 4 : 6 on this line.

Discuss how to show the ratio 4 : 6 on this line.
Use your second line to write another ratio equal to 4 : 6. What other ratios are equal
to 4 : 6? **Discuss**.

● **Discuss** how to shade a circle so that the ratio of the shaded part to the unshaded
part is 10 : 2. What other ratio is equal to 10 : 2? **Discuss**.

What other ratios could you show by shading parts of a circle? **Discuss**.

141

The ratio of **red** dots to **black** dots is 12 : 4.
There are three times as many red dots as black dots.
The ratio is also 3 : 1.

• • • • • • •   • •
• • • • • •   • •

12 : 4 and 3 : 1 are **equivalent ratios**.

To find an equivalent ratio we can divide both parts of a ratio by the same number **or** we can multiply both parts of a ratio by the same number.

Examples

$$\div 4 \left(\begin{array}{c} 12 : 8 \\ = 3 : 2 \end{array} \right) \div 4 \qquad \times 6 \left(\begin{array}{c} 2 : 5 \\ = 12 : 30 \end{array} \right) \times 6$$

Worked Example

What number goes in the gap to make these ratios equivalent?

a 1 : 5 = 3 : __   b 12 : 30 = __ : 5

Answer

a
$$\times 3 \left(\begin{array}{c} 1 : 5 \\ = 3 : 15 \end{array} \right) \times 3$$
b
$$\div 6 \left(\begin{array}{c} 12 : 30 \\ = 2 : 5 \end{array} \right) \div 6$$

We can **simplify a ratio** by cancelling.

Examples   15 : 3   Dividing both parts of the ratio by 3
             = 5 : 1

             20 : 36   Dividing both parts of the ratio by 4
             = 5 : 9

5 : 1 and 5 : 9 are in their simplest form.

A ratio is in its **simplest form** when both parts of the ratio have no common factor other than 1.

Exercise 3

1 What number goes in the gap to make these ratios equivalent?
   a 1 : 2 = 3 : __   b 3 : 9 = __ : 3   c 1 : 4 = 2 : __   d 4 : 16 = __ : 4
   e 1 : 5 = 5 : __   f 6 : 30 = 1 : __   g 2 : 3 = 4 : __   h 2 : 3 = __ : 12
   i 10 : 4 = 5 : __   j 3 : 15 = __ : 5

2 Which of these ratios is equivalent to 4 : 16?
   A 4 : 1   B 8 : 32   C 5 : 17   D 1 : 3

3 Give these ratios in their simplest form.
   a 4 : 12   b 5 : 20   c 3 : 18   d 12 : 20
   e 16 : 24   f 20 : 36   g 36 : 48

You can use cancelling to do this.

4 On a school camp there were 15 adults and 40 pupils.
   Write the ratio of adults to pupils in its simplest form.

kJ are a measure of energy.

*5 Cycling burns 2700 kJ an hour. Slow walking burns 300 kJ an hour.
    Write the ratio of kJ burnt cycling to kJ burnt slow walking as a ratio
    in its simplest form.

*6 This table shows the colour of hats sold.
    Write these as ratios in their simplest form.
    a red hats to green hats   b green hats to black hats
    c blue hats to black hats   d black hats to red hats

Red	24
Green	16
Blue	21
Black	30

Review 1 What number goes in the gap to make these ratios equivalent?
a 1 : 3 = 2 : __   b 1 : 3 = __ : 12   c 8 : 10 = 4 : __
d 2 : 3 = __ : 24   e 6 : 15 = 2 : __

Review 2 Give these ratios in their simplest form.
a 3 : 9   b 4 : 16   c 8 : 20   d 18 : 30

Practical

1 Use a recipe book to find ratios used in cooking.

You could find the ratio *quantity of flour : quantity of sugar* for cakes and biscuits. You could find some other ratios.

2 Look at food labels. Find ratios for *protein : fat* or *fat : carbohydrate* or *fat : sugar* ...
How many different ratios can you find?

3 **You will need** a tape measure.

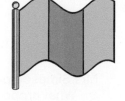

Find these ratios for the pupils in your class.
    handspan : height
    length of thumb : wrist circumference
    head circumference : neck circumference
Are the ratios different for boys and girls?

You might have to approximate the numbers to make the ratios easier.

Find the same ratios for some older people and some younger people.
Do the ratios change with age?

Ratio and proportion

Ratio compares part to part.
Proportion compares part to whole.

Example   This flag has three parts, two green and one red.
        The ratio of green to red is two parts to one part or 2 : 1

        Two of the three parts are green.
        The proportion of green in the whole flag is $\frac{2}{3}$.

Worked Example
There are 20 jelly beans in a jar. 14 are black and six are blue.
a Write the ratio of black jelly beans to blue jelly beans in its simplest form.
b Write the proportion of blue jelly beans as a percentage.

Answer
a ratio of black to blue = 14 : 6
                    = **7 : 3**
b Six out of 20 are blue.
   $\frac{6}{20} = \frac{3}{10}$
   30% are blue.

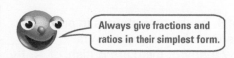

Always give fractions and ratios in their simplest form.

Number

1 The RSPCA had 12 kittens needing homes. Four are grey and eight are ginger.
   a What proportion of the kittens are grey? Give your answer as a fraction.
   b What proportion of the kittens are ginger? Give your answer as a fraction.
   c What is the ratio of grey to ginger kittens?

2 On a busy road, there were ten accidents last month. Three were injury and seven were non-injury accidents.
   a Give the proportion of injury accidents as a fraction.
   b Give the proportion of non-injury accidents as a percentage.
   c What is the ratio of injury to non-injury accidents?

3 This scale shows how much had been raised towards new computers.
   a What is the ratio of blue to red on this scale?
   b What proportion of the scale is blue? Give the answer as a fraction, decimal and percentage.
   c What proportion of the scale is red? Give the answer as a fraction, decimal and percentage.

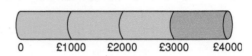

4 a What is the ratio of sad to happy faces on this banner?
   b What proportion of faces are sad?
   c What proportion of faces are happy?
   Give the answer to b and c as a fraction, decimal and percentage.

5 Of 20 people asked, 15 liked a summer holiday and five liked a winter holiday.
   a Write, in its simplest form, the ratio of people who liked a summer holiday to people who liked a winter holiday.
   b What proportion liked a summer holiday?
   c What proportion liked a winter holiday?
   Give the answers to b and c as a fraction, decimal and percentage.

6

   Mrs Simpson put these 20 pieces of fruit in a bowl.
   a What proportion of the fruit are bananas?
   b What proportion are pears?
   c What is the ratio of bananas to pears?

*7 One-fifth of Juliet's friends went rock climbing.
   a What is the ratio of friends who went rock climbing to those who didn't?
   b What proportion went rock climbing?

*8 Rajiv and Rosie each have some stamps in an album. The table shows how many stamps they have and how many of them are British. Who has the greater proportion of British stamps, Rajiv or Rosie?

	Number of stamps	Number of British stamps
Rajiv	165	15
Rosie	216	18

Review 1   In a class of 20 drama students, 11 were boys.
a   Give the proportion of boys as a fraction.
b   Give the proportion of girls as a percentage.
c   What is the ratio of boys to girls in the class?

Review 2
a   What is the ratio of purple to yellow squares?
b   What proportion of squares are purple?
    Give the answer as a fraction, decimal and percentage.
c   What proportion of squares are yellow?
    Give the answer as a fraction, decimal and percentage.

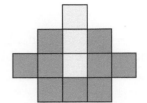

* **Review 3**
On Thursday and Friday Brett counted the
number of cars which stopped in the 'no
stopping' area outside the school.
This table shows the number of cars that
stopped outside the school and the number
that stopped in the 'no stopping' area.
On which day did the greater proportion of
cars stop in the 'no stopping' area?

	Number of cars stopping	**Number of cars in 'no stopping'**
Thursday	189	21
Friday	216	27

Practical

Gather data from your class to make a poster about ratio and proportion.

You could include things like

    ratio of boys to girls
    proportion of pupils with computers at home
    proportion of pupils with skate boards
    ratio of those who wear glasses to those who don't
    ratio of left-handed to right-handed pupils
    proportion of pupils who like a particular TV programme
    proportion of pupils who can roll their tongue
    ratio of those who walk to those who come by bus
    proportion of pupils who were not born in Britain.

Solving ratio and proportion problems

Worked Example
At a party there are six fruit bowls and five plates of cakes on every table.
The are 48 fruit bowls altogether.
How many plates of cakes are there?

Answer
48 fruit bowls = **8** lots of 6 fruit bowls.
So there must be **8** lots of 5 plates of cakes.
There are **40** plates of cakes.

Number

Worked Example
The ratio of boys to girls in a drama class is 2 : 5.
There are 30 girls in the class.
How many boys are there?

Answer
30 girls = **6** lots of 5 girls.
So there must be **6** lots of 2 boys or **12** boys.

Exercise 5

1 A recipe for a salad has two peppers for every seven tomatoes.
   Ruth makes the salad and uses six peppers. How many tomatoes
   does she use?

2 For every three girls in a class there are four boys.
   There are 15 girls in the class.
   How many boys are there?

3 For every US$16 you get £10.
   How many pounds would you get for US$48?

4 A class voted between going to the cinema and going to the beach.
   The result was 5 : 3 in favour of the cinema.
   If 20 pupils voted for the cinema, how many voted for the beach?

5 The ratio of dogs to ponies at a show was 2 : 5
   There were 18 dogs at the show.
   How many ponies were there?

6 The ratio of non-smokers to smokers in a café was 10 : 2.
   There were 120 non-smokers.
   How many smokers were there?

7 The ratio of protein to carbohydrate in a Big Mac burger is 2 : 3.
   There are 39 g of carbohydrate in a Big Mac.
   How many grams of protein are there?

8 Pizzas are sold in packs of 25.
   There are two vegetarian ones in every five.
   a What fraction are vegetarian?   b How many of the pack of 25 are vegetarian?

*9 One in every five families has a single parent.
   Of 15 million families, how many have a single parent?

Review 1 How much anti-freeze should be added to 6 litres of water?

Review 2 A school voted about changing the uniform.
The result was 9 : 1 in favour of changing it.
If 360 voted to change it, how many voted against changing it?

Review 3 Three out of every four pupils bus to school.
Of 800 pupils, how many bus to school?

Anti-Freeze

Mix in the ratio

3 parts water to

2 parts anti-freeze

Puzzle

1 Ellie has two china teddy bears.
She measures the shorter one with hair pins.
It is six hair pins high.
She measures them both with matches.
The shorter one is four matches high.
The taller one is six matches high.
How high would the taller one be if she measured it with hair pins?

2 Lena is aged between 20 and 50.

   a The ratio of Lena's age to her daughter Anthea's age is 4 : 1.
What are all the possible ages to the nearest year Lena and Anthea could be?

   b If in four years time the ratios of their ages will be 3 : 1, how old are Lena and Anthea now?
How old will Lena and Anthea be in four years time?

  *c How old will Lena and Anthea be when the ratio of their ages is 2 : 1?

Dividing in a given ratio

Discussion

Ashrad and Catherine shared 42 sweets in the ratio 4 : 3.
They did this by counting them out as follows.

    4 for Ashrad, 3 for Catherine   total 7
    4 for Ashrad, 3 for Catherine   total 14
    4 for Ashrad, 3 for Catherine   total 21

They kept doing this until all 42 sweets had been counted out.
How many sweets did each get?

Discuss Ashrad and Catherine's method of sharing.
Use their method to share these.

    54 sweets in the ratio 4 : 5   60 sweets in the ratio 2 : 3

How else could Ashrad and Catherine work out how to share the sweets? **Discuss**

Worked Example
A box of 40 chocolates had milk and dark chocolates in the ratio 2 : 3.
How many were milk chocolates?

Answer
Out of every 5 chocolates,
2 are milk and 3 are dark.
$\frac{2}{5}$ are milk chocolates.
$\frac{1}{5}$ of 40 = 40 ÷ 5
      = 8
$\frac{2}{5}$ of 40 = 2 × 8
      = 16
16 chocolates were milk chocolates.

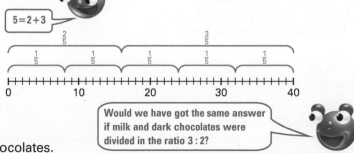

$5 = 2 + 3$

Would we have got the same answer if milk and dark chocolates were divided in the ratio 3 : 2?

Number

1 Melissa had £ 50.
   She spent it on CDs and clothes in the ratio 1 : 4.
   How much did she spend on CDs?

2 36 pupils are going to visit the museum.
   The boys and girls are in the ratio of four girls to five boys.
   How many boys are there?

3 A school has Indian and Asian students in the ratio 3 : 2.
   There are 500 Indian and Asian students altogether.
   How many of each are there?

4 Asad spent £120 on CDs and a stereo in the ratio 1 : 5.
   How much did he spend on each?

5 To make purple paint you mix red and blue paint in the ratio 2 : 7.
   How much of each colour do you need to make
   a 9 ℓ **b** 27 ℓ **c** 4·5 ℓ **d** 36 ℓ

6 A drink is made from juice and lemonade in the ratio 2 : 3.
   a How much juice is needed to make 500 mℓ of the drink?
   b How much lemonade is needed to make 1 ℓ of the drink?
   c Answer **a** and **b** again if the ratio of juice and lemonade is 3 : 2.
       Do you get the same answers? Explain.

7 To make Joanne's salad dressing you mix oil and vinegar in the ratio 3 : 1.
   How much of each do you need to make these amounts of salad dressing?
   a 400 mℓ **b** 800 mℓ **c** 200 mℓ **d** 1 ℓ
   e Todd made some salad dressing.
       He mixed the oil and vinegar in the ratio 1 : 3.
       Would it taste the same as Joanne's? Explain.

*8 Kate spent £44 in the holidays.
   She spent three times as much on clothes as she did on food.
   How much did she spend on food?

Review 1   Manzoor has 80 books.
The ratio of fiction to non-fiction is 3 : 1.
How many fiction books does he have?

Review 2   A 35 cm length of liquorice is cut into two pieces in the ratio of 2 : 3.
What is the length of the longer piece?

Review 3   Georgia has 35 tops.
She has four times as many short-sleeved tops as long-sleeved.
How many short-sleeved tops does she have?

Summary of key points

 A We can solve **proportion problems**.

Example   Four packets of Easter Eggs cost £2·80.

12 packets cost three times as much.

£2·80 × 3 = £8·40. 12 packets cost £8.40.

 B 5 : 2   4 : 7   12 : 19

These are all **ratios**.

The order in a ratio is important. 5 : 2 is different from 2 : 5.

Example   There are four black rabbits and three white rabbits.

The ratio of black rabbits to white rabbits is 4 : 3.

The ratio of white rabbits to black rabbits is 3 : 4.

 C We can make an **equivalent ratio** by multiplying or dividing both parts of the ratio by the same number.

Example   ×5 $\left(\begin{array}{c} 2 : 5 \\ = 10 : 25 \end{array}\right)$ ×5   ÷4 $\left(\begin{array}{c} 8 : 12 \\ = 2 : 3 \end{array}\right)$ ÷4

 D A ratio is in its **simplest form** when both numbers in the ratio have no common factors other than 1.

Example   In its simplest form the ratio 8 : 12 is 2 : 3. We divide both parts by 4.

 E **Proportion** compares part to whole.

Ratio compares part to part.

Example

The ratio of red to blue is one part to five parts or 1 : 5.

The proportion that is red is 1 part out of 6 parts or $\frac{1}{6}$.

 F We can solve **ratio and proportion** problems.

Example   The ratio of boys to girls in a cricket club is 5 : 3.

If there are 20 boys then we can work out the number of girls.

20 is **4** lots of 5

There must be **4** lots of 3 or **12** girls.

 G We can **divide in a given ratio**.

Example   There are 45 pupils in the badminton club.

The ratio of boys to girls is 4 : 5.

Out of every 9 pupils, 4 are boys.

$\frac{4}{9}$ are boys.

$\frac{1}{9}$ of 45 = 5

$\frac{4}{9}$ of 45 = 4 × 5

= 20

There are 20 boys in the badminton club.

Test yourself

1 Five plants cost £5.20.
How much will these cost?
   a 10 plants   **b** 15 plants   **c** 20 plants

5 Plants
£5.20   Ⓐ

2 On a school camp there were 31 students and four teachers.
What was the ratio of students to teachers on the camp?   Ⓑ

3 At a school fete 14 stalls were outside and nine were inside.
Find the ratio of
   a inside to outside stalls
   b outside to inside stalls.   Ⓑ

4 Charlotte has twice as many pairs of shoes as Olivia.
Find the ratio of the number of shoes Charlotte has to the number Olivia has.   Ⓑ

5 What goes in the gap to make these ratios equivalent?
   a $2 : 3 = 8 :$ ___   **b** $28 : 12 = 7 :$ ___   **c** $4 : 9 =$ ___ $: 45$ Ⓒ

6 **a** What is the ratio of red to yellow squares?
Give your answer in its simplest form.
   b What proportion of the squares are red?
Give you answer as a percentage.
   c What proportion of the squares are yellow?
Give your answer as a decimal.   Ⓓ Ⓔ

7 20 pupils ordered pizza. Seven ordered barbeque chicken and 13 ordered meat lovers.   Ⓔ
   a Give the proportion that ordered meat lovers as a fraction.
   b Give the proportion that ordered barbeque chicken as a percentage.
   c What is the ratio of those who ordered barbeque chicken to those who ordered meat lovers?

8 A recipe for fertiliser says to add 5 litres of water
to every 2 litres of fertiliser.
How much water should be added to 6 litres of
fertiliser?

Fertiliser
Mix 5 litres
of water to
every 2 litres
of fertiliser   Ⓕ

9 The ratio of nurses to children in a children's ward at a hospital is 2 : 5.
How many nurses are needed if there are 20 children?   Ⓕ

10 Natasha was given a 700g bag of sweets.
She shared this with her sister in the ratio 3 : 4.
Her sister got the larger share.
How many grams of sweets did Natasha and her sister each get?   Ⓖ

11 Rebecca and Lewis share a 56 hour job in the ratio 5 : 3.
How many hours does Rebecca work?   Ⓖ

12 Sam got 175 marks for his science project. He got 25 of these for presentation.
Victoria got 168 marks and 28 were for presentation.
Who got the greater proportion of marks for presentation?   Ⓔ

Algebra Support

Unknowns

In each of these, ▓ stands for an **unknown** number.

$$8 + ▓ = 12 \qquad 3 \times ▓ = 21 \qquad 24 + 36 = ▓$$

In $8 + ▓ = 12$, ▓ stands for 4 because $8 + \mathbf{4} = 12$.

Practice Questions 3, 5, 7

Counting on and back

Counting on from 4 in steps of 5 we get

$$4 \underset{+5}{} 9 \underset{+5}{} 14 \underset{+5}{} 19 \underset{+5}{} 24 \underset{+5}{} \cdots$$

Counting back from 72 in steps of 4 we get

$$72 \underset{-4}{} 68 \underset{-4}{} 64 \underset{-4}{} 60 \underset{-4}{} 56 \underset{-4}{} 52 \underset{-4}{} \cdots$$

We usually write the numbers with commas in between.

72, 68, 64, 60, 56, 52, ...

Counting back from 40 in steps of 10 we get

40, 30, 20, 10, 0, ⁻10, ⁻20, ...

Practice Questions 1, 2, 4, 6

Practice Questions

1 Write down the first six numbers you get when you
   a start at 0 and count on in threes
   b start at 0 and count on in nines
   c start at 60 and count back in fives
   d start at 200 and count back in tens
   e start at 200 and count back in steps of
   f start at 72 and count on in steps of six
   g start at 4 and count back in steps of two.

2 What is the next number?
   a 4, 8, 12, ...
   b 3, 6, 9, ...
   c 5, 10, 15, ...
   d 8, 16, 24, ...
   e 20, 40, 60, ...
   f 100, 90, 80, ...
   g 40, 35, 30, ...
   h 12, 19, 26, ...
   i 17, 26, 35, ...
   j 81, 72, 63, ...
   k 100, 93, 86, ...

3 What number does ▓ stand for in these?
   a $▓ + 4 = 6$
   b $▓ + 5 = 10$
   c $8 + ▓ = 14$
   d $▓ + 9 = 16$
   e $9 + ▓ = 20$
   f $▓ + 45 = 80$
   g $33 + ▓ = 50$
   h $4 \times ▓ = 12$
   i $5 \times ▓ = 20$
   j $7 \times ▓ = 21$
   k $4 \times ▓ = 36$
   l $4 \times ▓ - 5 = 15$
   m $3 \times ▓ + 6 = 12$
   n $\frac{▓}{4} = 10$
   o $\frac{▓}{3} = 6$

Algebra

4 Find the missing numbers in these sequences. Explain the rule for the sequence. For **a** the rule is 'start at 1 and count on in steps of 2'.

    a 1, 3, ☐, ☐, 9, ...
                             b 10, 20, ☐, ☐, 50, ...

    c 6, 10, ☐, ☐, 22, ...
                             d 50, 44, ☐, ☐, 26, 20, ☐, ☐, ...

    e ☐, 8, 11, 14, ☐, ☐, ...
                             f ☐, ☐, 19, 16, 13, ☐, ☐, ...

    g ☐, ☐, 45, 49, ☐, 57, 61, ☐, ...

5 What number does ▲ stand for in these?

    a $10 - ▲ = 4$ **b** $20 - ▲ = 10$ **c** $16 - ▲ = 11$ **d** $25 - ▲ = 16$

    e $100 - ▲ = 80$ **f** $37 - ▲ = 29$ **g** $50 - ▲ = 21$

6

+	1	2	3	4	5
1	2	3	4	5	6
2	3	4	5	6	7
3	4	5	6	7	8
4	5	6	7	8	9
5	6	7	8	9	10

There are many numbers patterns in this addition table.

In the blue 'box', the numbers on each diagonal (8 and 10, 9 and 9) add to the same number.
Is this true for all square 'boxes'?

In the pink 'box', the numbers read clockwise, beginning at the top left,
are 3, 4, 5, 6, 5, 4, 3. Do all rectangular 'boxes' have a similar number pattern?

What other number patterns can you find in this addition table?

7 Each letter from A to G is a code for one of these digits: 1, 3, 4, 5, 6, 8 or 9. Crack the code.

    $D + D = A$ $D \times D = BC$ $D + E = BF$ $G \times E = DG$

    $E + E = BA$ $E \times E = AB$ $D \times E = FC$

8 Expressions, Formulae, Equations

You need to know

 unknowns page 151

Key vocabulary

algebra, brackets, commutative, denominator, equation, evaluate, expression, formula, formulae, numerator, prove, simplest form, simplify, solution, solve, squared, substitute, symbol, therefore ($\therefore$), unknown, value, variable, verify

 Get the Message

This is the Semaphore Alphabet. Symbols stand for letters.

Decode this message.

Make up a message of your own using the Semaphore Alphabet.
You could draw stick people like this.
Give your message to a friend to decode.

In algebra, letters stand for numbers.

Understanding algebra

Brad has a number of borrowed videos plus 8 videos of his own.
We don't know how many borrowed ones he has.
We could say he has $n + 8$ videos.

$n + 8$ is an **expression**.

 We sometimes call an unknown a *variable*.

n is a letter **symbol** standing for an **unknown** number.
In an expression, n can have any **value**.

If we know Brad has 11 videos altogether we can write
$n + 8 = 11$.
$n + 8 = 11$ is an **equation**.

 You will learn more about equations later.

Expressions are usually written without a multiplication sign.

$2 \times n$ or $n \times 2$ is written as $2n$.
$1 \times n$ or $n \times 1$ is written as n.

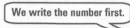 We write the number first.

Here are some more examples of expressions.

Examples		
	add 5 to a number	$n + 5$
	subtract 3 from a number	$n - 3$
	subtract a number from 6	$6 - n$
	divide a number by 3	$\frac{n}{3}$
	multiply a number by 5	$n \times 5$ or $5n$
	multiply a number by 3 then add 4	$n \times 3 + 4$ or $3n + 4$
	add 4 to a number then multiply by 3	$(n + 4) \times 3$ or $3(n + 4)$
	multiply a number by itself	$n \times n$ or n^2
*	multiply a number by 2 then multiply the answer by itself	$2n \times 2n$ or $(2n)^2$ or $4n^2$

Exercise 1

1 Write these without a multiplication sign.
 a $3 \times x$ **b** $4 \times b$ **c** $n \times 3$ **d** $h \times 5$ **e** $9 \times (x + 3)$ **f** $(n - 4) \times 3$
 g $5 \times (m - 3)$ **h** $n \times n$ **i** $p \times p$ **j** $5 \times b \times b$ **k** $3 \times a \times a$

2 Write these **with** a multiplication sign.
 a $4n$ **b** $5x$ **c** $12c$ **d** $8p$ **e** $4(x + 3)$ **f** $8(n - 2)$
 g $7(p + 3)$ **h** x^2 **i** q^2 **j** $5b^2$

3 Write an expression for these. Let the unknown number be n.
 a subtract 5 from a number   **b** add 6 to a number
 c multiply a number by 9   **d** divide a number by 8
 e multiply a number by 5 then add 4
 f multiply a number by 6 then subtract 3
 g add 4 to a number then multiply by 3
 h subtract 3 from a number then multiply by 4
 i multiply a number by itself
 j multiply a number by itself then multiply by 5
 * **k** multiply a number by 4 then multiply the answer by itself

 Write your expressions without a multiplication sign.

4 Write an expression for these. Let *a* be the unknown number.
   a a number plus 4   **b** a number minus 8
   c a number multiplied by 10   **d** a number divided by 6
   e a number multiplied by 3 then 2 added
   f a number multiplied by 7 and then 5 subtracted
   g a number plus 3 and then multiplied by 5
   h a number minus 7 and then multiplied by 3
   i a number multiplied by itself then multiplied by 4
  * j a number multiplied by 6 and then the answer multiplied by itself

5 Explain the difference between
   a $3p$ and $p + 3$ **b** n^2 and $2n$ **c** $5(n + 2)$ and $5n + 2$ ***d** $5m^2$ and $(5m)^2$.

Review 1   Write these without a multiplication sign.

a $5 \times n$ **b** $n \times 4$ **c** $p \times 3$ **d** $6 \times b$ **e** $4 \times (x + 3)$
f $(n - 4) \times 7$ **g** $(m + 3) \times 8$ **h** $m \times m$ **i** $3 \times n \times n$ **j** $7 \times c \times c$

Review 2   Write an expression for these. Let the unknown number be *n*.

a add 3 to a number   **b** a number minus 7
c multiply a number by 10   **d** a number divided by 4
e multiply a number by 3 and then add 7
f a number plus 3 and then multiplied by 6
g subtract 8 from a number and then multiply by 5
h multiply a number by itself then multiply by 4
* i multiply a number by 5 then multiply the answer by itself

---

Operations used in algebra follow the same rules as those used in arithmetic.

Arithmetic
$4 + 5 = 5 + 4$
$4 \times 5 = 5 \times 4$
$2 + (4 + 5) = (2 + 4) + 5$
$2 \times (4 \times 5) = (2 \times 4) \times 5$

Algebra
$a + b = b + a$
$a \times b = b \times a$ or $ab = ba$
$a + (b + c) = (a + b) + c$
$a \times (b \times c) = (a \times b) \times c$ or $a(bc) = (ab)c$

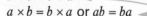

This is the *commutative* rule.

Exercise 2

Remember each letter stands for a number

1 Write true or false for these.
   a $3 + n = n + 3$ **b** $n + m = m + n$ **c** $x + y = xy$ **d** $3b = 3 + b$
   e $pq = qp$ **f** $(a + b) \times 3 = 3(a + b)$ **g** $3ab = \frac{ab}{3}$ **h** $4nm = 4(n + m)$
   i $ab \times 3 = 3ab$ **j** $st \times r = str$ **k** $3 + (x + y) = (3 + x) + y$ **l** $5 + c + d = 5cd$
  m $a \times b \times 4 = 4ab$ **n** $(x + y) \times 7 = 7xy$

Review   Write true or false for these.

a $5 + t = t + 5$ **b** $xy = yx$ **c** $abc = acb$ **d** $(4 + n) \times 5 = 5(4 + n)$
e $5 + m + n = mn + 5$ **f** $5(n + m) = 5nm$ **g** $x \times y \times 7 = 7xy$

Algebra

Collecting like terms

We can simplify expressions by adding and subtracting **like terms**.
This is called **collecting like terms**.

Arithmetic

$4 + 4 + 4 = 3 \times 4$
$8 + 8 + 8 + 8 = 4 \times 8$
$3 \times 6 + 4 \times 6 = 7 \times 6$
$5 \times 4 - 2 \times 4 = 3 \times 4$
$2 + 3 \times 2 + 2 = 5 \times 2$

Algebra

$a + a + a = 3a$
$n + n + n + n = 4n$
$3c + 4c = 7c$
$5d - 2d = 3d$
$b + 3b + b = 5b$

Examples

$b + b + b + b = 4b$
$2p + 4p + 8 = 6p + 8$

$a + b + b + c = a + 2b + c$

$5a + 2a - 6a = 7a - 6a$
$\qquad\qquad = a$

$1a = a.$

$2n + 4n = 6n$
$8x - 6x + 3 = 2x - 3$

$8y - 3y - y = 5y - y$
$\qquad\qquad = 4y$

It is best to write the 'like terms' next to each other.

$p + q + 3p + 4q = p + 3p + q + 4q$
$\qquad\qquad\qquad = 4p + 5q$

$7a + 6 - 4a - 2 = 7a - 4a + 6 - 2$
$\qquad\qquad\qquad = 3a + 4$

Worked Example
Write an expression for the perimeter of this rectangular pool. Simplify your expression.

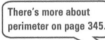

Answer
The perimeter is the distance around the outside of the rectangle.
An expression for the perimeter is $2a + b + 2a + b$.
$2a + b + 2a + b = 2a + 2a + b + b$
$\qquad\qquad\quad = \mathbf{4a + 2b}$

There's more about perimeter on page 345.

Exercise 3

1 Simplify these expressions.

a $a + a$	**b** $x + x + x$	**c** $n + n + n + n$	**d** $y + y + y + y + y$
e $5n + 3n$	**f** $4x + 3x$	**g** $2y + 7y$	**h** $3p + 7p$
i $5x + 2x$	**j** $3a + a$	**k** $7n + n$	**l** $3c + 5c$
m $8x + 4x$	**n** $7b + 4b$	**o** $8c + 12c$	

2 Simplify these expressions.

a $6a - 3a$	**b** $5n - 3n$	**c** $5n - n$	**d** $7p - p$
e $10b - 3b$	**f** $9x - 8x$	**g** $12m - 7m$	**h** $20t - 6t$
i $18f - 12f$	**j** $13c - 12c$	**k** $20a - 16a$	

3 Simplify.

a $2b + b + 3b$ b $4x + 2x + x$ c $3y + 2y + 2y$ d $3m + 2m + 5m$

e $4x + 3x + 5x$ f $2a + 3a + 6a$ g $7b + 2b + 4b$ h $4x + 3x - 2x$

i $7y + 2y - 3y$ j $8p + 6p - 13p$ k $10x - 8x + 2x$ l $8n - 3n + n$

m $5x - 3x + 6x$ n $3g + 2g - 4g$ o $7m - m - 3m$ p $15x - 10x - 3x$

q $15j + 2j - 13j$ r $20p - 12p + 3p$ s $14a + 6a - 7a$

4 Simplify these.

a $8x + x + 2$ b $4b + b + 2$ c $2m + 3m + 4$ d $5y + 3y + 7$

e $9a + 5 + a$ f $n + 8 + 3n$ g $4p + 7 + 3p$ h $6x + 4 - 3x$

i $8m + 7 - 2m$ j $4y + 8 - 2y$ k $6b + 10 + 5 + 3b$ l $6 + 4x + 4 + 5x$

*m $5 + 8x + 2 - 2x$ *n $4b + 7 - 3 - 2b$ *o $5p + 8 - 3 - 3p$ *p $9y + 12 - 6 - 3y$

*q $10m + 8 - 3m + 4$ *r $9p + 12 - 5 + 3p$ *s $11x + 13 - 10x - 5$

5 Simplify

a $b + b + a + a$ b $p + p + p + q + q$ c $2c + 3c + d$ d $f + g + 2g$

e $s + 2s + t$ f $2m + 3m + 4n + 3n$ g $5p + 4q + 2p + q$ h $3a + 2b + 2a + b$

i $3x + 8y + x + y$ j $4t + 3s + 2t + 7s$ k $3a + 2b + 4a + b$ l $4p + 3q + 7p + 8q$

m $14s + 11t + 5s + 8t$ n $12x + 10y + 7x + 13y$ o $11c + d + 12c + d$

6 Write an expression for the perimeter of these gardens.
Simplify your expression.

a
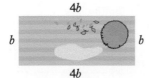

b
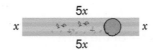

c
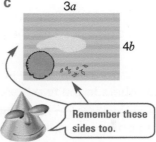

Remember these sides too.

d
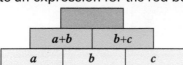

e
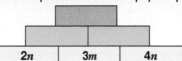

7 The number in each box is found by adding the numbers in the two boxes below it.
Write an expression for the red box. Write the expression as simply as possible.

a

b

8 Here are some algebra cards.

$a + 2$ $a \div 2$ $2a - 1$ $2a + a$ $2a$

$a - 2$ a^2 $a + a$ $2a + 2$ $3a + 3$

a Which card will always give the same answer as

i $\frac{a}{2}$ ii $2 + a$ iii $3a - a + 2$ iv $5a - 2a + 3$ v $a \times a$?

b Which two cards will always give the same answer as $2 \times a$?

c When the expressions on two cards are added they can be simplified to give $5a + 2$.
Which two cards are they?

d Write a new card that will always give the same answer as

i $4a + 2a$ ii $3a + 5a$ iii $8a - 3a$

*9 The answer is $3p + 4q$. What could the question have been?

* 10 Draw some shapes that have a perimeter of $8x + 12$.

T

Review 1

													E				

| $\overline{3a}$ | | $\overline{4a}$ $\overline{8a}$ $\overline{7a}$ $\overline{7a}$ $\overline{12a}^{E}$ $\overline{13a}$ | | $\overline{5a}$ $\overline{3a}$ $\overline{9a}$ $\overline{9a}$ | | $\overline{2a}$ $\overline{3a}$ $\overline{4a}$ |

$-$

| $\overline{11a}$ $\overline{2a}$ $\overline{18a}$ $\overline{13a}$ $\overline{11a}$ $\overline{10a}$ | | $\overline{11a}$ $\overline{6a}$ $\overline{8a}$ | | $\overline{a}$ $\overline{3a}$ $\overline{15a}$ $\overline{12a}^{E}$ $\overline{9a}$ $\overline{4a}$ |

Use a copy of the box.
Simplify these expressions. Write the letter that is beside each expression above the answer in the box.

E $9a + 3a = 12a$ **O** $9a - a$ **N** $4a + 11a$ **P** $7a - 6a$ **H** $10a - 8a$
B $8a - 3a$ **I** $9a + 2a + 7a$ **Y** $9a + 3a - 2a$ **T** $10a + 4a - 3a$ **W** $11a + 2a - 7a$
A $16a - 7a - 6a$ **R** $15a - a - a$ **S** $20a - 11a - 5a$ **C** $19a - 5a - 7a$ **L** $17a - 6a - 2a$

Review 2   Simplify these.

a $7b + 3b + 3$ **b** $5n + 2n - 3$ **c** $5x + x + 3$ **d** $8y - 2y + 4$ **e** $4a - 3a + 2$
f $5p - 4p + 1$ **g** $3h + 4 + 2h$ **h** $2x - 3 + 4x$ **i** $3n + 12 - n$

Review 3   This diagram shows a rectangular park.
Write an expression for the perimeter of the park.
Simplify your expression.

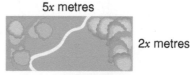

5x metres

2x metres

Review 4   The number in each box is found by adding the numbers in the two boxes below it.

Write an expression as simply as possible for the
a red box   **b** green box   **c** orange box.

p q r

*Review 5   Draw some shapes that have a perimeter of $4a + 6$.

Simplifying expressions by cancelling

We can **simplify expressions** by **cancelling**.

> There is more about cancelling fractions on page 110.

Arithmetic	**Algebra**	
$\dfrac{4}{4} = 1$	$\dfrac{n}{n} = 1$	
$\dfrac{3^1 \times 4}{3^1} = 4$	$\dfrac{3^1 n}{3^1} = n$	Divide numerator and denominator by the common factor 3.
$\dfrac{8^2 \times 5}{4^1} = 2 \times 5$	$\dfrac{8^2 n}{4^1} = 2n$	Divide numerator and denominator by the common factor 4.
$\dfrac{5 \times 6^1}{6^1} = 5$	$\dfrac{5a^1}{a^1} = 5$	Divide numerator and denominator by the common factor a (6 in the arithmetic example).
$\dfrac{10^5 \times 7}{6^3} = \dfrac{5 \times 7}{3}$	$\dfrac{10^5 a}{6^3} = \dfrac{5a}{3}$	Divide numerator and denominator by the common factor 2.

Exercise 4

1 Write these expressions in their simplest form.

a $\frac{m}{m}$ b $\frac{c}{c}$ c $\frac{y}{y}$ d $\frac{n}{n}$ e $\frac{3a}{3}$ f $\frac{6b}{6}$

g $\frac{7m}{7}$ h $\frac{9y}{9}$ i $\frac{5x}{x}$ j $\frac{7x}{x}$ k $\frac{8p}{p}$ l $\frac{5x}{x}$

m $\frac{8n}{4}$ n $\frac{10p}{2}$ o $\frac{14n}{7}$ p $\frac{21y}{7}$ q $\frac{24x}{8}$ r $\frac{54b}{9}$

s $\frac{42m}{7}$ t $\frac{35a}{7}$ u $\frac{21b}{b}$ v $\frac{64y}{8}$ w $\frac{72x}{9}$

∗2 Simplify these expressions by cancelling.

a $\frac{12a}{8}$ b $\frac{20x}{8}$ c $\frac{15y}{10}$ d $\frac{45m}{27}$ e $\frac{18b}{12}$ f $\frac{54y}{42}$

g $\frac{72x}{48}$ h $\frac{64m}{40}$ i $\frac{54p}{72}$ j $\frac{45r}{75}$ k $\frac{48n}{60}$

∗3 $\frac{\square}{\square}$ = 2x. What might go in the boxes to make this true?

Review

Use a copy of this box.

									A					
$\overline{2}$	$\overline{10}$	$\overline{12}$	$\overline{\frac{3x}{2}}$	$\overline{\frac{9x}{2}}$	$\overline{\frac{3x}{2}}$	$\overline{6x}$		$\overline{1}$	$\overline{12}$	$\overline{\frac{3x}{2}}$	$\overline{8x}$	$\overline{10x}$	$\overline{\frac{3x}{2}}$	

	A													
$\overline{6x}$	$\overline{1}$	$\overline{3x}$	$\overline{\frac{3x}{2}}$		$\overline{6x}$	$\overline{7x}$	$\overline{2x}$	$\overline{\frac{3x}{2}}$		$\overline{6}$ $\overline{12}$ $\overline{2}$ $\overline{3x}$	$\overline{\frac{5x}{2}}$ $\overline{7x}$ $\overline{12}$ $\overline{8x}$ $\overline{10x}$

Simplify these expressions by cancelling. Write the letter that is beside each expression above the answer in the box.

A $\frac{x}{x} = 1$ **O** $\frac{2w}{w}$ **F** $\frac{6x}{x}$ **U** $\frac{10x}{x}$ **R** $\frac{12x}{x}$

M $\frac{6x}{2}$ **Z** $\frac{20x}{10}$ **H** $\frac{20x}{2}$ **T** $\frac{24x}{3}$ **S** $\frac{30x}{5}$

I $\frac{42x}{6}$ **B** $\frac{20x}{8}$ **E** $\frac{24x}{16}$ **Y** $\frac{63x}{14}$

Brackets

We can write an **expression without brackets** by following the rules of arithmetic.

Arithmetic

$6 \times 52 = 6 \times (50 + 2)$

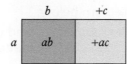

$6 \times (50 + 2) = (6 \times 50) + (6 \times 2)$
$= 300 + 12$

Algebra

$a(b + c) = a \times (b + c)$

$a \times (b + c) = (a \times b) + (a \times c)$
$= ab + ac$

Examples $2(a + 5) = \mathbf{2a + 10}$

$3(x + 7) = \mathbf{3x + 21}$

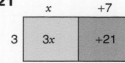

Algebra

1 Write these expressions without brackets. Use the grids to help.

a $3(a + 2)$

b $4(x + 2)$

c $5(n + 4)$

d $7(p + 1)$

e $3(m + 6)$

f $4(y + 3)$

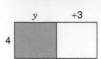

g $2(x + 5)$

h $3(m + 4)$

i $4(p + 10)$

j $6(a + 6)$

2 Write without brackets.

a $4(a + 2)$	**b** $3(x + 5)$	**c** $5(y + 3)$	**d** $7(b + 2)$
e $9(a + 4)$	**f** $8(m + 6)$	**g** $7(n + 1)$	**h** $4(p + 2)$
i $6(b + 5)$	**j** $17(c + 1)$	**k** $14(p + 2)$	**l** $16(m + 2)$
m $24(n + 1)$	**n** $2(a + 13)$	**o** $3(n + 12)$	

Review 1 Write these expressions without brackets.

a $5(x + 1)$

b $3(p + 4)$

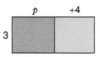

c $7(c + 8)$

d $24(x + 2)$

Use the grids to help.

Review 2 Write these without brackets.

a $4(x + 3)$	**b** $7(b + 5)$	**c** $9(n + 3)$	**d** $18(a + 1)$
e $2(x + 4)$	**f** $3(a + 6)$	**g** $5(m + 8)$	

Review 3 Use a copy of the diagram below.

Write each expression below without brackets. Find your answer in the box. Cross out the letter above it. What three word sentence do the letters that are left make?

$4(n + 3)$	$7(n + 5)$	$9(n + 3)$	$36(n + 1)$	$2(n + 4)$
$3(n + 6)$	$5(n + 8)$	$10(n + 4)$	$6(n + 6)$	$8(n + 1)$

C	R	A	A	T	Y	S	F	I	C	S	A	N	H	T	E	V	O	A	M	T	I	T
$3n{+}18$	$9n{+}12$	$5n{+}40$	$3n{+}9$	$4n{+}3$	$4n{+}12$	$3n{+}6$	$8n{+}8$	$7n{+}35$	$8n{+}1$	$10n{+}40$	$9n{+}3$	$7n{+}5$	$9n{+}27$	$5n{+}8$	$36n{+}36$	$10n{+}4$	$10n{+}14$	$6n{+}36$	$36n{+}1$	$2n{+}8$	$2n{+}4$	$4n{+}7$

Card Pairs – a game for 2 players

You will need a set of cards like this.

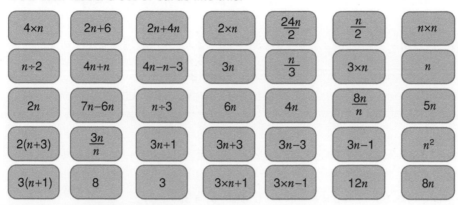

$4{\times}n$	$2n{+}6$	$2n{+}4n$	$2{\times}n$	$\dfrac{24n}{2}$	$\dfrac{n}{2}$	$n{\times}n$
$n{\div}2$	$4n{+}n$	$4n{-}n{-}3$	$3n$	$\dfrac{n}{3}$	$3{\times}n$	n
$2n$	$7n{-}6n$	$n{\div}3$	$6n$	$4n$	$\dfrac{8n}{n}$	$5n$
$2(n{+}3)$	$\dfrac{3n}{n}$	$3n{+}1$	$3n{+}3$	$3n{-}3$	$3n{-}1$	n^2
$3(n{+}1)$	8	3	$3{\times}n{+}1$	$3{\times}n{-}1$	$12n$	$8n$

To play
- Shuffle the cards.
- Deal half of them to each player.
- Find all of the pairs you have in your hand.
  A pair is made from two cards which match.

Example $n \times n$ and n^2 are a pair.

- Put your pairs down in a pile beside you.
- Take turns to choose a card from your partner's hand.
  Try and match it with a card in your hand to make a pair.
- The player who has the most pairs at the end is the winner.

Substituting

Lauren made some sandwiches.
Some were ham and 10 were egg.
We could say she made $n + 10$ sandwiches.
Lauren said 'I made 12 ham sandwiches.'
We can now work out the value of $n + 10$.

$n + 10 = \mathbf{12} + 10$ 12 has been put in place of n.
$\qquad\quad = 22$ We call this **substitution**.

We **evaluate** an expression by **substituting** values for the unknown into the expression.
When we evaluate an expression, the **order of operations** is the same as for arithmetic.

Brackets
Indices
Division and **M**ultiplication
Addition and **S**ubtraction

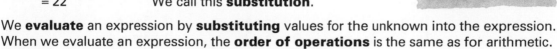

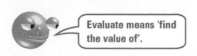
Evaluate means 'find the value of'.

Algebra

Example   In the expression $4 + 3a$, the multiplication is worked out before the addition.

Worked Example
If $n = 6$ find the value of   **a** $2n$ **b** $n - 2$ **c** $20 - 3n$ **d** $\frac{30}{n}$ **e** $\frac{n}{3} - 1$

Answer

a $2n = 2 \times 6$ **b** $n - 2 = 6 - 2$ **c** $20 - 3n = 20 - 3 \times n$ **d** $\frac{n}{2} = \frac{6}{2}$ **e** $\frac{n}{3} - 1 = \frac{6}{3} - 1$

$\quad\quad = 12$ $\quad\quad\quad = 4$ $\quad\quad\quad\quad = 20 - 3 \times 6$ $\quad\quad = 3$ $\quad\quad\quad\quad = 2 - 1$

$\quad\quad\quad\quad\quad\quad\quad\quad\quad\quad\quad\quad\quad\quad\quad = 20 - 18$ $\quad\quad\quad\quad\quad\quad\quad\quad\quad\quad = 1$

$\quad\quad\quad\quad\quad\quad\quad\quad\quad\quad\quad\quad\quad\quad\quad = 2$

Do $\times$ first.

 Do $\div$ first.

Worked Example
If $a = 4$, $b = 3$ and $c = 1$, evaluate   **a** $a + b$ **b** $9a - c$ **c** $3(a + b)$ **d** $b(12 - a)$

Answer

a $a + b = 4 + 3$ **b** $9a - c = 9 \times 4 - 1$ **c** $3(a + b) = 3(4 + 3)$

$\quad\quad\quad = 7$ $\quad\quad\quad\quad = 36 - 1$ $\quad\quad\quad\quad\quad = 3 \times 7$

$\quad\quad\quad\quad\quad\quad\quad\quad\quad = 35$ $\quad\quad\quad\quad\quad = 21$

d $b(12 - a) = 3(12 - 4)$

$\quad\quad\quad\quad = 3 \times 8$

$\quad\quad\quad\quad = 24$

Exercise 6

1   If $x = 1$ find the value of
   a $3x$ **b** $5x$ **c** $x + 7$ **d** $2x + 1$ **e** $3x + 2$.

2   If $b = 3$ find the value of
   a $3b$ **b** $5b$ **c** $2b + 5$ **d** $12 - b$ **e** $2b - 3$.

3   If $y = 4$ find the value of
   a $2y + 2$ **b** $12 - y$ **c** $3y + 4$ **d** $18 - 3y$
   e $20 - 2y$ **f** $3(y + 2)$ **g** $8(y + 1)$ **h** $4(y - 2)$
   i $7(8 - y)$ **j** $\frac{y}{2} + 3$.

4   If $p = 2$ and $q = 4$ evaluate
   a $p + q$ **b** $p + q - 1$ **c** $2p + q$ **d** $3q - p$
   e $2p + 3q$ **f** $20 - 2p - q$ **g** $2q - 3p$ **h** $3(p + q)$
   i $4(q - p)$.

5   If $a = 2$, $b = 3$, $c = 6$ and $d = 4$ evaluate
   a $a + b$ **b** $a + c - b$ **c** $\frac{16}{d}$ **d** $\frac{20}{a}$
   e $\frac{d}{2} + 1$ **f** $\frac{c}{2} - 1$ **g** $5(c - 2)$ **h** $3b + c$
   i $4d + 2$ **j** $2c - 10$ **k** $d - b + 2$ **l** $20 - 2d$
   m $2a + b - d$ **n** $3a - b$ **o** $2(b - a)$ **p** $5(c - b)$
   q $\frac{c}{2} + b$ **r** $\frac{b}{3} + a$ **s** $3b - a$ **t** $12 - 2d$
   u $c - 2a$ **v** $2c - 2d$ **w** $3a - 2b$ **x** $4(2b - d)$.

6 The expression $3d + 2$ gives the total number of dots needed to make a row with d blue dots.

How many dots in total are needed to make a row with 15 blue dots?

*7 $2n + 1$, $2n + 3$, $2n + 5$, $2n + 7$, $2n + 9$
    a Substitute $n = 1$ into each of the above expressions.
       You should get five consecutive odd numbers. Explain why.
    b What do you get if $n = 2$?
    c Which value of n would give the numbers, 13, 15, 17, 19, 21?

Review 1   If $y = 2$ find the value of
a $5y$ **b** $y + 6$ **c** $3y + 4$ **d** $2y - 1$ **e** $7y - 5$

Review 2   If $x = 3$, $y = 5$ and $z = 1$ find the value of
a $x + y$ **b** $y - x + z$ **c** $3(x + y)$ **d** $9y - z$ **e** $\frac{21}{x}$ **f** $\frac{y}{5} + 3$.

Review 3

$\overline{7}$	$\overline{5}$	$\overline{3}$	$\overline{22}$	$\overline{12}$	$\overline{8}$	$\overline{5}$	$\overline{8}$	$\overline{3}$		$\overline{18}$	$\overline{9}$	$\overline{0}$	$\overline{13}$	$\overline{8}$		$\overline{12}$
$\overline{13}$	$\overline{12}$	$\overline{2}$	$\overline{5}$	$\overline{10}$	$\overline{0}$		$\overline{11}$	$\overline{22}$	$\overline{0}$	$\overline{2}$	$\overline{0}$			$\overline{5}$	$\overline{12}$
$\overline{5}$	$\overline{13}$			$\overline{2}$	$\overset{A}{\overline{6}}$	$\overline{5}$	$\overline{8}$	$\overline{5}$	$\overline{8}$	$\overline{3}$					

Use a copy of this box.
If $p = 2$, $q = 5$, $r = 7$ evaluate these.
Write the letter that is beside each expression above the answer in the box.

A $3p = 6$ **I** $r - p$ **E** $p + q - r$ **W** $2p + r$ **H** $3q + r$

O $2r - q$ **D** $2(p + r)$ **S** $3q - p$ **N** $\frac{16}{p}$ **T** $2(p + 4)$

K $2(r - p)$ **G** $\frac{3q}{5}$ **L** $\frac{r}{7} + 6$ **R** $\frac{3p}{2} - 1$

Review 4   The expression $2t + 1$ gives the total number of coloured rods needed to make a row with t triangles.

How many rods are needed to make a row with 24 triangles?

Investigation

Substituting

1 Marcia wanted to know if $2n + 3n = 5n$ was a true statement.
She showed it was true when $n = 4$.

$$2n + 3n = 2 \times 4 + 3 \times 4 \qquad 5n = 5 \times 4$$
$$= 8 + 12 \qquad\qquad\qquad = 20$$
$$= 20$$

Is it true for other values of n? **Investigate**.

Are the statements below true for different values of the letters? **Investigate**.

$b + b + b = 36 \qquad\qquad 4n + 5n = 9n \qquad \frac{8n}{2} = 4n \qquad m + 3m - 2m = 2m$

$4(p + 3) = 4p + 12 \qquad 2(h + 4) = 2h + 8$

> Use a graphical calculator to help.

2 In arithmetic, if we are told that $56 + 83 = 139$,
then we know that

$83 + 56 = 139$
$139 - 83 = 56$
$139 - 56 = 83$.

$$56+83=139 \xleftrightarrow{\text{Inverse}} 139-83=56$$
$$\updownarrow \text{Commutative}$$
$$83+56=139 \xleftrightarrow{\text{Inverse}} 139-56=83$$

In algebra, if $m + n = 7$, does this mean these are true?

$n + m = 7 \qquad 7 - n = m \qquad 7 - m = n$ **Investigate**.

What if $m + n = 7$ is replaced with $m + n = 100$ or $m + n = 5 \cdot 4$ or $m + n = -2$?

3 If $ab = 48$, does this mean these are true?

$ba = 48 \qquad \frac{48}{b} = a \qquad \frac{48}{a} = b$ **Investigate**.

What if $ab = 48$ is replaced with $ab = 64$ or $ab = 4 \cdot 8$ or $ab = -8$?

Practical

1 You will need a spreadsheet.

	A	B	C
1	x	y	Expression
2			=3*A2+B2

> Put your own values for x and y here

Use a spreadsheet to find at least five different values of x and y that make the
expression $3x + y$ equal to   **a** 60   **b** 75.

T

2 **You will need** a graphical calculator.

Jodie used a graphical calculator to find the value of $4x^2 - 2$ for $x = 3$.

She keyed ③ ➡ ⒜LPHA ⓧ ⒠XE to give x the value 3

then ④ ⒜LPHA ⓧ ⓧ² ⊖ ② ⒠XE to find the value of $4x^2 - 2$ for $x = 3$.

The ⒜LPHA key is used to get letters on the screen.

The keying sequence on *your* calculator may be a little different. Find out how to do this on *your* calculator.

Use your calculator to find the value of these when $x = 1$, $x = 2$, $x = 3$, $x = 4$ and $x = 5$.

a $x^2 + 3$ b $x^2 - 1$ c $2x^2 + 3$ d $2x^2 - 4$
e $3x^2 + 4$ f $3x^2 - 7$ g $4x^2 - 5$

Formulae

age in years = age in months divided by 12 is a **formula**.
If we know the age in months we can find the age in years.
A **formula** is a rule for working something out.

The plural for *formula* is *formulae*.

Discussion

$$\text{cost for each pupil} = \frac{\text{total cost}}{\text{number of pupils}}$$

number of eggs = number of pupils $\times$ 6

Here are two formulae that might be used when planning the school camp.
Think of other formulae that might be used. **Discuss.**
Think of other formulae that might be used in a school. **Discuss.**
What about other places? **Discuss.**

Worked Example
Emily does jobs for her parents.
She uses this formula to work out how much she gets paid.

Amount paid = hours worked $\times$ £2 + £3

How much does she get paid for working

a 1 hour   b 3 hours   c 6 hours?

Answer
a Amount $= 1 \times 2 + 3$ b Amount $= 3 \times 2 + 3$ c Amount $= 6 \times 2 + 3$
        $= 2 + 3$ $= 6 + 3$ $= 12 + 3$
        $= £5$ $= £9$ $= £15$

Algebra

1

  1 m   2 m   3 m

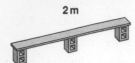

Shelves are made from a length of timber and bricks. The length (in metres) of the shelf is found from this formula.

length of shelf = number of bricks subtract one

Use this formula to find the length of the shelf that is on
a 5 bricks   **b** 9 bricks   **c** 16 bricks   **d** 30 bricks.

2 A recipe for cooking lamb gives the formula for cooking time as

time = number of kilograms × 40 minutes

What is the cooking time, in minutes, for a piece of lamb that weighs
a 2 kg   **b** 3 kg   **c** 4 kg   **d** 0·5 kg?

3 This formula gives the cost for an advertisement in a magazine.

cost = number of lines × £4

How much does it cost for an advertisement of
a   4 lines
b   8 lines
c   18 lines?

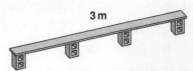

FOR THE PERFECT BRITISH HOLIDAY, JUST ADD WATER.

4

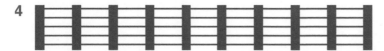

In this fence there are 10 posts and 45 spacers.
Pete used this formula to find the number of spacers needed.

number of spacers = 5 times the number of posts subtract 5

How many spacers are needed to build a fence with
a   6 posts   **b** 9 posts   **c** 25 posts   **d** 50 posts?

5 A formula for finding the area of a rectangle is

area = length times width

Use this formula to find the area of a rectangle if
a   length = 7 cm, width = 5 cm
b   length = 14 cm, width = 10 cm
c   length = 16 mm, width = 5 mm
d   length = 4·2 cm, width = 2 m.

width

length

6

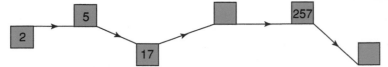

2   5   17   257

Rhian drew this number chain.
The formula for finding the next number is

next number = 4 times the previous number subtract 3

Use this formula to find the missing numbers.

7

Christine made these dot patterns.
She used this formula to find the number of green dots in each.

number of green dots = **2 times the number of red dots plus 2.**

How many green dots are there in a diagram which has
a 5 red dots   **b** 8 red dots   **c** 35 red dots   **d** 105 red dots?

*8   To change a temperature given in °C to °F we can use the formula

°F = multiply °C by 9, then divide by 5, then add 32.

There is one temperature where °F = °C. What is it?

Use this formula to change the following to °F.
a 100 °C   **b** 0 °C   **c** 50 °C   **d** 30 °C   **e** 10 °C

*9   The formula for how tall a boy is likely to grow, in cm, is

add mother's and father's heights (in cm), divide by 2 and add 7.

A boy is likely to grow to this height plus or minus 10 cm.
What is the greatest and least heights these boys will grow to?
a   Ryan: mother 168 cm father 178 cm
b   Jordan: mother 172 cm father 176 cm
c   Mark: mother 163 cm father 180 cm
d   Luke: mother 158 cm father 183 cm

Review 1   A formula for finding the cost of sweets is

cost = cost of one packet × number of packets.

Use this formula to find the cost if
a   cost of one packet = £2, number of packets = 5
b   cost of one packet = £1·50, number of packets = 4.

Review 2   A formula for changing pounds to kilograms is

kilograms = 5 times the number of pounds then divide by 11.

Use this formula to answer these.
a   Lin weighs 110 lb. How heavy is Lin in kilograms?
b   Lin's baby brother weighs 11 lb. How heavy is this in kilograms?

Review 3

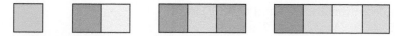

Kieran used this formula to find the perimeter, in cm, of each diagram.

perimeter = 2 times the number of squares plus 2

Use this formula to find the perimeter of a diagram which has
a 5 squares   **b** 10 squares   **c** 20 squares   **d** 100 squares.

Algebra

Practical

You will need to draw this diagram.

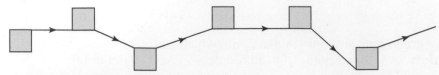

Write down a formula such as 'double each number and subtract five.'
Choose a starting number. Use your formula to find the next five or six numbers to make a chain.
Give your number chain to someone else to see if they can work out your formula.

Formulae are used by many people in their jobs.
Usually a **formula is written in symbols**.

Example
Max makes picture frames.
He uses the formula $P = 2l + 2w$ to work out the length of wood he needs to make a frame.
If $l = 12$ cm and $w = 10$ cm
$P = 2 \times 12 + 2 \times 10$
$\quad = 24 + 20$
$\quad = \mathbf{44\ cm}$

> Remember to do multiplication before addition.

Exercise 8

1 $P = 4l$ gives the perimeter of a square.
Find P if $l = 5$ cm.

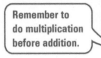
l

2 $A = lw$ gives the area of a rectangle.
Find A if $l = 7$ cm and $w = 5$ cm.

l
w

3 $y = \dfrac{f}{3}$ is a formula for changing feet, f, into yards, y.

Find y if f is   **a** 18   **b** 24   **c** 30   **d** 45

4 A formula for finding force from mass, m, and acceleration, a, is $F = ma$.
Find F if   **a** $m = 5$ and $a = 8$ **b** $m = 12$ and $a = 4$

5 A formula for finding speed, S, from distance, D and time, T, is $S = \dfrac{D}{T}$.
Find S if   **a** $D = 150$ and $T = 5$ **b** $D = 320$ and $T = 4$

6 $g = \dfrac{p}{8}$ is a formula for changing pints, p, into gallons, g.

Find g if   **a** $p = 40$ **b** $p = 160$ **c** $p = 192$

l
w

7 $P = 2l + 2w$ gives the perimeter of a rectangle.
Find P if
a $l = 5$ cm, $w = 2$ cm   **b** $l = 7$ cm, $w = 4$ cm   **c** $l = 10$ cm, $w = 6$ cm

8 A taxi company uses the formula $C = 2k + 3$ where k is the number of kilometres travelled and C is the cost in pounds.
Find the cost to travel
  a 1 km   **b** 5 km   **c** 8 km   **d** 20 km.

9 The cost, in pounds, of hiring a car from 'Have Wheels' is given by the formula
$C = 30d + 20$. d is the number of days hired. Find the cost of hiring a car for
  a 3 days   **b** a week   **c** a fortnight   **d** 16 days.

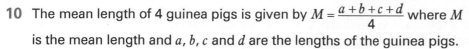

10 The mean length of 4 guinea pigs is given by $M = \dfrac{a+b+c+d}{4}$ where M

There is more about the mean (average) on page 376,

is the mean length and a, b, c and d are the lengths of the guinea pigs.
Find M if $a = 17$ cm, $b = 16$ cm, $c = 16.5$ cm, $d = 18.5$ cm.

11 The weekly wage of a car salesperson is worked out using $w = 200 + 100n$. n is the number of cars sold and w is the weekly wage, in pounds.
How much is Will's weekly wage if he sells
  a 4 cars   **b** 3 cars   **c** no cars   **d** 1 car   **e** 6 cars?

12 Jamie uses this formula to work out how much money, in pounds, he will get from his parents each week.
$p = 3 + 2h$. h is the number of hours he works doing jobs.
How much money will Jamie get if he works for
  a 5 hours   **b** 3 hours   **c** 10 hours   **d** 14 hours   **e** 8 hours?

*13 $h = \dfrac{28 - a}{2}$ is thought to be a formula which gives the

number of hours of sleep that a child needs. h is the number of hours of sleep. a is the age, in years, of the child.

Copy and complete this table.

a	2	4	6	8	10	12
h	13					

Review 1 $m = \dfrac{c}{100}$ where m is metres and c is centimetres.
Find m if
  a $c = 200$ **b** $c = 2600$ **c** $c = 6500$ **d** $c = 18\,000$

Review 2 $S = 20 + 4t$ is a formula which gives the distance, S metres, travelled in t seconds by a car going downhill.
What distance is travelled in   **a** 10 seconds   **b** 1 minute?

Review 3   A live-in nanny is paid £6 an hour. Each week £40 is taken for board. The nanny's weekly earnings are given by the formula $E = 6h - 40$ where h is the number of hours worked in that week.
Find this nanny's earnings for a week in which 45 hours were worked.

Puzzle

  * Replace each of the letters with one of the digits 0, 1, 2, 3, 4, 5, 6, 7, 8, 9 so that the additions are correct.

```
1     S O M E        2       S U N        3     F O U R
    +   C A N            + B U R N            + F I V E
    ─────────            ─────────            ─────────
      H E A R            F E V E R              N I N E
```

Is there more than one possible answer?

Practical

1 **You will need** a tape measure.
The formula for finding the area, in cm², of skin on your body is **S = 2ht** where
h is your height, in cm, and t is the distance around your thigh, in cm.
Use the formula to work out the area of your skin.

2 Choose a job such as dietician, chemist, builder, nurse, ...
Find out what formulae people in these jobs use.
Make a poster or booklet to show what you found out.
You could use the internet to help.
You could e-mail someone who works in the job you choose.

Writing expressions and formulae

Writing Expressions

Paul had x coins in his hand. He is given three more coins.
An expression for the number of coins Paul has now is $x + 3$.

Worked Example
The height of the small box is h.
Write an expression for the height of another box which is
three times higher.

Answer
The height of the other box is $3 \times h$ which we write as $3h$.

Exercise 9

1 Write an expression for the number of marbles Kishan has. The answer to **a** is $m - 5$.
   a Kishan had m marbles. He gave 5 away.
   b Kishan had p marbles. He was given 8 more.
   c Kishan had n marbles. He lost 6.
   d Kishan had x marbles. He gave 9 to a friend.

2 Aaron has x CDs. Helen has three times as many as Aaron.
   Write an expression for the number of CDs Helen has.

3 There are n seats in a row.
   Write an expression for the number of seats in 30 rows.

4 A box holds p books.
   How many books do 10 boxes hold?

5 Tickets cost £t.
   How much do 12 tickets cost?

6 Alex is a years old.
Write an expression for his age
   a in 8 years time   **b** in 10 years time   **c** 7 years ago.

7 There are n coins in this pile. Write an expression for the number of coins in a pile that has

   a two more coins   **b** five fewer coins
   c four times as many coins   **d** half as many coins

8 The pile of coins in question **7** is divided into three equal piles.
Write an expression for the number of coins in each pile.

9 You have d sweets.
   a Milly has four more than you.
       How many does Milly have?
   b You give away three sweets.
       How many do you have left?
   c Shabir has twice as many as you.
       How many does Shabir have?
  *d Shabir shares her sweets equally between herself and three other friends.
       How many do they each get?

10 Katie makes cheese, egg, ham and tomato sandwiches.
   s stands for the number of cheese sandwiches Katie makes.
   a Katie makes three times as many egg sandwiches as cheese ones.
       How many egg ones does she make?
   b Katie makes the same number of ham as cheese sandwiches.
       How many ham sandwiches does she make?
   c Katie makes ten more tomato than cheese sandwiches.
       How many tomato sandwiches does she make?
   d How many sandwiches does she make altogether?
       Write your answer as simply as possible.

*11 In a sale, a comic costs c pence. Hadley gives £3 to pay for four comics.
Which of these expressions gives his change in pence?
   A $4c - 3$ **B** $4c - 300$ **C** $2 - 4c$ **D** $300 - 4c$

*12 Ella has three bags of sweets.
Each bag has p sweets inside.
Ella takes out some sweets.
Now the total number of sweets in Ella's three bags
is $3p - 6$.
Some of these statements could be true.
Which ones?

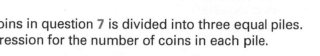

   A Ella took one sweet out of each bag.
   B Ella took two sweets out of each bag.
   C Ella took three sweets out of one bag and none out of the others.
   D Ella took six sweets out of one bag and none out of the others.
   E Ella took six sweets out of each bag.

Review 1   Sam had p trees in his garden.
Write an expression for the number of trees someone has if they have
  a three more than Sam   **b** ten fewer than Sam
  c four times as many as Sam   **d** half as many as Sam.

Review 2   Ben has blue, red, green and yellow model planes.
p stands for the number of blue planes he has.
  a   Ben has twice as many red planes as blue planes.
      How many red planes does he have?
  b   Ben has four more green planes than blue planes.
      How many green planes does he have?
  c   Ben has the same number of yellow planes as blue planes.
      How many blue planes does he have?
  d   How many planes does he have altogether?
      Write your answer as simply as possible.

*Review 3   A ride at a fair costs r pence. Matthew gives £8 to pay for six rides. Which of these expressions gives his change in pence?
  A $6r - 8$ **B** $800 - 6r$ **C** $8 - 6r$ **D** $6r - 800$

Writing formulae

There are 60 seconds in a minute.
We can **write a formula** for finding the number of seconds in any
number of minutes. In words, the formula is
  number of seconds = 60 times the number of minutes
using S for the number of seconds and m for the number of minutes we write
  $S = 60m$

Exercise 10

  1   Write a formula for these. The answer to **a** is $c = 35n$.
     a   the cost, c, of n chocolate bars at 35p each $c =$ ___
     b   the cost, c, of m pencils at 21p each $c =$ ___
     c   the cost, c, of l buttons at 17p each $c =$ ___
     d   the cost, c, of h hair ties at £2 each $c =$ ___
     e   the total money raised, r, in a sponsored walk at 20p per kilometre, k $r =$ ___
     f   the total number of cans of baked beans, b, eaten by a boy who eats 6 cans a week for
             w weeks $b =$ ___
     g   the total weight, w, of a cat weighing c kg and a dog weighing 20 kg $w =$ ___
     h   the total length, l of a caravan of length c m and a car of length 4 m $l =$ ___
     i   the cost, c, per person if a meal costing m pounds is shared by b people $c =$ ___
   ***j**   the final length, l, of a 4 m piece of wood with c cm cut off $l =$ ___
   ***k**   the final length, l, of an 8 m piece of wood with c cm cut off $l =$ ___

> Be careful with the
> units. 1 m = 100 cm.

  *2   ABCD is a square of side p cm.
       Two identical isosceles triangles are
       drawn in two of the corners.
       Express r in terms of m and p.

> Express r in terms of m and p
> means write a formula with r,
> m and p that begins $r =$ ____

Review   Write a formula for these.
  a   the cost, c, of p packets of biscuits at £2 each $c =$ ___
  b   the total height, h, of a man of height m cm and his hat of height 12 cm $h =$ ___
  c   the final length, l, of a 2 m piece of rope with r cm cut off $l =$ ___

Practical

1 **You will need** a spreadsheet.
Choose one of these formulae or a formula of your own.
Use a spreadsheet to print a table of values.

pints = gallons × 8

$$speed = \frac{distance}{time}$$

perimeter of rectangle = $2l + 2w$
area of rectangle = lw

Example   pints = gallons × 8

	A	B	C	D	E	F
1	gallons	1	= B1 + 1	⟶	copy across	
2	pints	= B1 * 8	⟶	copy across		

2 **You will need** a spreadsheet.
You are going on holiday overseas.
Choose a country you would like to visit.
Find out the conversion rate for the currency of that country.
Use a spreadsheet to find out how much of that currency you would get for
£10, £50, £100, £200, £500, £1000, ...

> You could change row
> 1 to 5, 10, 15 ... gallons.

Investigation

Finding Formulae

1

1 blue 2 red
2 lines

2 blue 3 red
6 lines

4 blue 2 red
8 lines

How many lines are needed to connect
5 blue dots to each of 3 red dots
4 blue dots to each of 4 red dots
8 blue dots to each of 2 red dots?
Copy and fill in this table.

number of blue dots (b)	1	2	4	5	4	8
number of red dots (r)	2	3	2	3	4	2
number of lines (n)	2	6	8			

Try to find a formula which gives the number of lines, n, needed to connect b blue
dots to r red dots. $n =$ ___

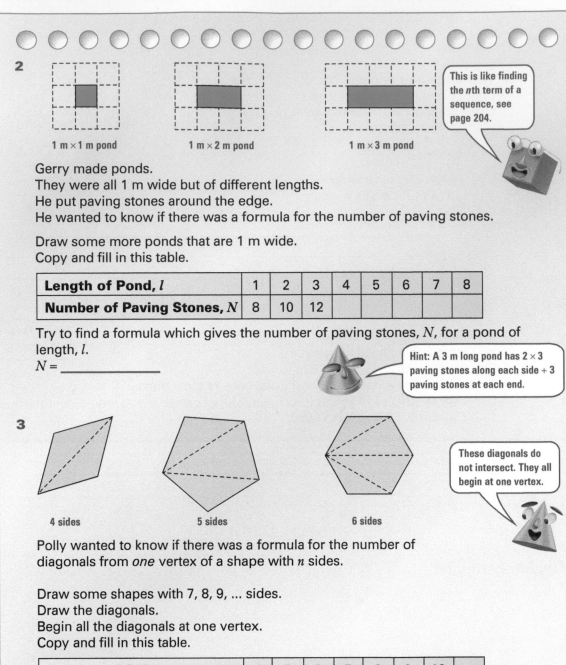

2

1 m × 1 m pond   1 m × 2 m pond   1 m × 3 m pond

This is like finding the *n*th term of a sequence, see page 204.

Gerry made ponds.
They were all 1 m wide but of different lengths.
He put paving stones around the edge.
He wanted to know if there was a formula for the number of paving stones.

Draw some more ponds that are 1 m wide.
Copy and fill in this table.

Length of Pond, l	1	2	3	4	5	6	7	8
Number of Paving Stones, N	8	10	12					

Try to find a formula which gives the number of paving stones, N, for a pond of length, l.

$N =$ _____

Hint: A 3 m long pond has 2×3 paving stones along each side + 3 paving stones at each end.

3

4 sides   5 sides   6 sides

These diagonals do not intersect. They all begin at one vertex.

Polly wanted to know if there was a formula for the number of diagonals from *one* vertex of a shape with n sides.

Draw some shapes with 7, 8, 9, ... sides.
Draw the diagonals.
Begin all the diagonals at one vertex.
Copy and fill in this table.

Number of Sides, s	4	5	6	7	8	9	10	...
Number of Diagonals, d	1	2	3					...

Try to find a formula which gives the number of diagonals, d, in a shape with s sides.

$d =$ _____

Writing equations

Yesterday Sarah went to the library. She got six books out for herself and some for her sister. We do not know how many she got for her sister.

We could write the **expression**, $n + 6$, for the number of books she borrowed. n is the number of books she got out for her sister.

If we are told that Sarah borrowed 14 books altogether we can write $n + 6 = 14$.

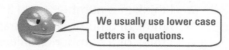

We could also write $14 = n + 6$.

$n + 6 = 14$ is an **equation**. n has a **particular value**.

$a + b = 8$ is also an **equation**. a and b can have many different values but they must always add to 8.

Example $a = 3, b = 5$.
An equation always has an equals sign.

Worked Example
I think of a number and multiply it by 3. The answer is 12.
Write an equation for this.

We usually use lower case letters in equations.

Answer
Let x be the unknown number.
x multiplied by 3 is $3x$.
The answer is 12.   **$3x = 12$ or $12 = 3x$**

Worked Example
Gregg bought some disco tickets.
He paid £2 for each of them. The total cost was £8.
Write an equation for this.

Answer
Let d be the unknown number of disco tickets
The cost of d tickets at £2 each is $2d$ pounds.
The tickets cost £8 altogether.   **$2d = 8$ or $8 = 2d$**

Worked Example
Louise had some pieces of pizza. She divided them equally onto two plates.
There were eight pieces on each plate.
Write an equation for this, using p for the number of pieces of pizza.

Answer
p is divided by 2 to get $\frac{p}{2}$.
The number of pieces on each plate is 8.   **$\frac{p}{2} = 8$ or $8 = \frac{p}{2}$**

Worked Example
I think of a number, multiply it by 2 and then add 3. The answer is 13.
Write an equation for this, using n as the unknown number.

We usually write the letters on the left hand side.

Answer
We multiply n by 2 to get $2n$.
We then add the 3 to get $2n + 3$.
So **$2n + 3 = 13$** is the equation.

Algebra

Exercise 11

1 Write an equation for each of the following. Use n as the unknown.
The answer to **a** is $n + 3 = \mathbf{15}$

Don't try to solve the equations.

  a I think of a number. I add 3. The answer is 15.
  b I think of a number. I subtract 4. The answer is 12.
  c I think of a number. I multiply it by 3. The answer is 21.
  d I think of a number. I divide it by 4. The answer is 5.
  e I think of a number. I subtract 7. The answer is 16.
  f I think of a number. I divide it by 9. The answer is 4.
  g I think of a number. I multiply it by 3. The answer is 18.
  h I think of a number. I add 12. The answer is 20.
  i I think of a number, multiply it by 3 and then add 7. The answer is 19.
  j I think of a number, multiply it by 4 and then add 8. The answer is 20.
  k I think of a number, multiply it by 6 and then subtract 5. The answer is 13.
  l I think of a number, multiply it by 8 and then subtract 20. The answer is 20.
  m When a number is multiplied by 3 the answer is 15.
  n Two less than a number is 7.
  o When a number is divided by 5 the answer is 2.
  p Three more than double a number is 15.
  q Twice the length of a hedge is 14 metres. n is the length of the hedge.
  r Russell ran round a track 24 times.
He ran 4800 m in total. n is the length of the track.
  s When Allanah doubles her lucky number and then adds 4,
she gets an answer of 30. (n is Allanah's lucky number.)
  ***t** Simon bought some peanut bars. They cost £2 each. The total cost was £16.
  ***u** Bianca bought some bags of sweets at £1 each and a drink for £2. The total cost was £6.
  ***v** 25 paving stones are stacked on top of one another. The total height of the stack is
88 cm. n is the thickness, in cm, of one paving stone.

*2 Write an equation for these.
  a Two numbers, a and b, add to 11.
  b The difference between two numbers, x and y, is 12.

Review Write an equation for each of the following. Use n as the unknown.
a I think of a number. I subtract 11. The answer is 9.
b I think of a number. I add 5. The answer is 23.
c I think of a number. I divide it by 4. The answer is 6.
d I think of a number. I multiply it by 5 and then add 4. The answer is 34.
e Rosalind's house number is four more than twice Rupert's house number. Rosalind's house
number is 36.
f When Fred divided his savings by 4 the answer was £42. (n is Fred's savings.)

Solving equations

We can use **inverse operations** to solve equations.

Discussion

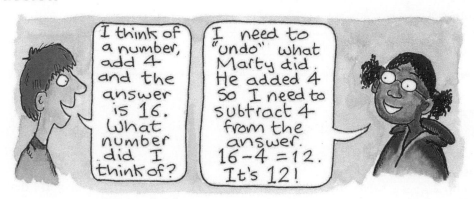

Marty

Rani

Is Rani correct? **Discuss**.

We can use **inverse operations** to **solve equations**.

Worked Example
I think of a number, add 8 and the answer is 24. What is the number?

Answer
Let n be the unknown number.
$$n + 8 = 24$$
$$\therefore n = 24 - 8 \qquad \text{The inverse of adding 8 is subtracting 8.}$$
$$n = \mathbf{16}$$
Verify the answer by substituting it back into the equation.
$$n + 8 = \mathbf{16} + 8 \qquad \text{Substitute 16 for } n.$$
$$= 24 \checkmark$$

$\therefore$ means therefore.

Verify means check.

Worked Example
Mick ran 25 times round a track. Altogether he ran 16 km. How long is the track?

Answer
Let t be the length of the track.
$$25t = 16$$
$$\therefore t = \frac{16}{25} \qquad \text{The inverse of multiplying by 25 is dividing by 25.}$$
$$t = \frac{64}{100} \qquad \text{Change to a denominator of 100 so we can write as a decimal.} \qquad \frac{16}{25} \overset{\times 4}{\underset{\times 4}{=}} \frac{64}{100}$$
$$t = 0{\cdot}64$$

The length of the track is 0.64 km.
Verify the answer.
$$25t = 25 \times \mathbf{0{\cdot}64} \qquad \text{Substitute } 0{\cdot}64 \text{ for } t.$$
$$= 16 \checkmark$$

Algebra

Worked Example

If I divide a number by 4 the answer is 6.
What is the number?

Answer

Let the number be n.

$\frac{n}{4} = 6$

$\therefore n = 6 \times 4$ **The inverse of dividing by 4 is multiplying by 4.**

$n = \mathbf{24}$

Verify the answer.

$\frac{n}{4} = \frac{24}{4}$

$= 6 \checkmark$

Exercise 12

1 What is the inverse of these?
   a adding 2 b multiplying by 3 c subtracting 4 d dividing by 7
   e adding 16 f dividing by 20 g dividing by 15

2 a I think of a number. b I think of a number.
      I add 7 to this number.   I multiply this number by 5.
      The answer is 12.   The answer is 30.
      What is the number?   What is the number?
   c I think of a number. d I think of a number.
      I divide this number by 3.   I subtract 6 from this number.
      The answer is 8.   The answer is 30.
      What is the number?   What is the number?

3 Solve these equations.
   a $y + 9 = 12$ b $x + 8 = 14$ c $7 + w = 21$ d $30 = 5 + p$
   e $m - 4 = 6$ f $n - 2 = 10$ g $y - 12 = 4$ h $m - 14 = 8$
   i $7x = 28$ j $6x = 54$ k $5a = 45$ l $24 = 3m$
   m $\frac{x}{4} = 2$ n $\frac{x}{7} = 3$ o $\frac{n}{4} = 6$ p $\frac{x}{5} = 12$
   q $m + 16 = 33$ r $\frac{w}{4} = 16$ s $4f = 64$ t $63 = p - 19$
   u $48 = n - 17$ v $\frac{a}{7} = 14$ w $16m = 80$

4 Write and solve an equation for these. Let the unknown be n.
   a Ross bought some CDs at 'Sounds'. He bought two more at 'CD Warehouse'.
      Altogether he bought eight CDs.
      How many did he buy at 'Sounds'?
   b Caroline bought some packets of muesli bars. They cost £3 each.
      The total cost was £12. How many did she buy?
   c 20 copies of the same book are stacked on top of one another. The stack is 35 cm high.
      What is the thickness of one book?

Review 1

a I think of a number. b I think of a number
   I add 6.   I divide by 7.
   The answer is 33.   The answer is 11.
   What is the number?   What is the number?

Review 2 Solve these equations.

a $x + 8 = 11$ b $y + 7 = 24$ c $9 + p = 23$ d $70 = 60 + p$
e $n - 3 = 7$ f $20 = a - 3$ g $6a = 54$ h $7a = 84$
i $\frac{n}{6} = 3$ j $\frac{n}{17} = 5$ k $72 = p - 27$

Discussion

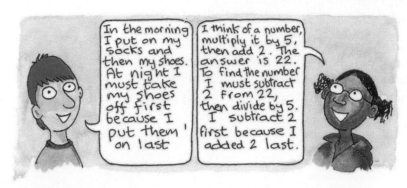

Marty   Rani

Discuss Marty's and Rani's statements.

To find the **solution** to equations like $3n + 6 = 18$, we can use inverse operations.
We undo the operations by working backwards doing the inverse.
To get $3n + 6 = 18$

$n \rightarrow$ [multiply by 3] $\rightarrow 3n \rightarrow$ [add 6] $\rightarrow 3n+6$

To solve the equation do the **inverse** operations in the reverse order.

$4 \leftarrow$ [divide by 3] $\leftarrow 12 \leftarrow$ [subtract 6] $\leftarrow 18$

$n = 4$

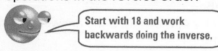

Start with 18 and work backwards doing the inverse.

Worked Example
Laura buys 4 Christmas presents for £n each and a card for £2.
The total cost is £34. Write and solve an equation to find the cost of each present.

Answer

$4n + 2 = 34$ n is multiplied by 4 then 2 is added
$4n = 34 - 2$ the inverse of adding 2 is subtracting 2
$4n = 32$
$n = \frac{32}{4}$ the inverse of multiplying by 4 is dividing by 4
$n = 8$

Each Christmas present cost £8.
Verify the answer.
$4n + 2 = 4 \times 8 + 2$ Substitute 8 for n.
$= 34$ ✓

Cost of card
$4 \times n + 2 = 34 \leftarrow$ Total cost
Number of presents   Cost of one present

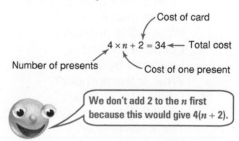

We don't add 2 to the n first because this would give $4(n + 2)$.

Algebra

Discussion

- Gemma keyed this on her calculator.
  $7 \times 8 = + 6 = - 6 = \div 8 =$.
  Without keying this, predict what the answer will be. **Discuss.**

  Would Gemma get the same answer as she did above if she keyed
  $7 \times 8 = + 6 = \div 8 = - 6 =$? **Discuss.**

  Look at the first keying sequence again.
  What would happen if Gemma missed keying the first $=$? **Discuss.**
  What if she missed keying the second $=$?
  What if she missed keying all of the $=$ except the last one? **Discuss.**

- Amy was asked to solve $3x + 2 = 14$.
  She wrote $3x + 2 = 14$
  $$= 14 - 2$$
  $$= \frac{12}{3}$$
  $$= 4$$
  What is wrong with this? **Discuss.**

Exercise 13

1 What is the inverse of
   a multiplying by 2 then adding 4   b multiplying by 6 then adding 7
   c multiplying by 4 then subtracting 2   d multiplying by 5 then subtracting 4
   e multiplying by 6 then adding 2   f adding 2 then multiplying by 5
   g adding 3 then multiplying by 4?

2 Find the answer to these using inverse operations.
   a I think of a number.
      When I multiply by 4, then add 2 the answer is 26.
      What is the number?
   b I think of a number.
      I divide by 2, then add 3. The answer is 8.
      What is the number?
   c I think of a number.
      I add 5 then multiply by 2.
      The answer is 22
      What is the number?
   d I think of a number.
      I multiply by 7 then add 3·5. The answer is 17·5
      What is the number?
  *e I think of a number.
      I add 2·7 then multiply it by 5
      The answer is 24·5.
      What is the number?

3 Solve these equations.
   a $2x + 5 = 13$ b $3y - 6 = 9$ c $5w + 3 = 23$ d $7y - 12 = 2$
   e $4p + 5 = 33$ f $6m - 7 = 29$ g $5m + 7 = 37$ h $4n - 9 = 27$
   i $8b - 6 = 66$ j $9a + 12 = 75$ k $7x + 20 = 76$ l $14 = 2 + 3n$
   m $11 = 4a - 1$ n $21 = 3t - 6$ o $17 = 3 + 2d$ p $66 = 3w + 6$
   q $5x + 7 = 67$ r $4p + 3 = 83$ s $6n + 3 = 75$

*4 Write and solve an equation for these.

a Jesse had saved £29. Her aunt gave her £5 of this. The rest she got from working for a neighbour for £3 an hour. How many hours work did she do?

b Sam bought some cakes for £3 each and one bag of biscuits for £2.
The total cost was £14.
How many cakes did Sam buy?

c Emma cycled four times around a track and then 800 m along the road.
The odometer on her bike showed she had cycled 2400 m altogether.
How long is the track?

d Ashad earned £5 for every car he cleaned.
His boss gave him an extra £10 for working hard.
Altogether Ashad earned £35.
How many cars did he clean?

Review 1 Solve these equations.

a $2n + 4 = 10$ b $3y + 4 = 22$ c $4x - 7 = 25$ d $9p - 7 = 56$
e $57 = 7m + 8$ f $31 = 5p + 6$ g $4x + 3 = 63$

Review 2 Write and solve equations for these.

a Melanie delivered seven packs of pamphlets minus the four pamphlets she had left over. Altogether she delivered 66 pamphlets. How many pamphlets were in each pack?

b Joseph measured the length he wanted his curtains. He multiplied this by 4 and then added 28 cm. His answer was 480 cm. What length does he want his curtains?

Worked Example
In this diagram the number in each box is found by adding the two numbers above it.
What are the missing numbers in this diagram?

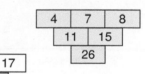

Answer
Let n be the number in the yellow box.
 red box $= n + 20$
blue box $= n + 17$
        $67 = $ red box $+$ blue box
        $67 = n + 20 + n + 17$
            $= 2n + 37$
$2n + 37 = 67$
    $\therefore 2n = 67 - 37$ the inverse of adding 37 is subtracting 37
    $2n = 30$
    $\therefore n = \frac{30}{2}$ the inverse of multiplying by 2 is dividing by 2
    $n = \mathbf{15}$

Verify the answer.
$2n + 37 = 2 \times \mathbf{15} + 37$ substitute 15 for n.
        $= 67$

The diagram can be filled in as shown.

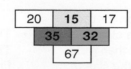

Algebra

1 Write and solve an equation to find the value of *n*. Use this to find the missing numbers. **a** and **b** have been started for you.

The number in each box is the sum of the two numbers above it.

a

b
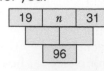

c
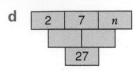

$12 + n + 8 + n = 32$ $16 + n + n + 20 = 60$

d
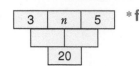

e
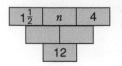

*f
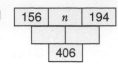

*g

*h

*i
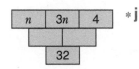

*j
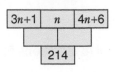

2 Write and solve an equation to find *x*.
   a is started for you.

a
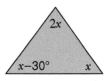

b
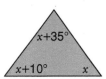

c
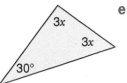

Remember:
The angles of a triangle add up to 180°.

$x + 70° + 40° = 180°$
$x + 110° = 180°$

d   **e**   *f

*g   *h   *i   *j

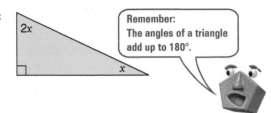

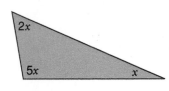

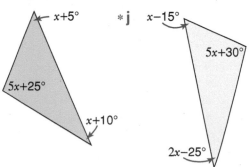

3 Write and solve an equation for these.

   a There are 32 pupils in a class.
   There are 6 more boys than girls.
   How many girls are in the class?
   Let *g* be the number of girls

   b There are 30 scones altogether on two plates.
   The second plate has 6 fewer scones than the first plate.
   How many scones are there on each plate?
   Let *n* be the number of scones on the first plate.

4 $m + n = 7$ is an equation with two unknowns.
If $m = 4$ and $n = 3$, the equation would be true.
What other values of m and n make this equation true?
Write down at least ten different values for m and n.

5 What could the numbers a and c be?
a $a + c = 32$ **b** $16 - a = c$ **c** $\frac{a}{c} = \frac{1}{2}$ **d** $ac = 24$

*6 The number in each box is the sum of the two numbers above it.
What positive values could x and y have?

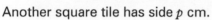

*7 The length of the side of this square tile is t cm.
a Write an expression for the area of the tile.
b The area of the tile is 64 cm².
What is the length of the side of the tile?
c What is the perimeter of the tile?

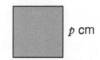

Another square tile has side p cm.
d Write an expression for the perimeter of the tile.
e If the perimeter is 40 cm, what is the area?

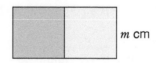

*8 This shape is made from two square tiles.
The length of a side of a tile is n cm.
a Write an expression for the area of the shape.
b The area of the shape is 50 cm².
What is the length of the side of each tile?
c What is the perimeter of the shape?
d Another shape is made with two different square tiles,
each of side m cm.
Write an expression for the perimeter of the shape.
If the perimeter is 24 cm, what is the area?

Review 1 The number in each box is the sum of the two numbers above it.
Find the value of n.

a 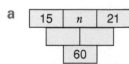   **b**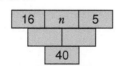

Review 2 Find x.

a 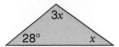   **b**    **c**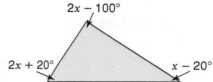

Investigation

Arithmagons

This is an arithmagon.
The number in each square is the sum of the numbers in the circles on either side of it.

What could the numbers A, B, C and D be?
Is there more than one answer? **Investigate**.

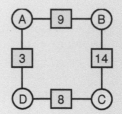

Manoli wrote   A + B = 9
                 A + B + C + D = 17
                 B = 9 – A

Write some more relationships between A, B, C and D.
Are any of the relationships you have written equivalent?

Manoli's first and third relationships are equivalent.

Check that the relationships are true for all of the sets of possible values for A, B, C and D you have found.
When writing this investigation we had to find numbers for the boxes that worked. How might we have found them? **Investigate**.

Summary of key points

 $p - 7$, $8n$, $7t + 4$ are all **expressions**.
The letter stands for an **unknown**.
We write expressions as simply as possible.
Example   Multiply a number by 2 then add 6 is written as $2n + 6$.

 The operations used in algebra **follow the same rules as in arithmetic**.
Examples

$x + y = y + x$ $pq = qp$ $m + (n + p) = (m + n) + p$ $a(bc) = (ab)c$
$3 + 4 = 4 + 3$ $8 \times 9 = 9 \times 8$ $5 + (6 + 4) = (5 + 6) + 4$ $3 \times (4 \times 5) = (3 \times 4) \times 5$

 We can **simplify** expressions by **collecting like terms**.
Examples $a + a + a = 3a$ $3n + 6n = 9n$ $8b - 3b = 5b$

$4m + 6 - 3m - 2 = 4m - 3m + 6 - 2$

$= m + 4$

$3x + 8y + 2x + y = 3x + 2x + 8y + y$

$= 5x + 9y$

Write the like terms next to one another.

We can **simplify** expressions by **cancelling**.

Examples $\frac{8b}{2} = 4b$ divide numerator and denominator by the common factor 2

$\frac{6a}{a} = 6$ divide numerator and denominator by the common factor a

 To **write an expression without brackets** we use the rules of arithmetic

Example $4(x + 7) = 4x + 28$

	x	$+7$
4	$4x$	$+28$

 We **evaluate an expression** by **substituting** values for the unknown.

Examples   If $a = 2$, $b = 3$ and $c = 5$ then

$3a = 3 \times 2$ $2c - b = 2 \times 5 - 3$ $\frac{c}{a} + 3 = \frac{5}{2} + 3$

$\quad\;\; = 6$ $\qquad\quad = 10 - 3$ $\qquad\;\; = 2 \cdot 5 + 3$

$\qquad\quad\;\; = 7$ $\qquad\;\; = 5 \cdot 5$

 A **formula** is a rule for working something out.

It can be written in words or symbols.

Examples   The cost in pounds of hiring a ladder = number of days × £3 then add £10.

This can be written in symbols as $c = 3d + 10$ where c is the cost and d is the number of days.

$T = \frac{D}{S}$ is the formula for finding the time taken to travel distance, D, at speed, S.

If $D = 20$ and $S = 4$ then $T = \frac{20}{4}$

$= 5$

 We can **write expressions and formulae** for practical situations.

 $n + 4 = 12$ is an **equation**. n has a particular value.

$a + b = 4$ is an **equation**. a and b can take many different values but they must always add to 4.

 We can **solve equations** using inverse operations.

Examples $x + 8 = 11$

$x = 11 - 8$ the inverse of adding 8 is subtracting 8

$x = 3$

> We start with the right-hand side of the equation and work backwards doing the inverse.

$2x - 3 = 15$

$2x = 15 + 3$ the inverse of subtracting 3 is adding 3

$2x = 18$

$x = \frac{18}{2}$ the inverse of multiplying by 2 is dividing by 2

$x = 9$

> Always verify your answer by substituting it back into the equation.

Test yourself

1 Simplify these.
   a $3 \times y$ **b** $b \times 4$ **c** $c \times c$ **d** $4 \times (x + 3)$ **e** $2 \times (a + 7)$
   f $(y - 4) \times 6$ **g** $p \times p$ **h** $3 \times b \times b$ **i** $5y \times 5y$

2 Write an expression for these.
   a subtract 4 from a number   **b** multiply a number by 7
   c multiply a number by 2 then add 3   **d** add 3 to a number then multiply by 2
   e multiply a number by 3 then multiply the result by itself

3 Simplify these.
   a $x + x + x + x$ **b** $8a + 3a$ **c** $10b - 5b$ **d** $10m + 3m + 2m$
   e $5n + 3n - n$ **f** $12p - 4p + 3p$ **g** $10q - 3q - 2q$ **h** $6p + 8p + 4$
   i $3y + 2y + 7 + 2$ **j** $3x + 5 + 2x - 1$ **k** $9m + 8 - 3m - 5$ **l** $3a + 4b + 2a + 2b$
   m $5x + 2y - 3x + 2y$ **n** $8p + 2q - 3p - q$ **o** $4m + 7n - 4m + 2n$

4 Holly's house is a rectangle.
Write an expression for the perimeter
of Holly's house.
Simplify your expression.

4x metres

16 metres

5 Simplify these expressions by cancelling.
   a $\frac{5w}{w}$ **b** $\frac{20n}{5}$ **c** $\frac{28x}{7}$ **d** $\frac{30m}{m}$ **e** $\frac{15r}{r}$ *f $\frac{21a}{14}$

6 Write these without brackets.
   a $5(x + 3)$ **b** $7(n + 4)$ **c** $8(a + 4)$ **d** $12(x + 2)$

7 Match each of these with an expression from the box.
   a $2x + x + 2x$ **b** $2x + 4 + x$ **c** $\frac{4x}{x}$
   d $2(x + 2)$ **e** $2 \times x \times x$ **f** $\frac{16x}{4}$
   g $(x + y) \times 3$ **h** $(x + y) + 2$

A $4x$	**B** $3(x + y)$
C $2x^2$	**D** $5x$
E $2 + (x + y)$	**F** $2x + 4$
G $3x + 4$	**H** 4

8 If $x = 3$, $y = 2$ and $z = 6$ find the value of
   a $x + 3$ **b** $y - 1$ **c** $3x$ **d** $5z$ **e** $4 - y$ **f** xy
   g $xy - z$ **h** $3z - y$ **i** $xy + z$ **j** $4(z - y)$ **k** $\frac{z}{y} + x$ **l** $4(2x + y)$

9 **a** If $x + y = 6$, write three other facts that must be true.
   b What values could x and y have?

10 **i** The cost of a child's ticket to a theatre is found by halving the cost of an
     adult ticket and then adding £1.
     How much does a child's ticket cost when an adult's ticket is
     a £10   **b** £16   **c** £15   **d** £9?
   ii C = cost of child's ticket A = cost of adult ticket
     Copy and complete this formula for the cost of a child's ticket.
     $C =$ _____

11 $D = \frac{M}{V}$ is a formula for finding density.
Find D if $M = 18$ and $V = 3$.

12 $c = 80 + 65h$ is the formula for the cost to hire a wedding car where c is the cost, in pounds, and h is the number of hours hired.
Find the cost to hire a car for
   a 4 hours   **b** 8 hours   **c** 5·5 hours.

13 There are n blocks in Rajiv's hand.
Write an expression for the number of blocks these people have.
   a Caitlin has two more blocks than Rajiv.
   b Richard has three fewer blocks than Rajiv.
   c Brian has three times as many blocks as Rajiv.
   d Nathan has one fifth as many blocks as Rajiv.
  *e Tejal has three less than four times as many blocks as Rajiv.

14 Write an equation for each of these. Let the unknown in each be n.
   a I think of a number. I subtract 5. The answer is 7.
   b I think of a number. I multiply by 4. The answer is 32.
   c When the length of a fence is divided by 6 the answer is 4 metres.
   d Three times James' height is 450 cm.
   e When the length of a pool was doubled and then 4 metres was added, the final length was 20 m.

15 Solve these equations
   a $b + 9 = 21$ **b** $x + 7 = 20$ **c** $5x = 50$ **d** $\frac{y}{9} = 4$
   e $2n + 6 = 16$ **f** $4y - 3 = 33$ **g** $8p + 12 = 76$

16 Write and solve an equation to find the answers to these.
   a I think of a number.
      When I multiply by 4 and add 9 the answer is 37.
      What is the number?
   b Jimmy bought 5 CDs at 'Sounds Right'.
      The total cost after the discount was £51.
      How much was each CD before the discount?
   c Find the value of x.

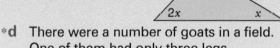

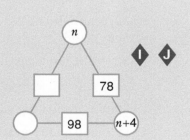

Sounds Right
Buy 5 CDs and get a £4 discount off the final price.

  *d There were a number of goats in a field.
      One of them had only three legs.
      The total number of legs was 127.
      How many goats were in the field?

*17 The number in the square is found by adding the numbers in the circles either side of it.
Write and solve an equation to find the value of n.
Copy the diagram and fill in the missing numbers.

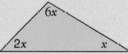

You need to know

✓ unknowns   page 151

✓ counting on and back   page 151

⋯ Key vocabulary ⋯⋯⋯⋯⋯⋯⋯⋯⋯⋯⋯⋯⋯⋯⋯⋯⋯

consecutive, continue, decrease, finite, function, function machine, generate, increase, infinite, input, mapping, *n*th term, output, predict, relationship, rule, sequence, term

▷▷ Wheel of Fortune

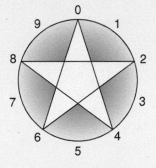

If we start at 0 and keep adding 4 we get this pattern.

0, 4, 8, 12, 16, 20, 24, 28, 32, 36, 40, 44, ...

The pattern formed by the last digit of these numbers is

0, 4, 8, 2, 6, 0, 4, 8, 2, 6, 0, 4, ...

If we join these last digits in order on a number wheel, we get the pattern shown.

What pattern do the last digits of these make on a number wheel?

1 Start at 0 and keep adding 2.

2 Start at 0 and keep adding 6.

3 Start at 0 and keep adding 3.

4 Start at 0 and keep adding 7.

Keep writing down numbers until you get a repeating pattern.

Sequences

2, 4, 6, 8, 10, 12

This is a **number sequence**. It is a set of numbers in a given order.
Each number in the sequence is called a **term**.

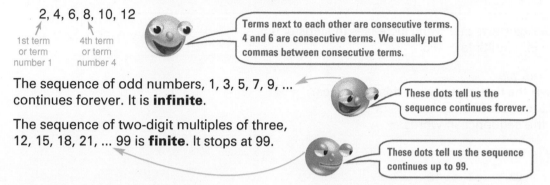

2, 4, 6, 8, 10, 12

1st term
or term
number 1

4th term
or term
number 4

Terms next to each other are consecutive terms.
4 and 6 are consecutive terms. We usually put
commas between consecutive terms.

The sequence of odd numbers, 1, 3, 5, 7, 9, ...
continues forever. It is **infinite**.

These dots tell us the
sequence continues forever.

The sequence of two-digit multiples of three,
12, 15, 18, 21, ... 99 is **finite**. It stops at 99.

These dots tell us the sequence
continues up to 99.

Discussion

● 7 MN were asked to write examples of sequences found in everyday life.

I wrote down
the numbers on
houses on my
side of the
street.
1,3,5,7,9...

I wrote down
the times the
sun sets
each day.
17·25, 17·26,
17·27, 17·29,
17·30.

The rule for Paul's sequence is 'start at 1 and count on in steps of 2'.
Sirah's sequence has a more complex rule (one that is harder to find).

Think of other sequences from everyday life. Think of some with simple rules and some
with a rule that is hard to find. **Discuss**.

●

I wrote down
the midday
temperatures
for May.
21°c, 24°c,
18°c, 19°c...

I wrote down
the winning
numbers in
the school
raffle.
3, 59, 26, 83,
92, 36...

James' sequence has an irregular pattern and no rule.
Maisy's sequence has no rule and no pattern. It is called a set of random numbers.
Think of other sequences from everyday life with an irregular pattern. **Discuss**.
Think of other sequences from everyday life that are a set of random numbers. **Discuss**.

● Is Paul's sequence finite or infinite? **Discuss**.
What about Sirah's, James and Maisy's?

Counting on or counting back

Declan is stacking books. Each book is 0·5 cm thick. He wrote down the total height of the pile after each book had been stacked.

0.5   1   1.5   2   2.5   3   3.5
   +0.5 +0.5 +0.5 +0.5 +0.5 +0.5

This sequence starts at 0·5 and increases in steps of 0·5.
Each term is a multiple of 0·5.

Worked Example
Write down the first six terms of this sequence.
Start at 32 and count back in steps of 5.
Describe the sequence in words.

Answer

32   27   22   17   12   7
   −5   −5   −5   −5   −5

The first six terms are **32, 27, 22, 17, 12, 7.**
We could describe the sequence by saying
**'the sequence begins at 32 and decreases in steps of 5.
Each term is 2 more than a multiple of 5'.**
How else could you describe each term?

We could say each term is 3 less than a multiple of 5.

Worked Example
Write down the next three terms of 4·2, 3·8, 3·4, 3·0, ...
Describe the sequence.

Answer

4.2   3.8   3.4   3.0   2.6   2.2   1.8
   −0.4 −0.4 −0.4 −0.4 −0.4 −0.4

The next 3 terms are **2·6, 2·2, 1·8**.
We could describe the sequence by saying
**'the sequence starts at 4·2 and decreases in steps of 0·4.
Each term is 0·2 more than a multiple of 0·4'.**

Exercise 1

1 Write down the first six terms of these sequences.
   a start at 5 and count on in steps of 3
   b start at 73 and count back in steps of 7
   c start at 77 and count back in steps of 8
   d start at 5 and count on in steps of 0·5
   e start at 2 and count back in steps of 0·2
   f start at 5·5 and count on in steps of 1·5
   g start at ⁻2 and count back in steps of 2
   h start at 2 and count back in steps of 5
   i start at 0·3 and count on in steps of 0·6

2 In which of the sequences you made in question **1** are the terms ascending (getter bigger)?

3 Jessica washed her hair every 3rd day. She wanted to mark the days on her calendar.
She washed it on 2nd July.
Write down the dates of the next five times she will wash her hair.

4 After a run, Patty took her pulse every 4 minutes. She took it the first time at 10·05 a.m. Write down the times of the next five times she took her pulse.

5 Write down the next 3 terms of these sequences.

a 6, 12, 18, 24, 30, ...   b 9, 18, 27, 36, 45, ...   c 7, 12, 17, 22, 27, ...
d 3, 11, 19, 27, 35, ...   e 0·3, 0·6, 0·9, 1·2, 1·5, ...   f 50, 48, 46, 44, 42, ...
g 5, 5·2, 5·4, 5·6, 5·8, ...   h 78, 70, 62, 54, 46, ...   i 98, 91, 84, 77, 70, ...
j 10, 5, 0, ⁻5, ⁻10, ...   k 2, ⁻1, ⁻4, ⁻7, ⁻10, ...   l 4, 0, ⁻4, ⁻8, ...
m ⁻64, ⁻55, ⁻46, ⁻37, ⁻28, ...   n 0·8, 0·4, 0, ⁻0·4, ⁻0·8, ...

6 Describe each of the sequences in question **5 a-f**.

7 Susannah used her graphical calculator to display the sequence 0·2, 0·6, 1, 1·4, 1·8, 2·2.
She keyed

Susannah then keyed
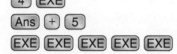
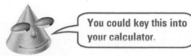

You could key this into your calculator.

a Describe the sequence Susannah will get on her screen.
b How many times would she have to press EXE after the 4 to get a number bigger than 50?

8 a Write down what you need to key on a graphical calculator to get a sequence which begins at 0·3 and goes up in steps of 0·3.
b Without using the calculator, predict which of these numbers will be in the sequence. How can you tell?

3·0   3·9   4·0   4·2   5·1

∗9 Jeff's calculator screen looked like this.
He described the sequence.
'The sequence begins at 0·3 and increases in steps of 0·3.'
a Describe the sequence that has terms 10 times larger than these.
b Describe the sequence that has terms 100 times larger than these.

Review 1 Write down the first six terms of these sequences.

a start at 4 and count on in steps of 8
b start at 98 and count back in steps of 7
c start at 2 and count on in steps of 1.5
d start at 5 and count back in steps of 0·8
e start at 3 and count back in steps of 5

Review 2 In which of the sequences you made in **Review 1** are the terms descending (getting smaller)?

Review 3 Write down the next three terms of these sequences.

a 3, 8, 13, 18, 23, ...   **b** 0·6, 1, 1·4, 1·8, 2·2, ...   **c** 2, ⁻6, ⁻14, ⁻22, ...

Review 4 Describe, in words, the sequence in **Review 3a.**

****Review 5** Predict if 3·9 will be a term of the sequence generated by this keying sequence.

⓪ · ⑤ EXE
Ans + ⓪ · ③
EXE EXE EXE EXE EXE ...

[T]

⭐ **Practical**

You will need a graphical calculator.

Use your graphical calculator to generate each of the sequences in question **5** of exercise **1**.

[T]

○ ○ ○ ○ ○ ○ ○ ○ ○ ○ ○ ○ ○ ○ ○ ○ ○ ○ ○

Investigation

Seating Numbers

This shows part of the seating at a football stadium.

17	18	19	20	21	22	23	24
9	10	11	12	13	14	15	16
1	2	3	4	5	6	7	8

The four seats in this 2 × 2 square of seating add to 22.
Call this square **1**.

9	10
1	2

What do the seats in this 2 × 2 square add to?
Call this square **2**.

10	11
2	3

What does square 3 add to?
What about square 4, square 5, ...
What do you notice? Why does this happen?

What if you added the six seat numbers in a 2 × 3 rectangle?
What if you added the nine seat numbers in a 3 × 3 square?
What if ...

Showing sequences with geometric patterns

The **numbers** in a sequence can be shown using diagrams.

Square Numbers

| 1 | 4 | 9 | 16 | 25 |

Triangular Numbers

| 1 | 3 | 6 | 10 | 15 |

Even Numbers

| 2 | 4 | 6 | 8 | 10 |

Multiples of 3

| 3 | 6 | 9 | 12 | 15 |

Discussion

Look at the geometric patterns above.

● The even numbers are called even because they can be arranged in matched pairs of dots. Can you draw the odd numbers in matched pairs? **Discuss.**
Draw diagrams to represent the first six odd numbers.
Describe what you notice.

● The multiples of 3 have been arranged in rectangles of dots.
Can multiples of any number be arranged in rectangles? **Discuss.**
Try to arrange the multiples of 4, 5, 6, 7, ... as rectangles.

● How else could the triangular numbers be drawn using dots? **Discuss.**

Writing sequences from rules

A sequence can be written down if we know the first term and the rule for finding the next term.

Example   **1st term   4   rule for finding the next term**   add 3

The sequence is   4,   7,   10,   13,   16,   19, ...

Each term is three more than the one before. The terms of this sequence increase in equal steps.

Example **1st term** 1000 **rule for finding the next term** halve

The sequence is 1000, 500, 250, 125, 62.5, ...   Each term is half of the one before.
           ÷2   ÷2   ÷2   ÷2   The terms of this sequence decrease in unequal steps.

Exercise 2

1 The first term and the rule for finding the next term are given.
Write down the first five terms of these sequences.
You could use a graphical calculator if you wish.

a **first term** 3
   rule add 4

b **first term** 1
   rule multiply by 2

c **first term** 100
   rule subtract 10

d **first term** 1
   rule add 5

e **first term** 1
   rule add 11

f **first term** 3
   rule double

g **first term** 2000
   rule halve

h **first term** 100 000
   rule divide by 10

i **first term** 1
   rule multiply by 4

j **first term** 50 000
   rule divide by 5

k **first term** 25
   rule subtract 2·5

l **first term** 0·1
   rule add 0·1

m **first term** 5
   rule subtract 0·25

n **first term** 1
   rule add 101

2 Which of the sequences in question **1** increase or decrease by unequal steps?

3 We could describe the sequence generated in question **1a** as 'the first term is 3 and each term is 4 more than the one before'.
Describe these sequences in this way.
Question **1 b, c, d** and **f**

4 Some of the terms of these sequences are smudged.
Find the missing terms.
a 3, ▢, ▢, ▢, 19 **rule** add 4
b 67, ▢, ▢, 46, ▢ **rule** subtract 7
c 1600, ▢, ▢, ▢, 6·25 **rule** divide by 4
d 1, ▢, 25, ▢, ▢ **rule** multiply by 5
e 2, ▢, ▢, ▢, 32, ▢ **rule** double

*5 Write down the first 6 terms of the sequences I am thinking of.

a I am thinking of a sequence.
The 3rd term is 8.
The rule is 'add 3'.

b I am thinking of a sequence.
The 4th term is 10.
The rule is 'subtract 4'.

c I am thinking of a sequence.
The 3rd term is 40,
The rule is 'multiply by 2'.

*6 The rule for a sequence is 'to find the next term, add 4'.
Two possible sequences with this rule are 1, 5, 9, 13, ... and 23, 27, 31, 35, ...
   a Write down another possible sequence with this rule.
   b Is it possible to find a sequence with the rule 'add 4' for which
      i   all terms are multiples of 4   ii   all terms are multiples of 8
      iii all terms are even numbers   iv   all terms are negative numbers
      v   none of the terms is a whole number?
      If it is possible, give an example.

Review 1   Write down the first five terms of these sequences.

a **first term** 4   b **first term** 4   c **first term** 80
   rule add 6   **rule** multiply by 2   **rule** subtract 5
d **first term** 5000   e **first term** 1   f **first term** 20 000
   rule divide by 5   **rule** add 9   **rule** divide by 2
g **first term** 2   h **first term** 1000
   rule multiply by 4   **rule** subtract 25

Review 2   Find the missing terms.

a 4, ▢, ▢, ▢, 48 **rule** add 11   b 80, ▢, ▢, 10, ▢, ▢ **rule** halve
c 243, 81, ▢, ▢, ▢ **rule** divide by 3   d 0·4, ▢, ▢, 0·1, ▢ **rule** subtract 0·1

*Review 3   The rule for finding the next term of a sequence is 'add 7'.

Is it possible to find a sequence with this rule for which
a all terms are multiples of 7   b all terms are multiples of 5
c all terms are even numbers   d every 7th term is a prime number?

If it is possible, give an example.

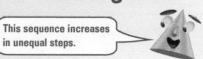

Investigation

Adding Terms

This sequence increases in unequal steps.

1, 3, 4, 7, 11, 18

The rule for this sequence is 'to find the next term, add the two previous terms'.
Use the same rule to find the next five terms of this sequence.

2, 4, ...

Use the same rule to find the missing terms in these.

3, ___ , 8, ___ , 21
8, ___ , ___ , ___ , 22
4, ___ , ___ , ___ , 26
___ , ___ , ___ , ___ , 24

* Is it possible to use the same rule to find the missing terms for these? If so, write down the missing terms.

4, ___ , ___ , 9

10, ___ , ___ , ___ , 24   Try fractions.

Algebra

Writing sequences given the *n*th term

We can show the sequence 3, 4, 5, 6 ... in a table.

Term number n	1	2	3	4	5	...
Term	3	4	5	6	7	...

One rule for this sequence is '**first term** 3 **rule** add 1'.

We can see that each term is the term number, n, plus **2**.

1st term ($n = $ **1**) = **1** + **2** = 3
2nd term ($n = $ **2**) = **2** + **2** = 4
3rd term ($n = $ **3**) = **3** + **2** = 5

$\vdots$

nth term (n) $= n + 2$

Another rule for this sequence is '**the *n*th term is $n + 2$**'.

Worked Example
The rule for the *n*th term of a sequence is $4n - 1$.
a Write the first four terms of this sequence.
b What is the 24th term?

Answer
a 1st term ($n = $ **1**) = $4 \times $ **1** $- 1$
                    $= 3$
   2nd term ($n = $ **2**) = $4 \times $ **2** $- 1$
                    $= 7$
   3rd term ($n = $ **3**) = $4 \times $ **3** $- 1$
                    $= 11$
   4th term ($n = $ **4**) = $4 \times $ **4** $- 1$
                    $= 15$
   The first 4 terms are **3, 7, 11, 15**.

b To find the 24th term, substitute $n = $ **24** into $4n - 1$.

   $4n - 1 = 4 \times $ **24** $- 1$
        $= 95$
   The 24th term is **95**.

For more on substituting look at page 161.

For more on substituting look at page 161.

Exercise 3

1 The *n*th term of a sequence is given. Write the first 5 terms.
   a $n + 1$ **b** $n + 3$ **c** $n + 4$ **d** $n - 1$ **e** $2n$
   f $2n + 1$ **g** $3n$ **h** $2n - 0.5$ **i** $20 - n$ **j** $3n - 4$
   k $100 - 5n$ **l** $4n - 2$ **m** $80 - 3n$ **n** $4 - 3n$

2 Write down the 20th term of each of the sequences in question **1**.

3 Describe each of the sequences in question **1** in words.
The answer to **1a** could be 'the integers greater than 1' or '**first term** 2   **rule** add 1'.

4

shape 1   shape 2   shape 3

The number of dots in the *n*th shape of this pattern is given by $5n + 1$.
Write down the number of dots in
a shape 5   **b** shape 10   **c** shape 20.

5 This table shows the sequence with *n*th term $2n + 1$.

Term number	1	2	3	4	5	6	...	n
Term	3	5	7	9	11	13	...	$2n + 1$

Use a copy of these tables. Fill in the missing terms.

a

Term number	1	2	3	4	5	6	...	n
Term	0	1					...	$n - 1$

b

Term number	1	2	3	4	5	6	...	n
Term	7	9					...	$2n + 5$

c

Term number	1	2	3	4	5	6	...	n
Term	4						...	$3n + 1$

d

Term number	1	2	3	4	5	6	...	n
Term	28						...	$30 - 2n$

*6 a Use a copy of this table.
Fill it in for the multiples of 3.
b Use another copy.
Fill it in for the multiples of 5.

Term number	1	2	3	4	5	6	...	n
Term							...	

Review 1 The *n*th term of a sequence is given. Write the first five terms.
a $n + 6$ **b** $6n$ **c** $2n - 2$ **d** $50 - 2n$ **e** $n - 5$

Review 2 Write down the 27th term of each of the sequences in **Review 1**.

Review 3 Describe each of the sequences in **Review 1** in words.

*Review 4 Use a copy of this table.
Fill it in for the multiples of 4.

Term number	1	2	3	4	5	6	...	n
Term							...	

Practical

You will need a spreadsheet and a graphical calculator.

1 4, 8, 12, 16, 20, ...
The terms of this sequence are the multiples of 4.
The rule for finding the next term is 'add 4'.
The nth term of the sequence is $4n$.

Using a spreadsheet we can find the first 20 terms of the sequence in two ways.

Using the rule for the next term

	A	B	C	D	E	
1	Term number	1	= B1+1	= C1+1	= D1+1	= E
2	Term	4	= B2+4	= C2+4	= D2+4	=

Using the expression for the nth term

	A	B	C	D	E	
1	Term number	1	= B1+1	= C1+1	= D1+1	=
2	Term	= B1*4	= C1*4	= D1*4	= E1*4	= F

Use a spreadsheet to find the first 20 multiples of 5 in two different ways.
Use a spreadsheet to find the multiples of these numbers in two different ways.

7   11   15   24

*2 Use a spreadsheet to find the first 20 terms of these.

100, 95, 90, 85, 80, ...
7, 11, 15, 19, 23, ...

Use the rule for finding the next term.

*3 Janice was asked to generate this sequence on her graphical calculator.

100 000, 50 000, 25 000, 12 500, ...

She keyed

(1)(0)(0)(0)(0)(0) [EXE] [Ans] (÷) (2) [EXE] [EXE] [EXE] [EXE]

Use a graphical calculator to generate these sequences.

128, 64, 32, 16, ...
1, 4, 16, 64, ...
1, ⁻1, 1, ⁻1, 1, ⁻1, ...

Write down the rules that you used.

Investigation

Shuffles

Put 16 chairs in 4 rows of 4.

Seat 15 students in the chairs as shown.

Move the student in seat 1 to seat 16 in as few moves as possible.

Follow these rules.

- The person in front, behind or next to the empty chair may move to it.
- No one else may move.
- No one may move diagonally.

1	2	3	4
5	6	7	8
9	10	11	12
	14	15	16

What is the fewest number of moves? **Investigate**.

What if the chairs were put in 5 rows of 5 and 24 students were seated. Leave seat 21 empty and move the student in seat 1 to seat 25.
What if ...

Sequences in practical situations

Discussion

Jamie made this pattern with matchsticks.

  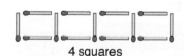

1 square   2 squares   3 squares   4 squares

Number of squares	1	2	3	4	...
Number of matchsticks	4	7	10	13	...

Jamie was asked how many matchsticks would be needed for 30 squares.

Is Jamie right?

Jamie worked out that for n squares the number of matchsticks needed is $n \times 3 + 1$ or $3n + 1$.
Is he right?

Each added square needs 3 matchsticks. But the first square needs an extra matchstick. That's $30 \times 3 + 1 = 91$.

Algebra

Worked Example
Jasmine made these shapes with floor tiles.

shape 1   shape 2   shape 3

Shape number	1	2	3	4	...
Number of tiles	3	5	7	?	...

a   Draw shape 4. How many tiles are in this shape?
b   Describe the pattern and how it continues.
c   Explain how you can find the number of tiles in the *n*th shape.
d   Write an expression for the number of tiles in the *n*th shape.

Look carefully at how the shapes are changing.

Answer
a   There are **9** tiles.

b   **The first L-shape has 3 tiles.**
    Two more tiles are added each time the next shape is drawn.
c   The first shape has two green tiles, the second four green tiles, the third six green tiles and so on. The number of green tiles in each shape is two times the shape number.
    Each shape also has 1 red tile. **The number of tiles altogether in each shape is 2 × shape number + 1.**
d   **$2n + 1$**

Exercise 4

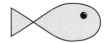

1   Ellen drew a fish. She made a pattern with the bubbles.

group 1   group 2   group 3   group 4

a   Copy and complete this table.

Group of bubbles	1	2	3	4	5
Number of bubbles	2	4	6		

b   Ellen drew ten groups of bubbles altogether. Explain how you could find the number of bubbles in the tenth group without drawing the whole pattern.
c   Explain how you could find the number of bubbles in the *n*th group.
d   Which of these expressions give the number of bubbles in the *n*th group?
    A $n + 1$ **B** $n + 2$ **C** $2n$ **D** $2n + 1$

2 Jess was building sheep pens.

**1 pen
4 rails**   **2 pens
8 rails**   **3 pens
12 rails**

a Draw 4 pens.

b Copy and finish filling in this table.

Number of pens	1	2	3	4	5
Number of rails	4	8	12		

c Describe the sequence made by the number of rails.

d Will one of the pens have 35 rails? Give a reason for your answer.

e Explain how you would find the number of rails needed to make n pens.

f Write an expression for the number of rails needed to make n pens.

g How many rails are needed for 16 pens?

3 'Vince's Mowing' has a V made of circles as its logo.

shape 1   **shape 2**   **shape 3**

a Copy and complete this table.

Size of V	1	2	3	4
Number of circles				

b Describe the sequence made by the number of circles.

c Will one of the V's have 29 circles? Give a reason for your answer.

d Explain how you would find the number of circles needed for the nth V.

e Which of these is the expression for the number of circles needed for the nth V?
   A $3n$ **B** $3n - 1$ **C** $2n$ **D** $2n + 1$

4

shape 1   **shape 2**   **shape 3**

Pritesh drew these shapes on her bedroom wall.

a Draw shape 4.
Copy and complete this table.

Shape number	1	2	3	4	5
Total number of circles	12	20			

b Describe the sequence made by the total number of circles.

c Explain how you could find the total number of circles needed for the nth shape.

*d Write an expression for the number of circles needed for the nth shape.

*e How many circles are needed for shape 20?

5 The Light Company make a wall panel with tiny blue and yellow lights.

model 1   model 2   model 3

a Copy and complete this table.

Model number	1	2	3	4
Number of blue lights				
Number of yellow lights				
Total number of lights				

b Describe the sequence made by the number of
   i blue lights   **ii** yellow lights   **iii** lights in total.
c Explain how you would find the number of blue lights there will be in the nth model.
∗**d** Write an expression for the number of blue lights in model n.
e Explain how you would find the total number of lights in model n.
∗**f** Write an expression for the total number of lights in model n.

6 Abigail made window hangers with coloured glass squares.

shape 1   shape 2   shape 3

a How many red and how many orange squares will there be in shape 8?
   Explain how you worked this out.
b Will one of the shapes have 29 orange squares? Give a reason for your answer.

7 Grace made these shapes for her bathroom wall with green and blue tiles.

   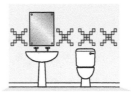

shape 1   shape 2   shape 3

a How many blue and how many green tiles will there be in shape 7?
   Explain how you worked this out.
b Will one of the shapes have 81 blue tiles? Give a reason for your answer.

Review 1 Paul made these matchstick squares.

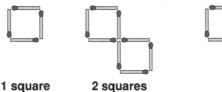 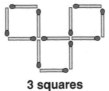 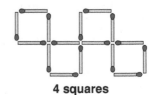

1 square   2 squares   3 squares   4 squares

a Draw the shape with 5 squares.
b What numbers go in the gaps in this table?
c How many matchsticks will be needed for 12 squares? Explain your answer.

Number of squares	1	2	3	4	5
Number of matchsticks	4	8			

d Will one of the shapes have 40 matchsticks? Give a reason for your answer.

Review 2

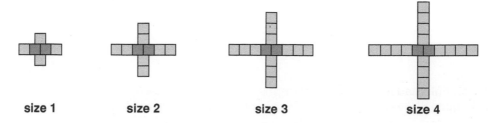

size 1   size 2   size 3   size 4

a Copy and complete this table.

Size of cross	1	2	3	4	5
Number of squares					

b Describe the sequence made by the number of squares.
c Will one of the crosses have 40 squares? Explain your answer.
d Explain how you could find the number of squares in cross size n.
∗e Write an expression for the number of squares in cross size n.

 Practical

You will need some matchsticks.

Danya made this matchstick pattern.

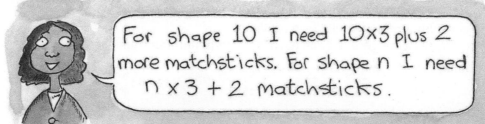

shape 1   shape 2   shape 3
5 matchsticks   8 matchsticks   11 matchsticks

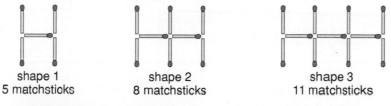

For shape 10 I need 10×3 plus 2 more matchsticks. For shape n I need n × 3 + 2 matchsticks.

Explore rows of other shapes made from matchsticks.
Find the rule for finding the number of matchsticks needed for the nth shape.

Investigation

Lego

Jacob only had lego blocks of lengths 1, 2 and 3 as shown in the picture.
This table shows the number of ways of making blocks of length 1 to 5.

Total block length	1	2	3	4	5
Blocks needed	1	1 + 1 or 2	1 + 1 + 1 or 1 + 2 or 3	1 + 1 + 1 + 1 or 2 + 2 or 1 + 3 or 1 + 1 + 2	1 + 1 + 1 + 1 + 1 or 1 + 2 + 2 or 1 + 1 + 1 + 2 or 2 + 3 or 1 + 1 + 3
Number of ways	1	2	3	4	5

Predict the number of ways of making a block of length 6.
Check your prediction.

Predict the number of ways of making blocks of lengths
7, 8, and 9.
Check your predictions.

What about a block of length 10?

2 1 is the same as 1 2

Hint: Sequences sometimes continue differently from what you might predict.

Finding the rule for the *n*th term

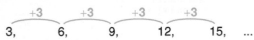

3,   6,   9,   12,   15,   ...
term 1   term 2   term 3   term 4   term 5

To find the 60th term we could keep adding 3.
Another way is to notice that each term is the term number multiplied by 3.

Term Number	1	2	3	4	5
Term	3	6	9	12	15

×3

If *n* is the term number, the *n*th term is $n \times 3$.

Using the rule for the *n*th term is much quicker when the term is a long way into the sequence, like the 60th term.

Discussion

● Mr James asked his class if anyone could explain how to find the 20th term of
5, 9, 13, 17, 21, ...

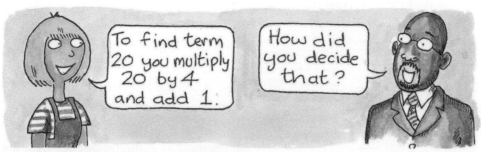

What might Rebecca's answer be? **Discuss**.
How could Rebecca find the *n*th term? **Discuss**.

Rebecca worked out the *n*th term of some other sequences.
This is what she wrote. **Discuss** how she might have got these answers.

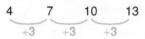
4   7   10   13   *n*th term is **3** times *n* plus 1

8   13   18   23   *n*th term is **5** times *n* plus 3

To find a **rule for the *n*th term** we look for a **relationship** between the term number and the
term itself.

Example   **Term number (*n*)**   1   2   3   4   5   ...
             Term   7   11   15   19   23   ...

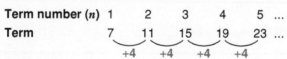

In the sequence 7, 11, 15, 19, 23, ...
Each term is **4** times the term number, *n*, plus **3**.

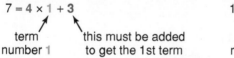

$7 = 4 \times 1 + 3$ $11 = 4 \times 2 + 3$

term   this must be added   term   this must be added
number 1   to get the 1st term   number 2   to get the 2nd term

---

Exercise 5    Try and write an expression
for the *n*th term for each.

1   Explain how to find the *n*th term of these.
    a   2, 4, 6, 8, 10, ...   **b**   3, 6, 9, 12, 15, ...   **c**   6, 12, 18, 24, 30, ...
    d   4, 9, 14, 19, 24, ...   **e**   3, 7, 11, 15, 19, ...   **f**   6, 11, 16, 21, 26, ...
   *g   100, 90, 80, 70, 60, ...   *h   30, 25, 20, 15, ...

Review   Explain how to find the *n*th term of these.
a 1, 3, 5, 7, 9, ...   **b** 4, 10, 16, 22, 28, ...   *c   60, 50, 40, 30, 20, ...

Algebra

Functions

We can find the **output** of a **function machine** if we are given the input.

Example   4 → [multiply by 3] → 12   If the input is 4, the output is
4 × 3 = 12.

x → [multiply by 3] → *y*   If the input is *x*, the output is
x × 3 = *y* or *y* = *x* × 3 or *y* = 3*x*.

Example   7 → [multiply by 2] → [add 4] → 18   If the input is 7, the output is 7 × 2 + 4 = 18.

x → [multiply by 2] → [add 4] → *y*   If the input is *x*, the output is *x* × 2 + 4 = *y*
or *y* = *x* × 2 + 4 or *y* = 2*x* + 4.

Sometimes we show the inputs and outputs on a **table**.

Example   *x* → [add 3] → [multiply by 2] → *y*

The outputs from the function machine form a sequence.

x	1	2	3	4	5	← inputs
y	8	10	12	14	16	← outputs

Sometimes the inputs and outputs are written like this.

3, 7, 2, 4 → [add 3] → [multiply by 2] → 12, 20, 10, 14

First number to be input.   First number to be output.

Exercise 6

1 Find the output for each of these if the input is 8.

a *x* → [add 2] → *y*   **b** *x* → [subtract 6] → *y*

c *x* → [multiply by 4] → *y*   **d** *x* → [divide by 2] → *y*

e *x* → [multiply by 2] → [add 1] → *y*   **f** *x* → [multiply by 3] → [subtract 4] → *y*

g *x* → [divide by 2] → [add 3] → *y*   **h** *x* → [divide by 4] → [add 5] → *y*

i *x* → [add 1] → [multiply by 2] → *y*   **j** *x* → [subtract 5] → [multiply by 3] → *y*

206

2 Find the output for each of the function machines in question **1** if the input is **12**.

3 Describe, in words, what each of the function machines in question **1** does to x to get y.
The answer to **a** is **2 is added to x to get y**.

4 Copy and fill in a table like this for each of these.

x	1	2	3	4	5
y					

a $x \rightarrow$ [multiply by 2] $\rightarrow$ [subtract 1] $\rightarrow y$

b $x \rightarrow$ [add 3] $\rightarrow$ [multiply by 4] $\rightarrow y$

c $x \rightarrow$ [multiply by 3] $\rightarrow$ [add 6] $\rightarrow y$

d $x \rightarrow$ [subtract 1] $\rightarrow$ [multiply by 3] $\rightarrow y$

5 Find the output for these.

a 3, 7, 8, 4.5 $\rightarrow$ [multiply by 2] $\rightarrow$ [subtract 4] $\rightarrow$ ___, ___, ___, ___

b 18, 6, 24, 2.1 $\rightarrow$ [divide by 3] $\rightarrow$ [add 1] $\rightarrow$ ___, ___, ___, ___

c 2, 1, 0, 6.5 $\rightarrow$ [add 4] $\rightarrow$ [multiply by 2] $\rightarrow$ ___, ___, ___, ___

d 7, 23, 49, 9.6 $\rightarrow$ [subtract 3] $\rightarrow$ [divide by 2] $\rightarrow$ ___, ___, ___, ___

Review 1 Find the output for each of these if 4 is the input.

a $x \rightarrow$ [divide by 2] $\rightarrow y$

b $x \rightarrow$ [multiply by 4] $\rightarrow$ [subtract 2] $\rightarrow y$

c $x \rightarrow$ [divide by 2] $\rightarrow$ [add 7] $\rightarrow y$

d $x \rightarrow$ [add 3] $\rightarrow$ [multiply by 4] $\rightarrow y$

Review 2 Copy and fill in a table like this for the function machines in **Review 1b** and **d**.

x	1	2	3	4	5
y					

Review 3 Describe, in words, what each of the function machines in **Review 1** does to x to get y.

Review 4 Find the output.

8, 9, 6, 14 $\rightarrow$ [subtract 4] $\rightarrow$ [multiply by 2] $\rightarrow$ ___, ___, ___, ___

$x \rightarrow$ [multiply by 3] $\rightarrow$ [add 2] $\rightarrow y$

The rule for this function machine is 'multiply by 3 then add 2'.
We can write this rule as $x \times 3 + 2 = y$ or $y = 3x + 2$,
or using a mapping arrow ($\longrightarrow$). $x \longrightarrow x \times 3 + 2$ or $x \longrightarrow 3x + 2$.
$y = 3x + 2$ and $x \longrightarrow 3x + 2$ are both **functions**.

We read this as x maps onto 3 times x plus 2'.

Algebra

Worked Example

a Write the rule for this function machine as $y = $ ___

$x \rightarrow \boxed{\substack{\text{subtract} \\ 3}} \rightarrow \boxed{\substack{\text{multiply} \\ \text{by } 4}} \rightarrow y$

b Write the rule for this function machine as $x \rightarrow$ ___

$x \rightarrow \boxed{\substack{\text{divide} \\ \text{by } 2}} \rightarrow \boxed{\substack{\text{add} \\ 4}} \rightarrow y$

Answer

a 3 is subtracted from x and the result is multipled by 4.

$y = (x - 3) \times 4$ or $y = \mathbf{4(x - 3)}$

b x is divided by 2 and then 4 is added.

$x \rightarrow \frac{x}{2} + \mathbf{4}$

We put a bracket around $x - 3$ because all of it must be multiplied by 4.

Exercise 7

1 Write the rules for these function machines as $y = $ ___.

a $x \rightarrow \boxed{\substack{\text{multiply} \\ \text{by } 2}} \rightarrow \boxed{\substack{\text{add} \\ 4}} \rightarrow y$

b $x \rightarrow \boxed{\substack{\text{multiply} \\ \text{by } 8}} \rightarrow \boxed{\substack{\text{subtract} \\ 7}} \rightarrow y$

c $x \rightarrow \boxed{\substack{\text{divide} \\ \text{by } 4}} \rightarrow \boxed{\substack{\text{subtract} \\ 3}} \rightarrow y$

d $x \rightarrow \boxed{\substack{\text{add} \\ 3}} \rightarrow \boxed{\substack{\text{multiply} \\ \text{by } 4}} \rightarrow y$

2 Write the rules for these functions machines as a mapping, $x \rightarrow$

a $x \rightarrow \boxed{\substack{\text{multiply} \\ \text{by } 3}} \rightarrow \boxed{\substack{\text{add} \\ 2}} \rightarrow y$

b $x \rightarrow \boxed{\substack{\text{multiply} \\ \text{by } 5}} \rightarrow \boxed{\substack{\text{subtract} \\ 2}} \rightarrow y$

c $x \rightarrow \boxed{\substack{\text{divide} \\ \text{by } 3}} \rightarrow \boxed{\substack{\text{add} \\ 2}} \rightarrow y$

d $x \rightarrow \boxed{\substack{\text{add} \\ 2}} \rightarrow \boxed{\substack{\text{multiply} \\ \text{by } 7}} \rightarrow y$

e $x \rightarrow \boxed{\substack{\text{subtract} \\ 3}} \rightarrow \boxed{\substack{\text{multiply} \\ \text{by } 5}} \rightarrow y$

3 a $x \rightarrow \boxed{\substack{\text{add} \\ 4}} \rightarrow \boxed{\substack{\text{multiply} \\ \text{by } 2}} \rightarrow y$

Copy and fill in this table.

x	1	2	3	4	5
y					

Write a function for this function machine.

b The order of the operations is changed.

$x \rightarrow \boxed{\substack{\text{multiply} \\ \text{by } 2}} \rightarrow \boxed{\substack{\text{add} \\ 4}} \rightarrow y$

Copy and fill in this table

x	1	2	3	4	5
y					

Write a function for this function machine.

c What happens when the order of the operations is changed?

4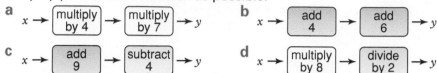

Mrs Patel asked her class to write the rule for this function machine as $x \longrightarrow$ ___.
Gemma wrote $x \longrightarrow x \times 4 \times 2$.
Anna wrote $x \longrightarrow 8x$.
Are both Gemma's and Anna's answers correct? Explain.

5 Write the function for these as $x \longrightarrow$ ___.
Simplify your answer as much as possible.

a $x \longrightarrow$ [multiply by 4] $\rightarrow$ [multiply by 7] $\rightarrow y$

b $x \longrightarrow$ [add 4] $\rightarrow$ [add 6] $\rightarrow y$

c $x \longrightarrow$ [add 9] $\rightarrow$ [subtract 4] $\rightarrow y$

d $x \longrightarrow$ [multiply by 8] $\rightarrow$ [divide by 2] $\rightarrow y$

*6 A function machine changes the number n to the number $2n - 3$. $n \longrightarrow 2n - 3$.
What number will it change these to?
a 4   **b** 5   **c** 7   **d** 23   **e** 0

Review 1   Write the rules for these function machines as $y =$ ___

a $x \longrightarrow$ [multiply by 4] $\rightarrow$ [add 10] $\rightarrow y$

b $x \longrightarrow$ [add 4] $\rightarrow$ [multiply by 3] $\rightarrow y$

Review 2   Write the rules for these as a mapping, $x \longrightarrow$ ___.

a $x \longrightarrow$ [divide by 3] $\rightarrow$ [add 7] $\rightarrow y$

b $x \longrightarrow$ [subtract 8] $\rightarrow$ [multiply by 4] $\rightarrow y$

Review 3   A function machine changes the number p to $3p + 2$. $p \longrightarrow 3p + 2$.
If this number is input, what is the output?
a 2   **b** 6   **c** 10   **d** 21   **e** 0

Review 4   Which function matches these function machines?

a $x \longrightarrow$ [multiply by 3] $\rightarrow$ [multiply by 5] $\rightarrow y$

b $x \longrightarrow$ [add 20] $\rightarrow$ [subtract 15] $\rightarrow y$

A $y = x + 15$
B $y = 6x$
C $y = 15x$
D $y = x + 5$

Review 5

$x \longrightarrow$ [multiply by 3] $\rightarrow$ [add 2] $\rightarrow y$ **machine 1**

$x \longrightarrow$ [add 2] $\rightarrow$ [multiply by 3] $\rightarrow y$ **machine 2**

a   Write a function for function machine 1.
b   Function machine 2 is the same as function machine 1 but the order of the operations has been changed.
If the input to both machines is the same, will the output be the same?
Explain your answer.

Finding the function given the input and output

 Guess My Rule – a game for a group

You will need pencil and paper and some copies of this function machine.

$x \rightarrow$ [ ] $\rightarrow y$ **or** $x \rightarrow$ [ ] $\rightarrow$ [ ] $\rightarrow y$

To play

- Choose a leader.
- This person fills in the box or boxes on the function machine with a rule, without showing anyone.
- Each person in the group takes turns to call out a number for x.
- The leader works out the value of y for each.
- The leader tells this to the group.
- The first person to work out the rule is the new leader.

Example Jill was the leader. She wrote $x \rightarrow \boxed{\text{multiply by 6}} \rightarrow y$

The numbers called were 2, 0, 4, 10.
The leader told the group the answers were 12, 0, 24, 60.

Worked Example

Find the rule for these function machines.

a 3, 5, 8, 2 $\rightarrow \boxed{?} \rightarrow$ 15, 25, 40, 10

b 4, 0, 7, 3 $\rightarrow \boxed{\text{multiply by 2}} \rightarrow \boxed{?} \rightarrow$ 11, 3, 17, 9

* c 0, 8, 6, 1 $\rightarrow \boxed{?} \rightarrow \boxed{?} \rightarrow$ 4, 8, 7, 4.5

Answer

a Each of the numbers has been multipled by 5. $x \rightarrow \mathbf{5x}$

b Put the input numbers in order first.

$0, 3, 4, 7 \rightarrow \boxed{\text{multiply by 2}} \rightarrow \boxed{?} \rightarrow$ 3, 9, 11, 17

 Remember to change the order of the output as well.

Find the output from the first box.

$0, 3, 4, 7 \rightarrow \boxed{\text{multiply by 2}} \rightarrow 0, 6, 8, 14 \rightarrow \boxed{?} \rightarrow$ 3, 9, 11, 17

We can now see that the rule for the second box is 'add 3'.

$0, 3, 4, 7 \rightarrow \boxed{\text{multiply by 2}} \rightarrow \boxed{\text{add 3}} \rightarrow$ 3, 9, 11, 17

The rule for the function machine is $x \rightarrow x \times 2 + 3$ or $x \rightarrow \mathbf{2x + 3}$.

c Put the numbers in order.

$$0, 1, 6, 8 \rightarrow \boxed{?} \rightarrow \boxed{?} \rightarrow 4, 4.5, 7, 8$$

Try ×2 $0, 1, 6, 8 \rightarrow \boxed{\text{multiply by 2}} \rightarrow 0, 2, 12, 16 \rightarrow \boxed{} \rightarrow 4, 4.5, 7, 8$

There is no rule that fits the 2nd box.

Try ×3 $0, 1, 6, 8 \rightarrow \boxed{\text{multiply by 3}} \rightarrow 0, 3, 18, 24 \rightarrow \boxed{} \rightarrow 4, 4.5, 7, 8$

There is no rule that fits the 2nd box.

Try ÷2 $0, 1, 6, 8 \rightarrow \boxed{\text{divide by 2}} \rightarrow 0, 0.5, 3, 4 \rightarrow \boxed{} \rightarrow 4, 4.5, 7, 8$

Putting 'add 4' in 2nd box will work.

The rule is $x \rightarrow \frac{x}{2} + 4$.

Exercise 8

1 Write down the rule for each of these.

a $2, 9, 3, 6 \rightarrow \boxed{?} \rightarrow 6, 27, 9, 18$ **b** $9, 1, 15, 5 \rightarrow \boxed{?} \rightarrow 13, 5, 19, 9$

c $33, 52, 21, 103 \rightarrow \boxed{?} \rightarrow 27, 46, 15, 97$ **d** $24, 16, 100, 10 \rightarrow \boxed{?} \rightarrow 6, 4, 25, 2.5$

e $3, 7, 9, 0.5 \rightarrow \boxed{?} \rightarrow 15, 35, 45, 2.5$ **f** $100, 35, 65, 45 \rightarrow \boxed{?} \rightarrow 80, 15, 45, 25$

2 Find the missing operations.

a $3, 2, 1, 4 \rightarrow \boxed{\text{multiply by 2}} \rightarrow \boxed{?} \rightarrow 5, 3, 1, 7$ **b** $0, 1, 5, 2 \rightarrow \boxed{\text{multiply by 3}} \rightarrow \boxed{?} \rightarrow 1, 4, 16, 7$

c $4, 1, 6, 2 \rightarrow \boxed{\text{multiply by 3}} \rightarrow \boxed{?} \rightarrow 10, 1, 16, 4$ ***d** $10, 3, 8, 0 \rightarrow \boxed{?} \rightarrow \boxed{?} \rightarrow 24, 10, 20, 4$

*e $1, 20, 3, 4 \rightarrow \boxed{?} \rightarrow \boxed{?} \rightarrow 4, 99, 14, 19$ ***f** $3, 10, 4, 6 \rightarrow \boxed{?} \rightarrow \boxed{?} \rightarrow 18, 53, 23, 33$

*g $1, 6, 8, 2 \rightarrow \boxed{?} \rightarrow \boxed{?} \rightarrow 7, 27, 35, 11$ ***h** $16, 20, 24, 7 \rightarrow \boxed{?} \rightarrow \boxed{?} \rightarrow 9, 11, 13, 4.5$

*i $12, 9, 15, 21 \rightarrow \boxed{?} \rightarrow \boxed{?} \rightarrow 6, 5, 7, 9$

3 a

$$4, 1, 8, 3 \rightarrow \boxed{?} \rightarrow \boxed{?} \rightarrow 7, 4, 11, 6$$

Joel was asked to fill in the missing operations.

He wrote $4, 1, 8, 3 \rightarrow \boxed{\text{add 2}} \rightarrow \boxed{\text{add 1}} \rightarrow 7, 4, 11, 6$

Jade said that the two operations could be replaced with one.

$$4, 1, 8, 3 \rightarrow \boxed{\text{add 3}} \rightarrow 7, 4, 11, 6$$

Is Jade correct?

b What could go in the boxes?

5, 3, 1, 4 → [?] → [?] → 30, 18, 6, 24

c What single operation could replace the two operations in **b**?

5, 3, 1, 4 → [?] → 30, 18, 6, 24

*4 Class 7T was asked to find the rule for this function machine.

5, 0, 3, 2 → [?] → [?] → 12, 2, 8, 6

Beth wrote $x \longrightarrow 2x + 2$

5, 0, 3, 2 → [multiply by 2] → [add 2] → 12, 2, 8, 6

Samuel wrote $x \longrightarrow 2(x + 1)$

5, 0, 3, 2 → [add 2] → [multiply by 2] → 12, 2, 8, 6

Who is correct? Explain your answer.

*5 Find two different ways of filling in these function machines.

a   4, 2, 0, 3 → [?] → [?] → 15, 9, 3, 12

b   5, 10, 1, 7 → [?] → [?] → 24, 44, 8, 32

Review 1   Find the rule for these.

a 24, 12, 18, 6 → [?] → 18, 6, 12, 0

b 3, 8, 4, 9 → [?] → 15, 40, 20, 45

c 6, 2, 1, 8 → [?] → [?] → 14, 2, −1, 20

d 24, 8, 2, 9 → [?] → [?] → 19, 11, 8, 11.5

Review 2

a What could go in the boxes?

5, 1, 7, 4 → [?] → [?] → 40, 8, 56, 32

b What single operation could replace these two operations?

*Review 3**   Find two different ways of filling in the function machines.

8, 3, 2, 1 → [?] → [?] → 45, 20, 15, 10

Investigation

Magic Squares

A is a magic square.
Each number in **A** has been mapped onto
a new number in **B** using the mapping
$x \longrightarrow x + 4$.

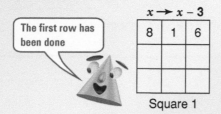

A

11	4	9
6	8	10
7	12	5

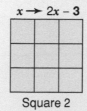 $x \longrightarrow x + 4$

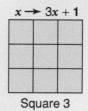

B

15	8	13
10	12	14
11	16	9

Copy these three diagrams.
Begin with the numbers in square **A** above and map each number onto a new number
using the mapping given.

The first row has
been done

$x \longrightarrow x - 3$

8	1	6

Square 1

$x \longrightarrow 2x - 3$

Square 2

$x \longrightarrow 3x + 1$

Square 3

Are squares 1, 2 and 3 magic squares?
What if the function is $x \longrightarrow \frac{x}{2} + 1$ or $x \longrightarrow 3x - 2$ or $x \longrightarrow 4x + 2$ or ...
Is the new square always a magic square? **Investigate**.

Finding the input given the output

Remember
To 'undo' an operation we use the **inverse operation**.
The inverse of adding is subtracting and the inverse of subtracting is adding.
The inverse of multiplying is dividing and the inverse of dividing is multiplying.

Worked Example

$x \longrightarrow \boxed{\begin{array}{c} \text{add} \\ 6 \end{array}} \longrightarrow y$

What was the input if 8 is the output?

Answer
We find the input using inverse operations.
The inverse of 'add 6' is 'subtract 6'.
We draw an inverse function machine.

$x \longleftarrow \boxed{\begin{array}{c} \text{subtract} \\ 6 \end{array}} \longleftarrow y$

 The direction of the
arrows changes.

We use this to find the input.

$2 \longleftarrow \boxed{\begin{array}{c} \text{subtract} \\ 6 \end{array}} \longleftarrow 8$

Begin with the output and work in
the direction of the arrows.

2 was the input.

Algebra

Worked Example

$x \rightarrow$ [multiply by 3] $\rightarrow$ [add 6] $\rightarrow y$

What was the input if 21 is the output?

Answer

Draw the inverse function machines.

$x \leftarrow$ [divide by 3] $\leftarrow$ [subtract 6] $\leftarrow y$

> The inverse of multiply by 3 is divide by 3.
> The inverse of add 6 is subtract 6.

Use the inverse function machine to find the input.

$5 \leftarrow$ [divide by 3] $\overset{15}{\leftarrow}$ [subtract 6] $\leftarrow 21$ $21 - 6 = 15$
$15 \div 3 = 5$

5 was the input.   Start with the output and work in the direction of the arrows.

Exercise 9

1 Draw the inverse function machines for these.

a $x \rightarrow$ [add 4] $\rightarrow y$ **b** $x \rightarrow$ [multiply by 6] $\rightarrow y$ **c** $x \rightarrow$ [divide by 2] $\rightarrow y$

> Remember:
> The arrows go in the opposite direction.

d $x \rightarrow$ [subtract 7] $\rightarrow y$ **e** $x \rightarrow$ [divide by 5] $\rightarrow y$ **f** $x \rightarrow$ [multiply by 10] $\rightarrow y$

2 Find the input.

a ? $\rightarrow$ [add 2] $\rightarrow 11$ **b** ? $\rightarrow$ [subtract 4] $\rightarrow 8$ **c** ? $\rightarrow$ [multiply by 3] $\rightarrow 24$

> Draw an inverse function machine to help.

d ? $\rightarrow$ [divide by 3] $\rightarrow 9$ **e** ? $\rightarrow$ [add 12] $\rightarrow 36$

3 Find the input for these.

a __, __, __, __ $\rightarrow$ [add 5] $\rightarrow$ 12, 20, 36, 72   **b** __, __, __, __ $\rightarrow$ [subtract 4] $\rightarrow$ 7, 24, 42, 63

c __, __, __, __ $\rightarrow$ [multiply by 4] $\rightarrow$ 28, 40, 100, 240   **d** __, __, __, __ $\rightarrow$ [divide by 3] $\rightarrow$ 9, 4, 6, 0

4 Find the input.
Draw the inverse function machines to help.

a __ $\rightarrow$ [multiply by 2] $\rightarrow$ [add 3] $\rightarrow 13$ **b** __ $\rightarrow$ [subtract 2] $\rightarrow$ [multiply by 6] $\rightarrow 60$

c __ $\rightarrow$ [divide by 2] $\rightarrow$ [add 5] $\rightarrow 13$ **d** __ $\rightarrow$ [multiply by 3] $\rightarrow$ [add 8] $\rightarrow 23$

e __ $\rightarrow$ [divide by 6] $\rightarrow$ [subtract 4] $\rightarrow 5$

5 Find the input for these.

a __, __, __ → [multiply by 3] → [add 4] → 34, 7, 16

b __, __, __ → [divide by 4] → [add 3] → 13, 6, 9

c __, __, __ → [add 8] → [divide by 2] → 6, 10, 7.5

d __, __, __ → [multiply by 4] → [subtract 7] → 5, 9, 33

e __, __, __ → [divide by 2] → [add 5] → 9, 15, 55

f __, __, __ → [subtract 3] → [multiply by 5] → 10, 20, 45

6 A function machine changes the number n to the number $3n + 2$.
What numbers must be input to get these output numbers?
a 11   **b** 5   **c** 20   **d** 62

7 1, 2, 3, 4, 5 → [multiply by 4] → [?] → 1, 2, 3, 4, 5

The numbers have stayed the same.
The inverse of 'multiply by 4' is 'divide by 4'.
The missing operation must be 'divide by 4'.
What are the missing operations in these?

a 1, 2, 3, 4, 5 → [add 2] → [?] → 1, 2, 3, 4, 5

b 1, 2, 3, 4, 5 → [subtract 6] → [?] → 1, 2, 3, 4, 5

c 1, 2, 3, 4, 5 → [multiply by 2] → [?] → 1, 2, 3, 4, 5

d 1, 2, 3, 4, 5 → [divide by 6] → [?] → 1, 2, 3, 4, 5

8 i

The numbers have stayed the same.

3, 4, 8, 12 → [add 4] → [multiply by 2] → [ ] → [ ] → 3, 4, 8, 12

The missing operations must be the inverse of 'add 4' and 'multiply by 2'.
These are 'subtract 4' and 'divide by 2'.
Which of these is correct?

A 3, 4, 8, 12 → [add 4] → [multiply by 2] → [divide by 2] → [subtract 4] → 3, 4, 8, 12

B 3, 4, 8, 12 → [add 4] → [multiply by 2] → [subtract 4] → [divide by 2] → 3, 4, 8, 12

ii

What are the missing operations in these?
Check to make sure you have the operations in the right order.

a 8, 4, 7, 2 → [multiply by 3] → [add 2] → [ ] → [ ] → 8, 4, 7, 2

b 9, 6, 12, 18 → [divide by 3] → [subtract 4] → [ ] → [ ] → 9, 6, 12, 18

c 4, 8, 1, 7 → [add 3] → [divide by 2] → [ ] → [ ] → 4, 8, 1, 7

d 7, 8, 4, 2 → [subtract 6] → [multiply by 3] → [ ] → [ ] → 7, 8, 4, 2

Algebra

*

9 In a game, orange numbers were found by multiplying a green number by 2 then adding 1. Carrie wrote this function.

green number × 2 + 1 = orange number

To find a green number from an orange number the function must be inverted.

green number = (orange number − 1) ÷ 2

Invert these functions to show how to get a green number from an orange number.

a (green number − 2) × 3 = orange number
b green number × 4 − 2 = orange number
c green number ÷ 2 + 3 = orange number

Review 1   Write down the input for each of these.

a ___, ___, ___ → [subtract 7] → 16, 24, 33   b ___, ___, ___ → [divide by 4] → 3, 8, 1

c ___, ___, ___ → [add 8] → 23, 42, 106   d ___, ___, ___ → [multiply by 5] → 25, 45, 100

Review 2   Find the input.

Draw the inverse function machines to help.

a ___ → [multiply by 2] → [add 7] → 19   b ___ → [subtract 3] → [divide by 4] → 9

c ___, ___, ___ → [subtract 8] → [multiply by 3] → 15, 21, 30

d ___, ___, ___ → [divide by 2] → [add 4] → 12, 54, 40

Review 3   What are the missing operations?

a 5, 1, 6, 10 → [add 4] → [   ] → 5, 1, 6, 10

b 3, 10, 7, 0 → [multiply by 3] → [add 2] → [   ] → [   ] → 3, 10, 7, 0

c 8, 9, 4, 3 → [divide by 4] → [add 1] → [   ] → [   ] → 8, 9, 4, 3

d 5, 14, 7, 24 → [subtract 4] → [multiply by 3] → [   ] → [   ] → 5, 14, 7, 24

Summary of key points

A A **sequence** is a set of numbers in a given order.
Each number is called a **term**.

Examples   8, 15, 22, 29, 36, ...   ← These 3 dots mean the sequence continues forever. It is *infinite*.
    ↑   ↑
    1st term   5th term

    4, 8, 12, 16, ... 100   ← This sequence is *finite*. It starts at 4 and ends at 100.

B We can **write a sequence** by **counting on** or **counting back**.

Examples   Starting at 7 and counting on in steps of 20 we get 7, 27, 47, 67, 87, ...
Starting at 2 and counting back in steps of 0·5 we get 2, 1·5, 1, 0·5, 0, ⁻0·5, ...

−0·5, ...

We can **write a sequence** if we know the **first term** and the **rule for finding the next term**.

Example   **1st term** ⁻6   **rule for finding the next term**   add 2

gives ⁻6, ⁻4, ⁻2, 0, 2, 4, ...

Each term is 2 more than the one before.

 We can **write a sequence** if we know the **rule for the *n*th term**.

Example   If the rule for the *n*th term is 3*n* – 1 the sequence is 2, 5, 8, 11, 14, 17, 20, ...

 Sequences in practical situations

shape 1

shape 2

shape 3

shape 4

Shape number	1	2	3	4
Number of squares	4	7	10	13

In the *n*th shape there are 3*n* + 1 squares.

There are **3** arms on each shape. The *n*th shape has 3 × *n* squares on the arms plus one in the middle.

 We can find a **rule for the *n*th term** by looking for the relationship between the term number and the term.

Example   In the sequence 5, 8, 11, 14, 17, ...

term number(*n*)   1   2   3   4   5

term   5   8   11   14   17

+3   +3   +3   +3

Each term is **3** times the term number, *n*, plus 2

 $x \rightarrow \boxed{\text{add } 3} \rightarrow \boxed{\text{multiply by 2}} \rightarrow y$ This is a **function machine**.

If 5 is the input, the output is (5 + 3) × 2 = 16.

We can write the **function** for this machine as $y = (x + 3) \times 2$ or $y = 2(x + 3)$.

It can also be written as a mapping. $x \rightarrow 2(x + 3)$.

 If we are given the input and output we can **find the rule** for the machine.

Example   3, 5, 10, 20 → ☐ → 9, 15, 30, 60

The rule for this machine is 'multiply by 3'.

 We use **inverse operations** to **find the input** if we are given the output.

Example   — → $\boxed{\text{multiply by 3}}$ → $\boxed{\text{subtract } 4}$ → 20

We draw an inverse function machine.

8 ← $\boxed{\text{divide by 3}}$ ← 24 $\boxed{\text{add } 4}$ ← 20 ← **Start with the output**

The input was 8.

Algebra

Test yourself

1 Write down the first six terms of these sequences.
   a Start at 4 and count on in steps of 2·5.
   b Start at 1 and count back in steps of 3.

2 Match the descriptions with the sequences given below.
   a The odd numbers from 9 to 21.
   b The sequence begins at 9 and increases in steps of 3.
   c Each term is a multiple of 4, plus 1.
   d The sequence begins at 9 and decreases in steps of 2.
   A 9, 12, 15, 18, ...   **B** 9, 13, 17, 21, ...
   C 9, 7, 5, 3, ...   **D** 9, 11, 13, 15, ... 21

3 Write down the next three terms of the sequence 0·8, 1·2, 1·6, 2, 2·4, ...
   Describe the sequence.

4 Show the first five multiples of 5 using dots. The first diagram is ⋮

5 Find the missing terms.
   a 26, ▢, ▢, ▢, 14   **rule** subtract 3   **b** 5000, ▢, ▢, 40, ▢   **rule** divide by 5
   c 0·5, ▢, ▢, ▢, ▢   **rule** double

6 The rule for finding the next term of a sequence is 'to find the next term, add 5'.
   a Write down two possible sequences with this rule.
   b Is it possible to find a sequence with this rule for which all terms are odd numbers? Explain.

7 The nth term of a sequence is given. Find the first five terms.
   a $n + 5$ **b** $2n - 3$ **c** $20 - 2n$

8 Write down the 15th term of each of the sequences in question **7**.

9 Use a copy of this table.
   Fill it in for the multiples of 6.

Term number	1	2	3	4	5	6	...	n
Term	6							

10 Dylan made these dot diagrams.

diagram 1   diagram 2   diagram 3

Diagram number	1	2	3	4	...
Number of dots	3	5	7	?	...

   a Draw diagram 4. How many dots are in this shape?
   b Describe the pattern and how it continues.
   c Will one of Dylan's diagram's have 24 dots? Give a reason for your answer.
   d Explain how you can find the number of dots in the nth diagram.
   e Write an expression for the number of dots in the nth diagram.

11 Explain how to find the nth term of the sequence 5, 8, 11, 14, 17, ...

12 Find the output for these.

a 4, 7, 9, 16 → [add 12] → __, __, __, __   **b** 12, 9, 27, 36 → [divide by 3] → __, __, __, __

c 5, 8, 3, 2 → [multiply by 3] → [add 4] → __, __, __, __

d 12, 20, 9, 17 → [subtract 4] → [multiply by 2] → __, __, __, __

13 Copy and fill in the table for this function machine.

x → [multiply by 2] → [add 7] → y

x	1	2	3	4	5
y					

14 Write the rules for these function machines as $y =$ ____.

a x → [multiply by 3] → [add 4] → y **b** x → [subtract 3] → [multiply by 4] → y

15 Write the rules for these function machines as a mapping, x → ____.

a x → [divide by 4] → [add 3] → y **b** x → [add 2] → [multiply by 5] → y

16 a The rule for Gemma's function machine is x → $2x - 4$.

Copy and finish Gemma's function machine.

x → [multiply by 2] → [ ] → y

b Copy and finish Ricky's function machine.

x → [ ] → [ ] → y

My function machine multiplies a number by 2 then subtracts 4.

Gemma

My function machine subtracts 4 from a number then multiplies the result by 2.

Ricky

c If 10 was put into both Gemma's and Ricky's function machines would they give the same output? Explain.

17 Match these functions and function machines.

a $y = 14x$ **b** $y = 7x$ **A** x → [add 6] → [add 8] → y **B** x → [add 14] → [subtract 7] → y

c $y = x + 14$ **d** $y = x + 7$ **C** x → [multiply by 2] → [multiply by 7] → y **D** x → [multiply by 28] → [divide by 4] → y

18 Write down the rule for each of these.

Find two different ways for **d**

a 3, 8, 9, 4 → [ ] → 7, 12, 13, 8   **b** 27, 6, 24, 18 → [ ] → 9, 2, 8, 6

c 5, 1, 0, 8 → [ ] → [ ] → 17, 5, 2, 26   **d** 1, 8, 6, 3 → [ ] → [ ] → 6, 27, 21, 12

19 Find the input. Draw inverse function machines to help.

a __, __ → [multiply by 3] → [add 4] → 22, 34   **b** __, __ → [divide by 3] → [subtract 5] → 5, 3

20 A function machine changes n to the number $3n - 4$.
   a What output would the function machine give if 7 was the input?
   b What number must be input to get 11 as the output?

21 What are the missing operations?

7, 2, 1, 6 → [add 7] → [divide by 2] → [?] → [?] → 7, 2, 1, 6

10 Graphs

Key vocabulary

axes, axis, coordinate pair, coordinate point, coordinates, equation (of a graph), graph, intersect, origin, relationship, straight-line graph, *x*-axis, *x*-coordinate, *y*-axis, *y*-coordinate

⏵⏵ **Working out the plot**

1 These two 'pictures' were made by joining the dots at (2, 2) (2, 10) (10, 2) (10, 10) (4, 4) (4, 8) (8, 4) (8, 8) in different ways.
What other 'pictures' can be made by joining these points? You will need to draw some copies of the grids and plot the given points.

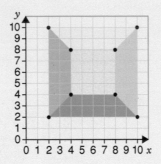 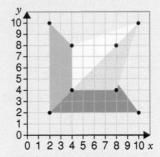

2

Gareth drew this picture of some of the buildings in his town. He plotted one point on this grid for each building.
Match each building with a point on the grid.

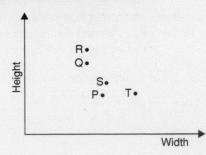

Coordinate pairs

(1, 3) is a **coordinate pair.**

x-coordinate   *y*-coordinate

(1, 3)   (2, 4)   (3, 5)
   +2   +2   +2

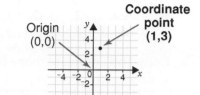

Origin (0,0)   Coordinate point (1,3)

In each of these coordinate pairs, the *y*-**coordinate** is two more than the *x*-**coordinate**.
The rule is $y = x + 2$.

We say 'the coordinate pairs *satisfy* the rule $y = x + 2$.

Worked Example
These coordinate pairs satisfy the rule $y = 12 - x$.
Fill in the missing numbers.
(1, ___), (2, ___), (3, ___), (4, ___)

Answer
We substitute the given *x*-coordinates into $y = 12 - x$.
When $x = 1, y = 12 - 1$ $x = 2, y = 12 - 2$ $x = 3, y = 12 - 3$ $x = 4, y = 12 - 4$
                = 11   = 10   = 9   = 8
We now fill in the missing numbers.
(1, **11**), (2, **10**), (3, **9**), (4, **8**)

We can list the coordinate pairs in a table.

Example

x	1	2	3	4
y	11	10	9	8

Exercise 1

1 Copy and complete these to find coordinate pairs that satisfy the rule $y = x - 1$.
   a When $x = 2, y = 2 - 1$ b When $x = 4, y = 4 - 1$
                    = ___   = ___
      The coordinate pair is (2, ___).   The coordinate pair is (4, ___).
   c When $x = 6, y = $ ___   d When $x = 10, y = $ ___
                    = ___   = ___
      The coordinate pair is (6, ___).   The coordinate pair is (10, ___).

2 For the rule $y = x + 3$, find the *y*-value when the *x*-value is
   a 1   b 2   c 4   d 8   e 10.

3 Copy and complete these coordinate pairs so they satisfy the rule $y = 3x$.
   a (1, ___)   b (3, ___)   c (5, ___)   d (___, 21)   e (___, 12)

Algebra

4 Copy and complete these coordinate pairs so they satisfy the given rule.

a $y = x + 1$ (0,), (1,), (2,), (3,), (4,)

b $y = 10 - x$ (0,), (1,), (2,), (3,), (4,)

c $y = 2x$ (1,), (2,), (3,), (4,)

d $y = x$ (3,), (2,), (1,), (0,), (⁻1,), (⁻2,), (⁻3,)

e $y = x + 5$ (3,), (2,), (1,), (0,), (⁻1,), (⁻2,)

f $y = x - 3$ (2,), (1,), (0,), (⁻1,), (⁻2,), (⁻3,)

5 Copy and complete these tables of coordinate pairs for the given rules.

a $y = x - 1$

x	⁻2	⁻1	0	1	2
y					

b $y = 5 - x$

x	3	4	5	10
y				

c $y = \frac{1}{2}x$

x	2	6	8	10	12
y					

6 The coordinate pairs (3, 6), (4, 8), (10, 20) all satisfy the rule $y = 2x$. Does (8, 16) satisfy the rule? Explain your answer.

7 Which of these coordinate pairs satisfy the rule $y = x + 4$?

A (1, 5)　**B** (10, 6)　**C** (20, 24)　**D** (⁻4, 0)　**E** (⁻1, ⁻5)　**F** ($\frac{1}{2}$, $4\frac{1}{2}$)

*8 Which of these coordinate pairs satisfy the rule $y = 2x - 3$?

A (2, 1)　**B** (4, 1)　**C** (10, 23)　**D** (6, 9)　**E** ($\frac{1}{2}$, ⁻2)

Review 1 Copy and complete these coordinate pairs so they satisfy the given rule.

a $y = x + 4$ (0,), (1,), (2,), (3,), (4,)

b $y = 8 - x$ (0,), (2,), (4,), (6,), (8,)

c $y = x - 4$ (6,), (4,), (0,), (⁻2,), (⁻4,), (⁻6,)

Review 2 Copy and complete these tables of coordinate pairs for the given rules.

a $y = 5x$

x	1	2	3	4	5
y					

b $y = 20 - x$

x	0	2	4	6	8
y					

c $y = x + 5$

x	⁻4	⁻2	0	2	4
y					

Review 3 Which of these coordinate pairs satisfy the rule $y = x - 3$?

A (3, 0)　**B** (0, ⁻3)　**C** (7, 10)　**D** (1, ⁻2)　**E** ($1\frac{1}{2}$, ⁻$1\frac{1}{2}$)

Drawing straight-line graphs

Coordinate pairs can be **plotted** on a grid.

Worked Example

The values on this table satisfy the rule $y = x + 2$.

x	⁻3	0	2
y	⁻1	2	4

a Write down the coordinate pairs.

c Choose another point on the line. Write down the coordinate pair of your point. Check that this coordinate pair satisfies the rule $y = x + 2$.

b Plot the coordinate points on a grid. Draw a straight line through the points. Label the line $y = x + 2$.

d Does $y = x + 2$ go through the point (⁻4, ⁻2)?

Answer

a (⁻3, ⁻1), (0, 2), (2, 4)

b The points are shown plotted.
A straight line is drawn through them.
The line is labelled $y = x + 2$.

c Choose a point on the line, say (**1**, 3).
$y = x + 2$.   If $x = $ **1**, $y = 1 + 2$
               $= 3$

(1, 3) **does satisfy the rule**.

d Check that (⁻4, ⁻2) satisfies $y = x + 2$.
If $x = -4, y = ⁻4 + 2$
           $= ⁻2$

$y = x + $ **2 does go through the point** (⁻4, ⁻2).

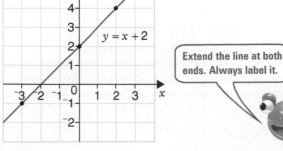

Extend the line at both ends. Always label it.

In the above worked example, $y = x + 2$ is the **equation of the line**. The coordinate pairs of *all* points on the line satisfy the equation of the line.

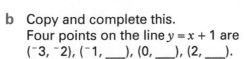

Exercise 2

Use a copy of the grids in this exercise.

1 a Copy and complete these to find three points on the line $y = x - 2$.

When $x = 2, y = 2 - 2$
              $= $ ___
A point is (2, ___)

When $x = 4, y = 4 - 2$
              $= $ ___
A point is (4, ___)

When $x = 6, y = $ ___
              $= $ ___
A point is (6, ___)

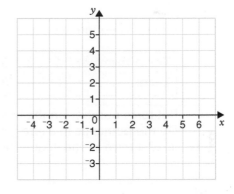

b On this grid, plot the three points found in **a**.

c Draw and label the line $y = x - 2$.

2 a $y = x + 1$
Copy and complete the table of values for this rule.

x	⁻3	⁻1	0	2
y	⁻2			

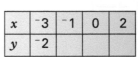

Draw your line carefully.

b Copy and complete this.
Four points on the line $y = x + 1$ are
(⁻3, ⁻2), (⁻1, ___), (0, ___), (2, ___).

c On this grid, plot these four points.
Draw and label the graph of $y = x + 1$.

d Does the point (⁻2, ⁻1) lie on this line?

e Should $y = x + 1$ go through the point (⁻3, ⁻4)?

f Should the point ($\frac{1}{2}$, $1\frac{1}{2}$) lie on this line?

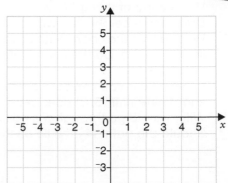

T

3 a (0,   ), (2,   ) (4,   )
Copy and complete these coordinate pairs for the rule $y = 1 - x$.
b Draw and label the line with equation $y = 1 - x$.
c Which of these points lie on the line?
(3, 4), (0, 1) (4, ⁻3)
d Will $y = 1 - x$ go through the point (7, ⁻6)?
e Choose two other points which lie on the line. Write down the coordinate pairs for each. Do these coordinate pairs satisfy the rule $y = 1 - x$?

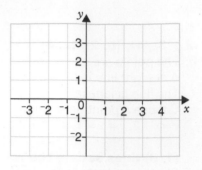

4 $x \rightarrow \boxed{\times 3} \rightarrow y$

a Copy and complete this table for the given function machine.

x	0	1	2	3	4
y					

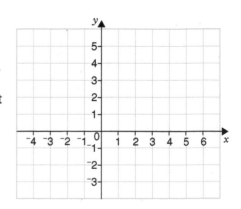

There is more about function machines on page 206.

b Write down the coordinate pairs from the table.
c On a grid, plot the five points.
Do the five points lie in a straight line?

5 a Plot this sequence on a grid.

Term number	1	2	3	4	5
Term	2	4	6	8	10

Put term number on the horizontal axis.
Do the points lie in a straight line?
b Is 22 a term of the sequence? Explain.

T

Review

a Copy and complete the table for $y = x - 3$.

x	⁻2	0	2
y			

b Write down the coordinate pairs for the points.
c Use a copy of this grid. On the grid, plot the three points.
Draw and label the line with equation $y = x - 3$.
d Write down the coordinate pairs for two other points that lie on the line. Do the coordinate pairs satisfy the rule $y = x - 3$?
e Explain how you can tell from the graph that the coordinate pair (5, 2) does satisfy the rule $y = x - 3$.
f Will $y = x - 3$ go through the point (20, 17)?
g Should the point $(4\frac{1}{2}, 1\frac{1}{2})$ lie on the line?

Investigation

T

Families of Graphs

You will need graph paper.

a Use a copy of these tables.

$y = x$

x	⁻4	0	4
y			

$y = 2x$

x	⁻3	0	3
y			

$y = 4x$

x	⁻2	0	2
y			

Draw x and y axes with x-values from ⁻5 to 5 and y-values from ⁻12 to 12.
On this set of axes draw and label the graphs of $y = x$, $y = 2x$ and $y = 4x$.
Describe what happens to the graph as the number multiplying x increases.
Which point do all the graphs have in common?

b Draw tables like the ones above for $y = 3x$ and $y = 5x$.
On the same set of axes, draw and label the graphs of $y = 3x$ and $y = 5x$.
Before you draw them, predict what they will look like.

2 a Use a copy of these tables.

$y = x + 1$

x	⁻3	0	3
y			

$y = x + 4$

x	⁻6	0	3
y			

$y = x + 5$

x	⁻2	0	2
y			

Draw x and y axes with x-values from ⁻8 to 8 and y-values from ⁻6 to 10.
On this set of axes draw and label the graphs of $y = x + 1$, $y = x + 4$, $y = x + 5$.
What do you notice about the slopes of these graphs?

b Look at the graphs you drew in **a**.
What do you notice about where each intersects
the y-axis and the equation of its line?

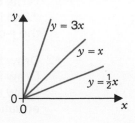

Intersects means crosses.

c Draw tables like the ones above for $y = x + 3$ and $y = x + 7$.
On the same set of axes, draw and label the graphs of $y = x + 3$ and $y = x + 7$.
Before you draw them, predict what they will look like.

3 Repeat questions **2a** and **2b** but replace the tables and equations with

$y = 8 - x$

x	0	2	5
y			

$y = 5 - x$

x	0	1	4
y			

$y = 4 - x$

x	0	1	3
y			

$y = mx$ is the equation of a **straight line** which goes **through the origin**.
m can have any value.
$y = 3x$ has a steeper slope than $y = x$.
$y = \frac{1}{2}x$ has a slope which is not as steep as $y = x$.
The bigger the value of m, the number in front of the x, the steeper
the slope.

Algebra

These lines have a **positive** slope. The values shown in green tell you where each crosses the y-axis.

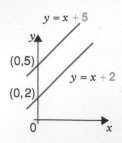

$y = x + 5$

$(0,5)$

$y = x + 2$

$(0,2)$

These lines have a **negative** slope. The values in red tell you where each crosses the y-axis.

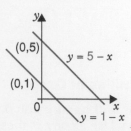

$(0,5)$

$y = 5 - x$

$(0,1)$

$y = 1 - x$

Exercise 3

1 Which graph will have the steepest slope? Explain why.
   A $y = 2x$ **B** $y = 3x$ **C** $y = \frac{1}{2}x$ **D** $y = 4x$

2 Will these have a positive or negative slope?
   a $y = x + 3$ **b** $y = 2x$ **c** $y = x - 4$ **d** $y = 6 - x$ **e** $y = 8 - x$

*3 Which of these graphs will cut the y-axis at (0, 3)?
   A $y = x$ **B** $y = x - 3$ **C** $y = 3 - x$ **D** $y = 3x$

*4 Which of these graphs will cut the y-axis at (0, ⁻5)?
   A $y = {}^{-}5x$ **B** $y = 5 - x$ **C** $y = x - 5$ **D** $y = x + 5$

*5 Will $y = 2x - 4$ have a steeper slope than $y = x$?

Review   Choose from the box.

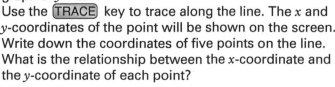

$y = 2x$ $y = 3 - x$ $y = x - 3$ $y = \frac{1}{2}x$ $y = 3x$

Which graph will   **a** have the second steepest slope
                   b have a negative slope
                   *c cut the y-axis at (0, ⁻3)?

★ Practical

You will need a graphical calculator.

1 Use your graphical calculator to draw the straight-line graph of $y = 3x$.
Use the [TRACE] key to trace along the line. The x and y-coordinates of the point will be shown on the screen. Write down the coordinates of five points on the line. What is the relationship between the x-coordinate and the y-coordinate of each point?

$Y_1 = 3x$

$x = {}^{-}1.2$ $y = {}^{-}3.6$

*2 Draw the graph of $y = x$.
Draw the graph of a line which is steeper than $y = x$. Write down the equation.
Draw the graph of a line which is not as steep as $y = x$. Write down the equation of the line.

T

Lines parallel to the x and y-axes

$y = a$ is the equation of a straight line **parallel to the x-axis**.
It cuts the y-axis at a. The x-coordinate can be any value at all.

Examples $y = 3$ is parallel to the x-axis and cuts the y-axis at 3.
   $y = {}^-5$ is parallel to the x-axis and cuts the y-axis at $^-5$.

$x = b$ is the equation of a straight line **parallel to the y-axis**.
It cuts the x-axis at b. The y-coordinate can be any value at all.

Examples $x = 2$ is parallel to the y-axis and cuts the x-axis at 2.
   $x = {}^-2$ is parallel to the y-axis and cuts the x-axis at $^-2$.

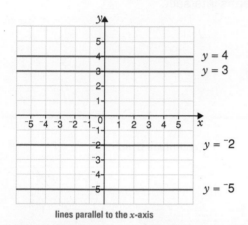

lines parallel to the x-axis

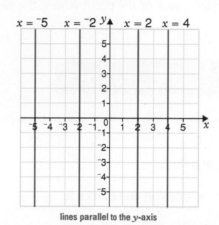

lines parallel to the y-axis

Exercise 4

1 Match these equations to the lines.
   a $y = {}^-1$
   b $x = 2$
   c $y = 1\frac{1}{2}$
   d $x = {}^-2$
   e $x = 1$
   f $x = {}^-1$
   g $y = 2$
   h $y = {}^-2$

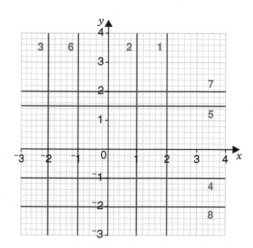

2 Draw a grid with x and y-values from $^-6$ to 6.
   On your grid, draw and label these graphs.
   a $x = 4$ b $y = {}^-2$ c $x = {}^-4$
   d $y = 2$ e $x = 1{\cdot}5$ f $y = {}^-3{\cdot}5$

3 Write down the equation of a line which is
   a parallel to the y-axis
   b parallel to the x-axis.

Algebra

*4 Draw a grid with x and y-values from $^-6$ to 6.
Copy and complete this table of values for $y = x - 2$.
Draw the graph of $y = x - 2$ on your grid.
Draw the graph of $y = 4$ on the same grid.
Write down the coordinates of the point where the two graphs meet.

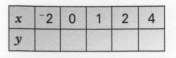

x	$^-2$	0	1	2	4
y					

Review 1 Which of these are equations of lines

a parallel to the x-axis
b parallel to the y-axis?

$$y = 4 \qquad x = 3 \qquad y = ^-1\tfrac{1}{2} \qquad x = 2\tfrac{1}{2} \qquad y = ^-4 \qquad x = ^-3$$

Review 2 Draw a grid with x and y values from $^-6$ to 6.

On your grid draw the graphs of $y = x - 4$ and $y = 3$.
Write down the coordinates of the point where the two graphs intersect.

Reading real-life graphs

Worked Example
Mr Taylor's class sat a test.
It was out of 40.
He drew this graph to convert the marks to
percentages.
a What percentage does each small division
on the vertical axis represent?
b Kyle got 32 out of 40. What percentage is
this?
c Todd got 52%. What mark out of 40 did
Todd get?

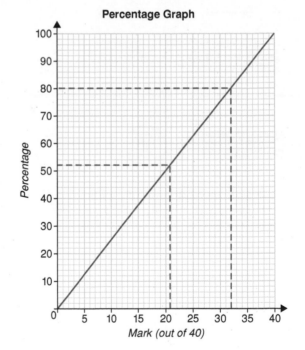

Percentage Graph

Answer
a Five small divisions represent 10%.
One small division must represent $\frac{10}{5} = $ **2%**.
b Find 32 on the horizontal axis.
Draw a dotted line from 32 vertically up to
the graph.
From there draw a horizontal dotted line out
to the other axis.
The percentage can then be read. It is **80%**.
c Find 52% on the vertical axis.
Draw a horizontal dotted line from 52 to the graph.
From there, draw a vertical dotted line down to the other axis.
The mark out of 40 can then be read. It is about **21**.

Exercise 5

1 Mia drew this graph to show the charge for
  swimming lessons.
  a Do the values between the points have any
    meaning? Explain.
  b How much does it cost for the first lesson?
  c How much does it cost for
    i 2 lessons   ii 4 lessons?

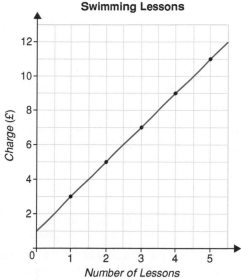

Swimming Lessons

2 a How many °C does each small division on the horizontal axis represent?
  b How many °F does each small division
    on the vertical axis represent?
  c Change these to °F.
    20 °C, 60 °C, 35 °C, 28 °C.
  d Change these to °C.
    50 °F, 65 °F, 142 °F, 161 °F.
  e Amanda is running a temperature. Her
    mother measured it as 100 °F. Her
    nurse measured it in °C. What did the
    nurse get for Amanda's temperature?

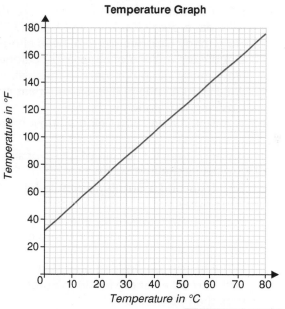

Temperature Graph

  f The hottest day recorded in the 19th century was 58 °C. This was in
    Mexico. What temperature is this in °F?
  g The normal temperature for the human body is about 98·4 °F.
    Approximately what is this in °C?

Give the answer to the
nearest unit.

3

Energy Conversions

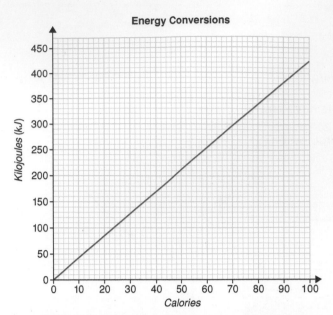

a How many Calories does each small division on the horizontal axis represent?
b How many kilojoules does each small division on the vertical axis represent?
c Convert these to Calories.
   i 50 kJ   ii 150 kJ   iii 210 kJ   iv 320 kJ
d Convert these to kilojoules.
   i 60 Calories   ii 74 Calories   iii 46 Calories   iv 92 Calories
e Allaf ate a pear (250 kJ) and a banana (350 kJ).
   How many Calories were in these altogether?
f Debbie ate a piece of cheese (50 Calories) and a sandwich (100 Calories).
   How many kilojoules were in these altogether?
g Jake is on a special diet. He must eat 420 kJ for breakfast and 350 kJ for lunch.
   How many Calories is this?
*h A milkshake has 800 Calories. How many kilojoules is this?

> Give the answer to the nearest division.

4 This graph gives the length of two types of candles as they burn.
   a Sarah burned a Burnbrite candle for 20 minutes.
      What length would it be after that?
   b Mrs Chen burned a Longlife candle for 40 minutes.
      What length would it be after that?
   c Jeff burned a candle for 30 minutes. It was then 20 mm long.
      What sort of candle was it?
   d Peter burned a Longlife candle until it was 25 mm long.
      For how long did he burn it?

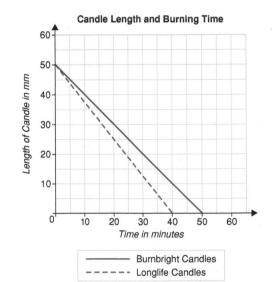

Candle Length and Burning Time

——— Burnbright Candles
– – – – Longlife Candles

5 This graph gives the minimum distance that should be between cars at different speeds.

   a Ed is driving in bad weather at 40 miles per hour.
   What is the minimum distance he should be from the car in front?

   b Fay is driving in good weather at 55 miles per hour.
   What is the minimum distance she should be from the car in front?

   *c Mr Shaw is driving 70 metres behind another car. The weather is bad.
   What is the maximum speed at which he should be driving?

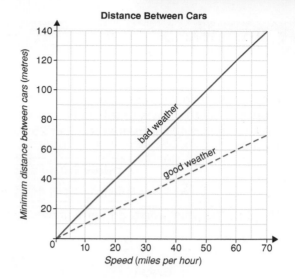

Review 1

a Work out what each small division on each of the axes represents.

b Convert these to pounds.
   i 35 kg   **ii** 10 kg

c Convert these to kilograms.
   i 60 lb   **ii** 100 lb

d Anita weighed 55 kg.
   How many lbs is this?

e Anita stands on the scales holding her cat. The cat weighs 10 lb.
   What do the scales read, in kg?

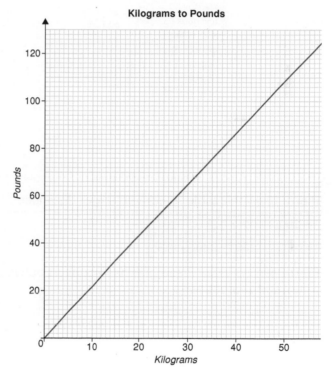

Review 2 This graph shows how much petrol two cars use.

- - - - - - Tom's Car
─────── Kim's Car

a Tom's car went 60 km.
   How many litres of petrol did it use?

b Kim's car went 100 km.
   How many litres of petrol did it use?

c Tom's car used 6 litres of petrol one day.
   How far did it go?

d How much more petrol does Tom's car use than Kim's car when each car travels 60 km?

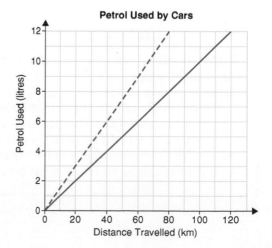

Plotting real-life graphs

Pierre comes from France. In France all the road signs are in kilometres.
Pierre wants to draw a graph so he can quickly convert miles to kilometres.
He knows that 40 kilometres is about 25 miles and so by doubling he knows that 80 km is about 50 miles.
He draws this table.

miles	0	25	50
km	0	40	80

Two points are enough to draw an accurate line graph. It is a good idea to plot at least three.

He plots the points on a grid and draws a line though them.

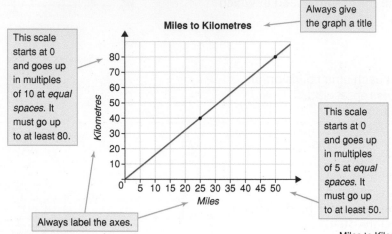

This scale starts at 0 and goes up in multiples of 10 at *equal spaces*. It must go up to at least 80.

Always give the graph a title

This scale starts at 0 and goes up in multiples of 5 at *equal spaces*. It must go up to at least 50.

Always label the axes.

Pierre drew this table to convert larger distances.

miles	125	200	250
km	200	320	400

Pierre can now use these graphs to convert between miles and kilometres.

This shows that the scale does not start at zero.

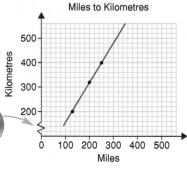

Discussion

Mr Reynolds wants to draw a graph so he can work out his charges for 0 to 20 hours work.
Discuss these questions.

> How many points should Mr Reynolds plot?
> What scales should he use on his axes?
> What title should he give his graph?

As part of your discussion draw a graph for Mr Reynolds.
Use your graph to work out what he should charge for $3\frac{1}{2}$ hours work.

REYNOLDS PLUMBING
£35 call out
plus
£10 per hour

Exercise 6

T 1 Use a copy of this.
Mary is buying velvet to make curtains.
This table shows the cost of 2, 5 and 15 metres of fabric.
  a Plot the points on your grid.
  b Draw a straight line through the points.
  c Use your graph to find the cost of 10 m of material.
  d Joe paid £165 for a length of velvet.
     How many metres, to the nearest m, did he buy?

Length (m)	2	5	15
Cost (£)	24	60	180

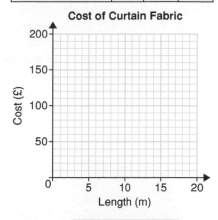

T 2 Shannon wants to draw a graph to convert pounds to kilograms.
She knows that 5 kg is about 11 lbs and that 20 kg is about 44 lb.
  a Copy and fill in this table.
  b Use a copy of this grid.
     Plot the points from the table on it.
     Draw a straight line through the points.
     Label the axes and give the graph a
     title.
  c Shannon bought a 40 lb bag of chaf for
     her horse.
     About how many kg is this?
  d Shannon paid £8.25 for a bag of pony
     nuts at 55p per kilogram.
     How many lbs of nuts were in the bag?

kg	0	5	20
lb	0		

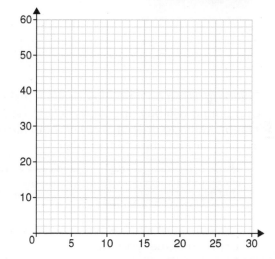

T 3 'Jones Electrical' charge a call out fee of £25 and then an hourly
rate of £15.
  a Copy and complete this table.
  b Use a copy of this grid. Plot the points
     from the table on the grid.
  c Draw a straight line through the points.
  d Use the graph to find the charge for 9
     hours work.
  e Mrs Cassidy was
     charged £175 for a
     job. About how
     many hours did it
     take?

Hours	0	4	12
Charge (£)	25	.	

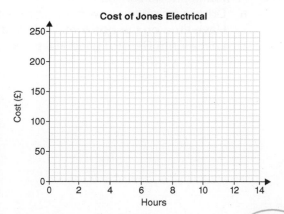

Algebra

T

4 Thomas wants to draw a graph to convert °C to °F.
He knows that 0 °C is 32 °F, 20 °C is 68 °F, 40° is 104 °F.

°C	0°	20°	40°
°F	0		

  a Copy and fill in this table.

  b Use a copy of this grid.
  Choose suitable scales for the
  axes. Have °C on the horizontal
  axis.

 * **c** Plot the points from the table
  on the grid. Draw a straight
  line through the points.

  d The temperature at sunset in
  Brighton was 11°C.
  What is this in °F?

Remember to label
the axes and give
the graph a title

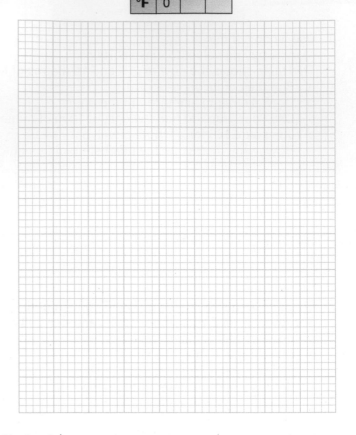

5 Trevor is going on a holiday to New Zealand.
One pound will buy approximately three New Zealand dollars.
Trevor wants to draw a graph to help him convert £0 to £50 into New Zealand dollars.

  a Copy and complete this table.

£	10		50
NZ$			

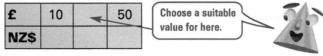

Choose a suitable
value for here.

  b Draw a set of axes. Have pounds on the horizontal axis and NZ$ on the vertical axis.
  c Draw and label your graph.
  d Draw another graph to convert from £200 to £300 to NZ$.
  Do not begin at (0, 0).

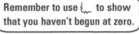

Remember to use ⌇ to show
that you haven't begun at zero.

 * **6** A mobile phone company charges £15 per month plus 15p per minute for peak time calls.
  a Draw a graph to show the cost of between 0 and 200 minutes of peak time calls.
  Have time on the horizontal axis and cost on the vertical axis.
  b Last month Gerrard used 180 peak time minutes. What will the charge be for this?

T **Review 1** Ingrid travelled to Scotland in her car. She used 5 ℓ of petrol for every 55 km she
travelled.

a Copy and complete this table.

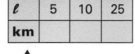

ℓ	5	10	25
km			

b Use a copy of this grid.
Plot the points from the table
on it.
Draw a straight line through
the points.
Finish labelling the graph.

c Ingrid drove to see her aunt
who lived 175 km away.
About how much petrol did
Ingrid's car use?

d One week Ingrid's car used
28 ℓ of petrol. About how far
did it travel?

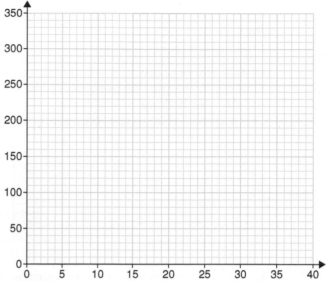

T **Review 2** A shop hires stereos. The cost is £50 plus £5 for every hour.

a Copy and complete this table.

Hours	0	4	8
Charge (£)			

b Use a copy of this grid.
Choose a suitable scale for the vertical axis.
Plot the points from the table on the grid.
Draw a straight line through the points.
Label the vertical axis and give the graph a title.

c Raewyn hired a stereo for 9 hours. Use your graph
to find the cost of this.

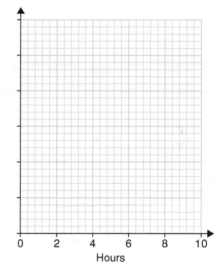

Review 3 There are about 2·5 acres in every hectare.

a Copy and complete this table.

Area in hectares	2	6	10
Area in acres			25

b Draw a graph to convert between acres and hectares.
Have area in hectares on the horizontal axis.

c One of Jim's field's is 12 acres. How many hectares is this?

* Practical

You will need a spreadsheet.

Crocodile hunting in Australia *Skiing in Switzerland*
Rodeo riding in Canada *Trekking in Nepal*
Cruising the Greek Islands

Decide where you would like to go for an overseas holiday.
Find the exchange rate between British pounds (£) and the currency you would use on your holiday.
Use a spreadsheet to convert pounds to the currency of the country.

Example

	A	B	C	D
1	Pounds	£10	= B1+10	= C1+10
2	A$	= B1*2.5	= C1*2.5	= D1*2.5

> This spreadsheet changes £ to Australian dollars.
> The exchange rate is £1 = A$2.5.

Use the graph function of the spreadsheet to draw a graph to change pounds into the currency you chose.

Interpreting real-life graphs

Worked Example

A pot of rice filled with water is cooked. This graph shows the amount of water left as the rice cooks.

a Explain this graph.
b What is the relationship between the amount of water left and the time the rice has been cooking?

Answer

a **As the rice cooks it absorbs water. The longer it cooks the less water is left.**
b **As cooking time increases, the amount of water left decreases.**

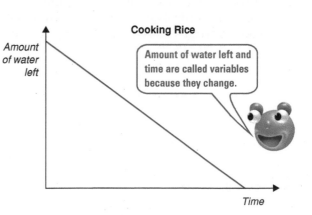

Cooking Rice

> Amount of water left and time are called variables because they change.

Exercise 7

1 Explain these graphs.
For each, say how the variable on the vertical axis is related to the variable on the horizontal axis. Use words like increases or decreases.

a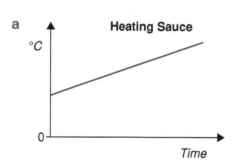

b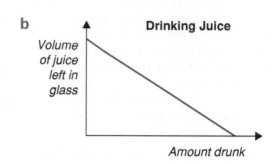

c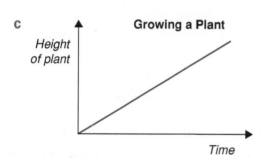

d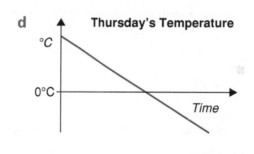

Review   Explain these graphs.
For each, say how the variables on the axes are related.
Use words like increases, decreases.

a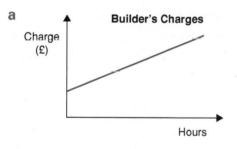

b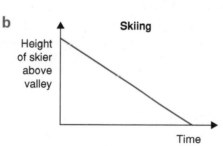

❋Discussion

Some pupils at a science camp heated four different containers of water over a burner.
The volumes of the water varied from 200 mℓ to 500 mℓ.
The pupils measured the temperature of the water in each container after 2 minutes.
The graph on the next page shows their results.
Discuss these questions.

1 Should the points have been joined?
2 Should the points be joined by a straight line?
3 How many different volumes do you think should have been heated to get an accurate picture of the trend?
4 What temperature do you think these volumes of water would be after 2 minutes?   350 mℓ   600 mℓ
5 If you wanted some water to be 63 °C after 2 minutes, how much water should you heat?
6 Describe the relationship between the temperature after 2 minutes and the volume of water.

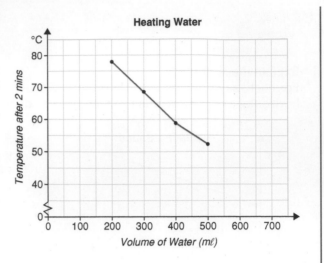

Practical

Find some real-life straight-line graphs.
You could look in science books, on the internet, in the newspaper, in magazines.
Make a poster of your graphs with an explanation of each graph underneath it.

Summary of key points

  (2, 3) is a **coordinate pair**.

Each of the coordinate pairs (⁻1, 0), (0, 1), (1, 2), (2, 3) satisfy the rule $y = x + 1$.

Example   (⁻**1, 0**) satisfies the rule $y = x + 1$ because **0** = ⁻**1** + 1.
$$\uparrow_y \qquad \uparrow_x$$

  To draw **a straight line graph**:

find some coordinate pairs that satisfy a rule

plot them on a grid

draw a straight line through the points.

Example $y = x - 1$

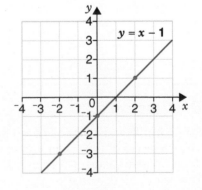

The coordinate pairs (⁻2, ⁻3), (0, ⁻1) and (2, 1) are shown plotted.

The **equation of the line** in the example is $y = x - 1$.

The coordinate pairs of all points on the line satisfy the equation of the line.

C $y = mx$ is the equation of a straight line through the origin. m can have any value. The greater the value of m, the steeper the slope.

 D $y = x + 2$
This line has a **positive** slope

$y = 4 - x$
This line has a **negative** slope

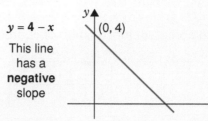

 E Lines **parallel to the x-axis** all have equation $y = a$.
a is any number.
Examples $y = 7$ $y = -1\frac{1}{2}$

Lines **parallel to the y-axis** all have equation $x = b$.
b is any number.
Examples $x = 4$, $x = -3\frac{1}{2}$.

 F We often draw straight-line graphs to represent **real-life situations**. We can read information from these graphs.
Example This graph shows monthly internet charges.
The points (0, £10), (10, £17.50) and (20, £25) have been plotted and joined with a straight line.

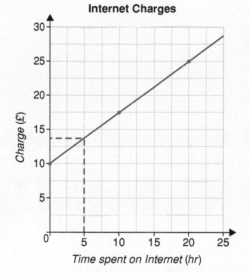

We can read other values from the graph.
Example If I spend 5 hours on the internet, the charge is about £13.75.

 G We can **interpret** straight-line graphs.

Test yourself

1 ($^-$2, ___) satisfies the rule $y = x - 5$. What is the missing coordinate?

2 Copy and complete these tables of coordinate pairs for the given rules.
   a $y = 4x$

x	1	3	5
y			

   b $y = x - 4$

x	$^-$1	2	5
y			

   c $y = 9 - x$

x	0	1	5
y			

T 3 Use a copy of this.

x	$^-$2	0	3
y			

   a Complete the table for $y = x - 1$.
   b Write down the coordinate pairs for the points.
   c On the grid, plot these three points.
      Draw and label the line with equation $y = x - 1$.
   d Does the point ($^-$4, $^-$5) lie on the line?
   e Write down the coordinates of two other points
      that lie on the line $y = x - 1$.

4 **a** $y = $___ x. What number could go in the gap to give the graph of a line which
      is steeper than $y = 2x$?
   b Which of these graphs has a positive gradient, $y = x + 3$ or $y = 5 - x$?
 ***c** Which of these graphs cuts the y-axis at (0, $^-$3)?
      A $y = x + 3$ **B** $y = 3 - x$ **C** $y = x - 3$ **D** $y = 3x$

5 Which of these is the equation of a line parallel to the x-axis?
   a $y = 4$ **b** $x = ^-4$ **c** $y = -3$ **d** $y = x$

T 6 The cost to hire a computer is £100 + £20 per week.
   a Copy and complete this table.

Weeks	2	6	10
Charge (£)			

   b Use a copy of this grid.
      Plot the points from the table on the grid.
      Draw a straight line through the points.
   c Explain what the symbol ⌇ on the vertical
      axis means.
   Show how to use your graph to answer these.
   d How much does it cost to hire a computer
      for 9 weeks?
   e Alison hired a computer and it cost her
      £240. For how many weeks did she hire it?

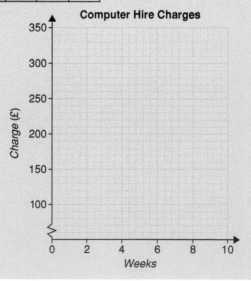

Shape, Space and Measures Support

Angles

Angle is a measure of turn. We measure angles in degrees. There are 360° in a full turn.

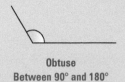

Acute	Obtuse	Right angle	Straight Angle
Less than 90°	Between 90° and 180°	90°	180°

We use a **protractor** or **angle measurer** to measure and draw angles.

Clockwise is the direction clock hands move.

These are the **arms** of the angle

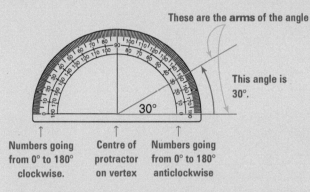

This angle is 30°.

30°

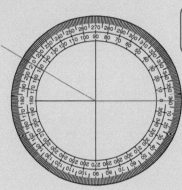

This angle is 150°.

↑ Numbers going from 0° to 180° clockwise.

↑ Centre of protractor on vertex

↑ Numbers going from 0° to 180° anticlockwise

We can **estimate angle size** by comparing with a right angle.
Angle A is about ⅓ of a right angle. It is about 30°.

The corner of a page is a right angle.

A

Practice Questions 7, 31, 32

Lines

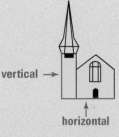

vertical →

↑ horizontal

Parallel lines are always the same distance apart.

Perpendicular lines intersect (cross) at right angles.

We can **measure lines** using a ruler.
This line is 4·4 cm or 44 mm long.

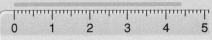

Practice Questions 1, 3, 4, 53

2-D Shapes

Triangles

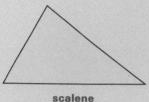
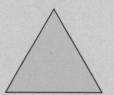
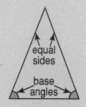

scalene
no equal sides
no equal angles

equilateral
3 equal sides
3 equal angles

isosceles
2 equal sides
2 base angles equal

right-angled

Quadrilaterals

A **quadrilateral** has four sides.

quadrilateral

These are some **special quadrilaterals**.

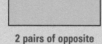

square
4 equal sides
4 right angles

rectangle
2 pairs of opposite
sides equal
4 right angles

parallelogram
Opposite sides equal
and parallel

rhombus
a parallelogram
with 4 equal sides

trapezium
1 pair of opposite
sides parallel

kite
2 pairs of adjacent
sides equal
1 pair of equal angles

Polygons

A 3-sided polygon is a **triangle**.
A 5-sided polygon is a **pentagon**.
A 7-sided polygon is a **heptagon**.
A 4-sided polygon is a **quadrilateral**.
A 6-sided polygon is a **hexagon**.
An 8-sided polygon is an **octagon**.
A **regular polygon** has equal sides and equal angles.

Examples

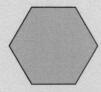
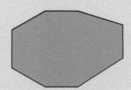

regular pentagon   regular hexagon   octagon (irregular)

A **diagonal** is a straight line drawn from one corner of a polygon
to another. These corners must not be next to each other.

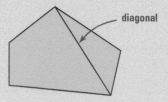

diagonal

Congruent shapes have identical shape and size.

Practice Questions 2, 10, 15, 16, 43, 48

3-D Shapes

 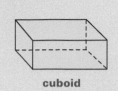 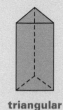

| cube | cuboid (rectangular prism) | triangular prism | pyramids | cylinder |

    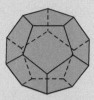

| sphere | cone | hemisphere | tetrahedron (triangular-based pyramid) | octahedron | dodecahedron |

Faces, Edges, Vertices

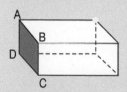

A **face** is a flat surface.
The above shape has six faces; One of the side faces is shown in red.

An **edge** is a line where two faces meet.
The above shape has 12 edges. One edge is shown in blue.

A **vertex** is a corner where edges meet. (**Vertices** is the plural of vertex.)
The above shape has eight vertices. One of the vertices has a yellow dot.

Practice Questions 2, 9, 11, 41

Nets

A **net** is a 2-D shape which folds to make a 3-D shape.
This net folds to make an open box.
The base is shown in yellow.

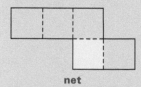

| net | open box |

There are other nets which fold to make an open box.

Practice Questions 45, 49

Coordinates

The **coordinates** of P are (5, 3).
The x-coordinate is 5.
The y-coordinate is 3.
We always write the x-coordinate first.
(x, y) is in alphabetical order.

Practice Question 19

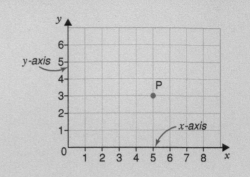

Symmetry

A shape has **reflective symmetry** if it can be folded so that one half fits exactly onto the other half.

Example   Shape **A** is **symmetrical** about the dotted line.
The dotted line is called a line of symmetry or an
axis of symmetry.

Shape **B** is **not** symmetrical.

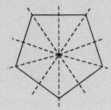

A **regular** polygon has the same number of lines of symmetry as it has sides.

Example   A regular pentagon has five sides and five lines of symmetry.

Practice Questions 17, 26, 51, 52

Transformations

The blue shape has been **translated** 3 units to the right and 1 unit down.

translation

The red shape has been **reflected** in the mirror line, m, to get the purple shape.

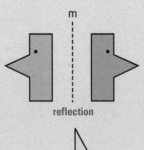
reflection

We can make a pattern by **rotating** a shape.
This pattern has been made by rotating a triangle through 90° angles.

$\frac{1}{4}$ turn = 90° $\frac{1}{2}$ turn = 180° $\frac{3}{4}$ turn = 270°

Practice Questions 13, 21, 38, 40

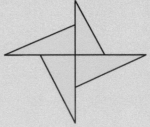
rotations of 90°

Measures

Time

1 minute = 60 seconds	1 millennium = 1000 years
1 hour = 60 minutes	1 century = 100 years
1 day = 24 hours	1 decade = 10 years
1 year = 12 months or 52 weeks and 1 day	
1 year = 365 days (or 366 days in a leap year)	

April, June, September, November have 30 days.
January, March, May, July, August, October, December have 31 days.
February has 28 days except in a leap year when it has 29 days.

All years that are divisible by 4 are leap years, except centuries which are leap years only if they are divisible by 400.

a.m. time is from midnight until noon; **p.m.** time is from noon until midnight.

24-hour clocks begin at midnight. The hours are numbered as shown.

00:00	01:00	02:00	...	12:00	13:00	...	22:00	23:00
midnight	1 a.m.	2 a.m.		12 p.m.	1 p.m.		10 p.m.	11 p.m.

24-hour clocks always have four digits.
1:20 p.m. is written as 13:20 or 1320 or 13·20.
4:35 a.m. is written as 04:35 or 0435 or 04·35.

Metric Measures

Length

1 kilometre (km) = 1000 metres
1 metre (m) = 1000 millimetres (mm)
1 metre (m) = 100 centimetres (cm)
1 centimetre (cm) = 10 millimetres (mm)

Capacity

1 litre (ℓ) = 1000 millilitres (mℓ)

Mass

1 kilogram (kg) = 1000 grams (g)

Examples

5 km = 5 × 1000 m   3 m = 3 × 100 cm
      = 5000 m   = 300 cm

Imperial Measures

The **imperial measures** you might still meet are:

length	capacity	mass
mile	pint	pound (lb)
yard, foot, inch	gallon	ounce (oz)

Reading scales

To **read a scale** we must work out what each space on the scale stands for.

Example   On these scales, each space stands for 10 kg.
The pointer is at about 92 kg or 90 kg to the nearest 10 kg.

Practice Questions 5, 6, 8, 12, 14, 18, 20, 22, 23, 24, 25, 27, 28, 29, 30, 34, 36, 37, 39, 47, 50

Perimeter and area

The distance right around the outside of a shape is called the **perimeter**.
Perimeter is measured in mm, cm, m or km.

The amount of surface a shape covers is called the **area**. Area is measured in square millimetres (mm^2), square centimetres (cm^2), square metres (m^2) or square kilometres (km^2).

We can count squares to find area. The area of each of these squares is 1 square centimetre. Since there are 12 squares in this rectangle, its area is 12 square centimetres or 12 cm^2.

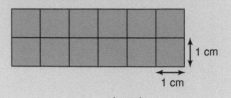

1 cm

1 cm

The **area of a rectangle** is found by multiplying the length by the width.

length

width

Example   Perimeter = $2 \times 8 + 2 \times 4$

$= 16 + 8$
$= 24$ cm

Area = 8×4
$= 32$ cm^2

8 cm

4 cm

Practice Questions 33, 35, 42, 44, 46

Practice Questions

1

T H E

Which lines are longer, the horizontal or vertical?
Check your answer by measuring.

2   Name these shapes.

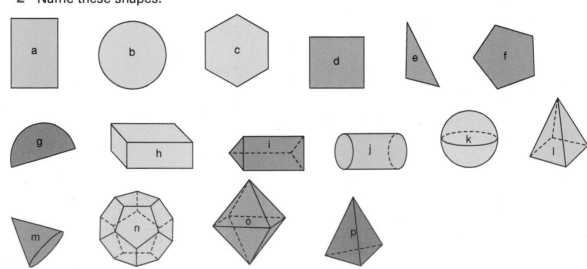

3

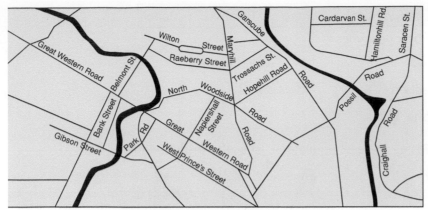

a Which street is parallel to Hopehill Road?
b Which road is perpendicular to West Prince's Street?
    A Great Western Road   **B** Hopehill Road   **C** Napiershall Street
c Is Gibson Street parallel to Great Western Road?
d Write down two other streets that are parallel.

4 Estimate the lengths of these lines to the nearest centimetre.
Then measure them accurately.

a
b
c
d

5 a How many days are there between March 1st and May 3rd, not including these days?
  b How many days are there between July 30th and September 20th, including both these days?
  c How many days were there between February 15th and March 15th in 1988, not including these days?
  d How many days were there between February 15th and March 15th in 2001, not including these days?

6 What goes in the gap?
  a 1 hour 25 minutes = ____ minutes   b 115 minutes = ____ hour ____ minutes
  c 124 seconds = ____ minutes ____ seconds   d 30 days = ____ weeks ____ days
  e 5 weeks 6 days = ____ days   f 2 centuries 10 years = ____ years
  g 1 millennium, 3 centuries, 2 decades, 4 years = ____ years

T

7 Use a copy of this.
  a Mark a right angle with a small square.
  b Put a cross (x) in an acute angle.
  c Put a dot (•) in an obtuse angle.
  d How many triangles are there altogether?

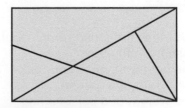

8

min	cm	mℓ	kg	m	mm	ℓ	g	km	sec

The abbreviation for each of the following is given in the box: kilometre, metre, centimetre, millimetre, gram, kilogram, litre, millilitre, minute, second.
Match the words with their abbreviations.

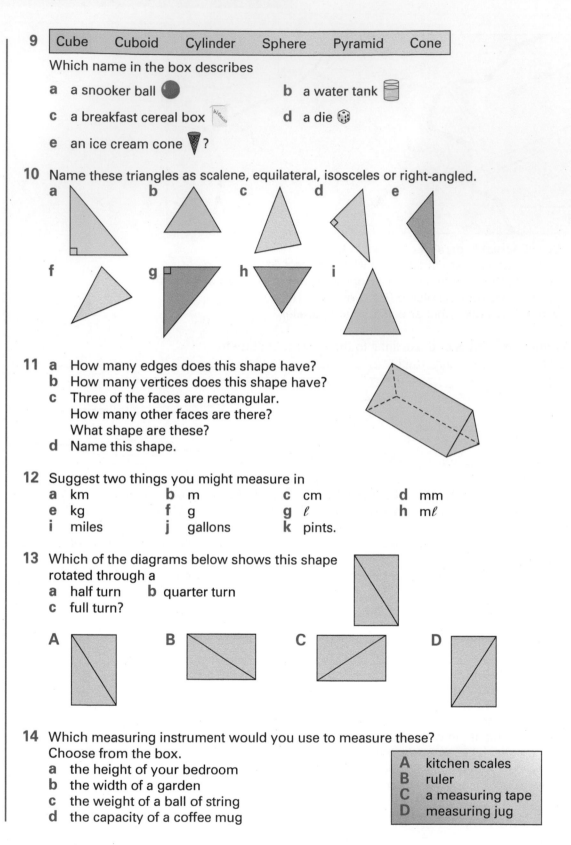

9 | Cube | Cuboid | Cylinder | Sphere | Pyramid | Cone |

Which name in the box describes

a a snooker ball

b a water tank

c a breakfast cereal box

d a die

e an ice cream cone ?

10 Name these triangles as scalene, equilateral, isosceles or right-angled.

a   b   c   d   e

f   g   h   i

11 **a** How many edges does this shape have?
    b How many vertices does this shape have?
    c Three of the faces are rectangular.
        How many other faces are there?
        What shape are these?
    d Name this shape.

12 Suggest two things you might measure in
    a km   **b** m   **c** cm   **d** mm
    e kg   **f** g   **g** ℓ   **h** mℓ
    i miles   **j** gallons **k** pints.

13 Which of the diagrams below shows this shape
    rotated through a
    a half turn   **b** quarter turn
    c full turn?

    A   B   C   D

14 Which measuring instrument would you use to measure these?
    Choose from the box.
    a the height of your bedroom
    b the width of a garden
    c the weight of a ball of string
    d the capacity of a coffee mug

    | A | kitchen scales |
    | B | ruler |
    | C | a measuring tape |
    | D | measuring jug |

15 Name these polygons.

16 Which of these shapes are regular?

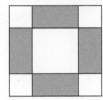

17 How many lines of symmetry do these shapes have?

a   **b**   **c**

18 The High Street in Wadestown is 2 km long.
How many metres is this?

19 Use a copy of this.

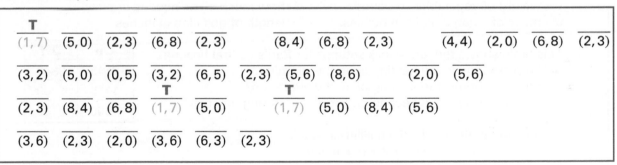

T
$\overline{(1,7)}$ $\overline{(5,0)}$ $\overline{(2,3)}$ $\overline{(6,8)}$ $\overline{(2,3)}$ $\overline{(8,4)}$ $\overline{(6,8)}$ $\overline{(2,3)}$ $\overline{(4,4)}$ $\overline{(2,0)}$ $\overline{(6,8)}$ $\overline{(2,3)}$

$\overline{(3,2)}$ $\overline{(5,0)}$ $\overline{(0,5)}$ $\overline{(3,2)}$ $\overline{(6,5)}$ $\overline{(2,3)}$ $\overline{(5,6)}$ $\overline{(8,6)}$ $\overline{(2,0)}$ $\overline{(5,6)}$

$\overline{(2,3)}$ $\overline{(8,4)}$ $\overline{(6,8)}$ **T** $\overline{(1,7)}$ $\overline{(5,0)}$ **T** $\overline{(1,7)}$ $\overline{(5,0)}$ $\overline{(8,4)}$ $\overline{(5,6)}$

$\overline{(3,6)}$ $\overline{(2,3)}$ $\overline{(2,0)}$ $\overline{(3,6)}$ $\overline{(6,3)}$ $\overline{(2,3)}$

Fill in the words by finding the letter beside each coordinate.
T is filled in for you.

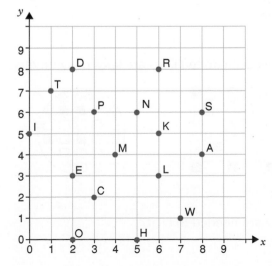

20 What do each of these read?

a   b   c   d

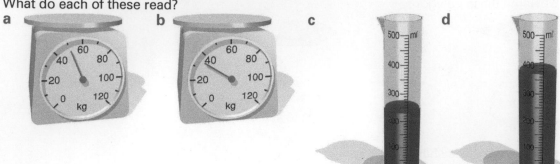

21 Describe how each of these shapes has been translated.

a   b

 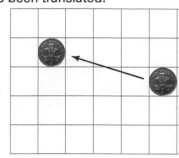

22 Alice's grandfather measured most things in imperial units. The imperial unit he used is given. Which metric unit could he have used instead?

a   distance to his brother's in miles   b   length of bedroom in feet
c   weight of vegetables in pounds   d   how much oil he needed in pints
e   mass of a bag of nails in ounces   f   length of golf club in inches

23 This table shows the time-clock settings for Rari's central heating.

a   For how long is the heating on in the morning?
b   For how long is the heating on in the afternoon?
c   For how long is the heating on altogether during one day?

On	Off
0645	0840
1115	1300
1500	2130

24 A medicine bottle holds 300 millilitres.
Robert is to have 15 millilitres twice a day.
How many days will the medicine last?

25 Last night the sun set at 19:30. It rose this morning at 06:45.
How many hours of darkness were there?

T   26 Use a copy of this.
Shade eight **more squares** so that both dotted lines are lines of symmetry.

a   b   c

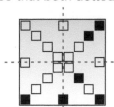 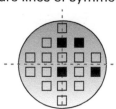

27 A sink holds $6\frac{1}{2}$ litres of water.
How many $\frac{1}{2}$ litre jugs full of water are needed to fill it?

28 What is the missing unit of measurement?
    a   7 cm = 70 ____   **b**   3 m = 3000 ____   **c**   3 m = 300 ____
    d   2 km = 2000 ____   **e**   6 cm = 60 ____   **f**   6 ℓ = 6000 ____
    g   5 kg = 5000 ____

29 Write these times as they would appear on a 24-hour clock.
    a 9·24 a.m.   **b** 2 a.m.   **c** 3·45 p.m.   **d** five minutes to midnight

30 Write these 24-hour clock times as a.m. or p.m. times.
    a 0400   **b** 0823   **c** 1415   **d** 2340   **e** 1700   **f** 2238   **g** 0016

31 Estimate the size of these angles.
Use a protractor to check your estimate.

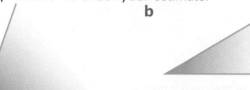

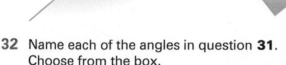

32 Name each of the angles in question **31**.
Choose from the box.

33 The distance between dots is 1 cm.
Without measuring, find the perimeters of the shapes.

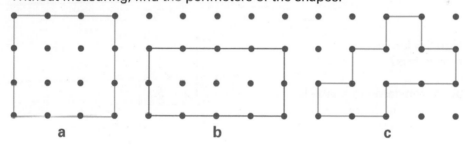

34 Which of these would you expect to be
    a   the length of a table
        A 3 cm   **B** 2 m   **C** 25 cm   **D** 30 m
    b   the weight of an orange
        A $\frac{1}{2}$ kg   **B** 250 g   **C** 760 g   **D** 25 g
    c   the amount of milk in a glass
        A 20 mℓ **B** 600 mℓ **C** 200 mℓ **D** 2 ℓ
    d   the distance between two cities
        A 150 mm   **B** 150 km   **C** 150 m   **D** 150 cm
    e   the length of a pencil
        A 15 cm   **B** 35 cm   **C** 3 cm   **D** 2 m
    f   the weight of a 3-year old
        A 450 g   **B** 5 kg   **C** 200 kg   **D** 18 kg
    g   the amount of water in a full bucket?
        A 2 ℓ **B** 750 mℓ **C** 10 ℓ **D** 50 ℓ

35 Lucy used this pattern to make Christmas decorations.
Count squares to estimate the area of each decoration.

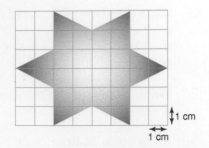

1 cm
1 cm

36 What reading is given by each of the pointers?

a

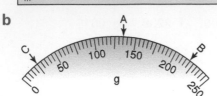

b

37 A cycle race began at 09:45 and finished at 15:35.
How long did the race last?

T

38 Use a copy of these. Draw the reflection in the mirror line, m.

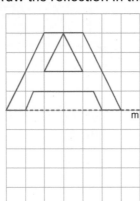

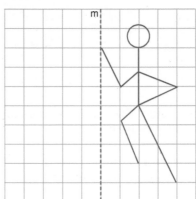

39 Ten apples weigh 2·5 kg.
How many grams is this?

40 Shape A has been translated to shape B.
Copy this sentence and fill in the gaps.
Shape A has been translated ____ units to the ____
and ____ units ____ .

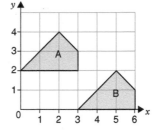

41 What is the smallest number of cubes you would need to add to these shapes to make
them into cuboids?

a

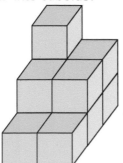

b

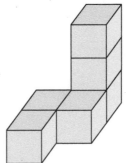

42 Find the perimeter of these.

a 12 m / 5 m

b 6 mm / 4 mm / 6 mm / 7 mm

c regular hexagon / 3 cm

T

43 Use a copy of this.
a Draw all the diagonals on the hexagon.
b How many diagonals does a regular hexagon have?
c How many diagonals does an octagon have?

Draw an octagon to check.

44 What is the area of these rectangles?

a 3 m / 2 m

b 16 cm / 2cm

c 5 mm / 20 mm

45 a Which of these nets will fold to make an open box?

There is more than one answer.

A   B   C   D   E

b Draw another two nets that fold to make an open box.

46 Choose a suitable unit to estimate the area of these.
Choose from the box.
a area of a classroom
b area of a book cover
c area of a calculator button
d area of a football field

mm²   cm²   m²

47 Mrs Sansom bought 2 kg of sugar.
She used 400 g of sugar to make some fudge.
How many grams of sugar were left?

48 Nishi had 12 squares. She made three different rectangles with them.
a How many different rectangles could she make with 24 squares?
b How many squares does she need to make exactly five different rectangles?

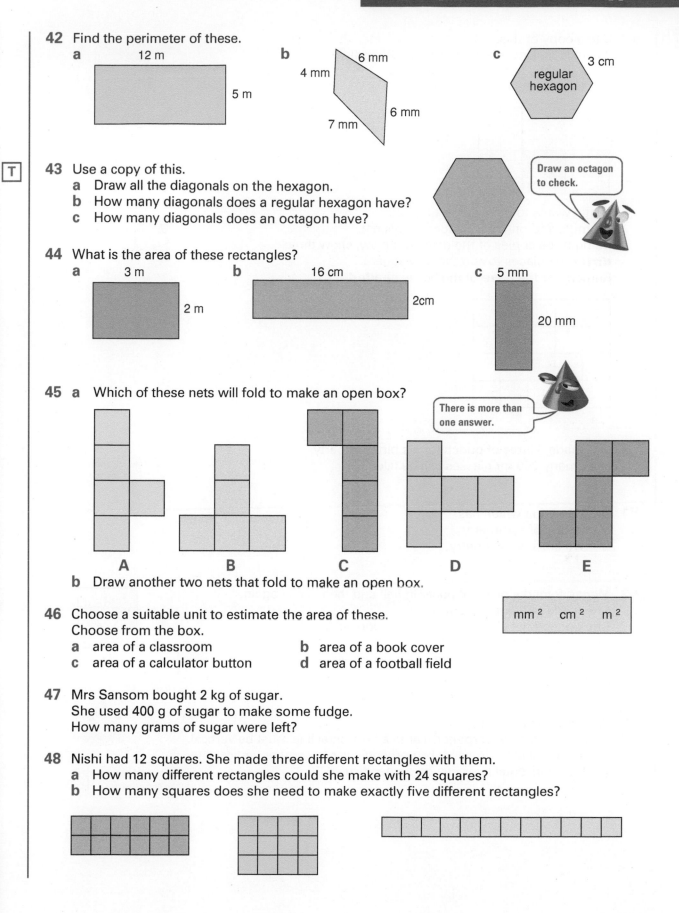

T

49 Use a copy of this.
Justin is making a box to display a butterfly.
He draws the net of the box like this.
The base of the box is shaded.

Justin wants to put a lid on the box.
He must add one more square to his net.
Using three copies of the diagram below, show three
different places to add the new square.
Remember the **base** of the box is **shaded**.

50 Jeff made 4 litres of punch for his birthday party.
How many 250 mℓ glasses can be filled?

51 Draw a shape which has
a one line of symmetry
b two lines of symmetry.

T

52 Megan folded a piece of paper in half and then in half again.
She made one straight cut.
She opened the sheet of paper and it looked like this.
Use a copy of this.
Draw on this folded sheet of paper where Megan
made her cut.

53 Which of these is true?
   A A line which is perpendicular to a horizontal line must be vertical.
   B A vertical line can never be perpendicular to another vertical line.
   C Two horizontal lines can be perpendicular.

There could be more
than one answer.

11 Lines and Angles

Key vocabulary

adjacent, angle, angles at a point, angles on a straight line, base angles, compasses, construction lines, degree(°), draw, intersect, intersection, isosceles triangle, line, line segment, measure, parallel, perpendicular, plane, point, proof, protractor (angle measure), prove, reflex, ruler, set square, vertex, vertically opposite angles, vertices

 Straight or Curved

1 Draw a right angle using a set square.
   Draw each 'arm' of the angle 10 cm long.
   Divide each 'arm' into cm lengths.
   Number the cm lengths as shown.
   Join 1 to 1, 2 to 2, 3 to 3, ...

2 Draw another right angle with arms 10 cm long.
   Divide each arm into $\frac{1}{2}$ cm lengths.
   Number the $\frac{1}{2}$ cm lengths 1 to 20 in the same
   order as the diagram in 1.
   Join 1 to 1, 2 to 2, 3 to 3, ...

3 Try and draw these.

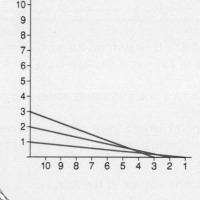

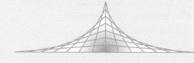

Naming lines and angles

This is a **line**. It goes on forever in both directions.
We say it has infinite length.

MN is a **line segment**.
It has end points M and N.
We say it has a finite length.

M   N

We think of lines and line segments as having no measurable width.
However when you **draw** lines they do have some width.

Two straight lines must either
be **parallel**
or cross once (**intersect**).

Is the horizon a line?

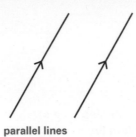

A is the point where the lines intersect.

parallel lines   intersecting lines

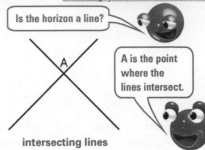

Discussion

Above we said that two straight lines must either be parallel or cross once.

What if one of the lines was drawn on the wall and the other on the ceiling? Would this still be true? **Discuss**.

If one line is drawn on the ceiling and the other on the wall the lines are on different **planes**.
Lines in the same plane are on the same flat surface.

When two line segments meet at a point they make an angle.
The point where they meet is the **vertex**.
The angle is a measure of turn from one arm to the other.

vertex   F

G   **arms or rays of angle**

                                          H

We name angles in two ways.

1 Using the letter at the vertex.
   The angle above is ∠G.

2 Using three letters, the middle one being the vertex.
   The angle above is ∠FGH or ∠HGF.
   We could write this as FĜH or HĜF.

We **must** use three letters if there is more than one angle at the vertex.

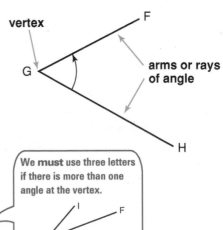

Exercise 1

1 Name the red line segment.

a P ——————— Q **b** C E B A D **c** R S T **d** G F H

2 Use a single letter to name the marked angles.

a X A P **b** Q A R **c** Q Y C **d** D E F

3 Use three letters to name the marked angles in question **2**.

4

SAL	M_	H_	O_	D_	A_	H_	T_	F_	T_
a	b	c	d	e	f	g	h	i	j

Copy this chart.
Name the marked angles on the diagrams below and fill in this chart.
The angle marked in **a** is ∠SAL so SAL is filled in as shown.

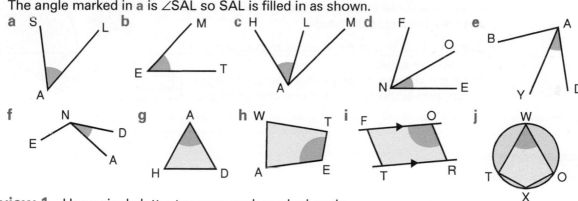

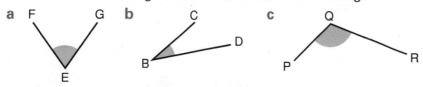

Review 1 Use a single letter to name each marked angle.

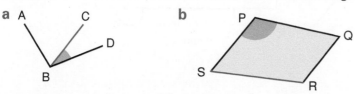

Review 2 Use three letters to name the marked angles in Review **1**.

Review 3 Use three letters to name each marked angle.

a A C D B **b** P Q S R

Review 4 Name the red line segments in Review **3**.

Special angles

Remember

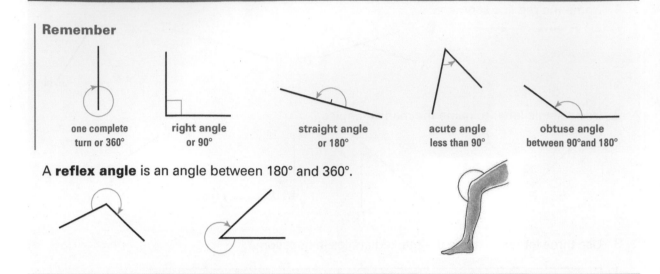

one complete turn or 360°	right angle or 90°	straight angle or 180°	acute angle less than 90°	obtuse angle between 90° and 180°

A **reflex angle** is an angle between 180° and 360°.

Practical

There is one acute angle, one right angle and three obtuse angles inside this shape.

1 Sketch a shape with three acute angles and one obtuse angle.
2 Sketch a shape with one reflex angle, one right angle, two acute angles and one obtuse angle.

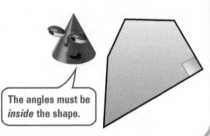

The angles must be *inside* the shape.

Exercise 2

1 Write down the colour of the   **a** acute   **b** obtuse   **c** right   **d** reflex
angles in each of these.

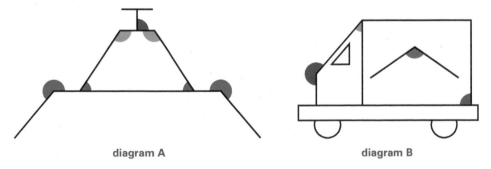

diagram A   diagram B

Review Copy this chart.

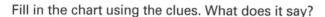

a   b   c   d   e

Fill in the chart using the clues. What does it say?

1 The last letter of the name of the angle in **a**.
2 The first letter of the name of the angle in **d**.
3 The second letter of the name of the angle in **c**.
4 The first letter of the name of the angle in **b**.
5 The first letter of the name of the angle in **e**.

Measuring angles

Remember
To measure an angle to the nearest degree, use a **protractor**.

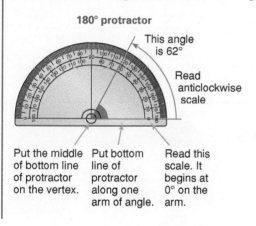

180° protractor

This angle is 62°

Read anticlockwise scale

Put the middle of bottom line of protractor on the vertex.

Put bottom line of protractor along one arm of angle.

Read this scale. It begins at 0° on the arm.

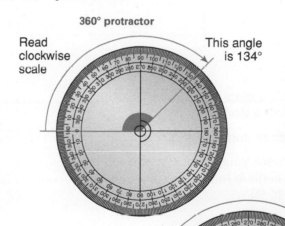

360° protractor

Read clockwise scale

This angle is 134°

Measuring reflex angles

The easiest way to **measure a reflex angle** is with a 360° protractor (angle measurer).

This reflex angle is 317°.

To measure a reflex angle with a 180° protractor measure the acute or obtuse angle then subtract it from 360°.

Example   Acute angle is 43°.
Reflex angle = 360° − 43°
= 317°

Measure this angle as 43°

Shape, Space and Measures

Before measuring, always **estimate** the size of the angle.
Do this by comparing it with a right angle.

The corner of a sheet of paper is a right angle.

Example   Estimate the size of the acute angle. It is about 80°.
The reflex angle is about 360° − 80° = 280°.

Exercise 3

1 Estimate the size of these angles. Check by measuring.

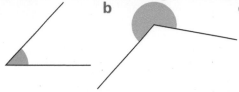

2 This is a diagram of a ladder leaning against a wall.

Measure angles **a** and **b**.

To check your accuracy, add angle **a** and angle **b** together. They should add to 90°.

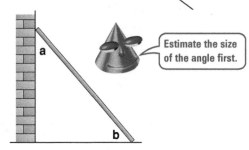

Estimate the size of the angle first.

3 The diagram below shows three angles, *a*, *b* and *c*.
   a Put the angles in order from largest to smallest.
   b Measure the size of each angle. Give your answer to the nearest degree.
      Add all three angles together. They should total 360°.

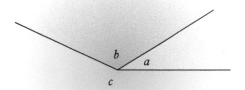

4 Measure the size of each of the marked angles.
As a check on the accuracy of your measurement, add all four angles together. They should total 1080°.

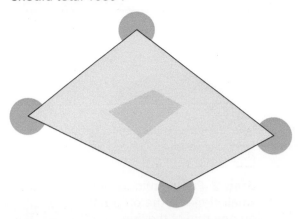

Review 1 Estimate the size of each of the marked angles.
Check your estimate by using a protractor to measure them.

a   b   c

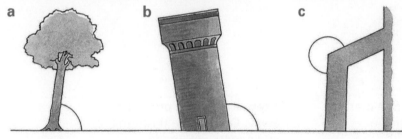

Review 2 Draw some large triangles. Make each triangle a different shape.
Measure the three angles in each triangle.
No matter what the shape of your triangle, the three angles should total 180°.

Review 3 Draw some large four-sided shapes. Measure the outside angles as you did in question **4**. No matter what shape you draw, the four outside angles should always total 1080°.

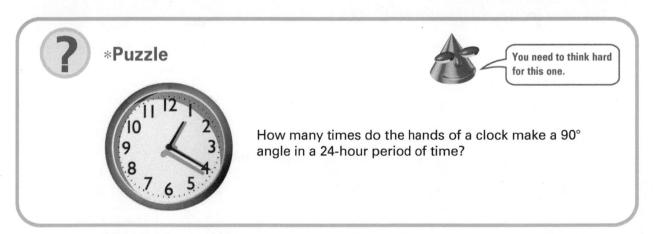

?   *Puzzle

You need to think hard for this one.

How many times do the hands of a clock make a 90° angle in a 24-hour period of time?

Drawing angles

To draw an angle of 70°, follow these steps.

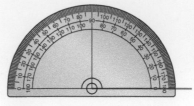

Step 1 Draw a straight line.

Step 2 Put the middle of the small circle on the protractor on one end of the line.

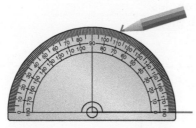

Step 3 Read the scale that begins at 0° on your line. Put a mark beside 70°.

Step 4 Take the protractor away. Draw a line through the small mark you made.

Using a 360° protractor, angles of any size between 0° and 360° may be drawn. Using a 180° protractor, reflex angles are drawn as follows.

1 Subtract the size of the angle to be drawn from 360°.

Example To draw 235°, subtract 235° from 360° to get 125°.

2 Draw this smaller angle.
The other angle in the diagram will be 235°.

Draw this angle of 125°

235°

Exercise 4

1 Draw these angles.
Ask someone else to check your answers by measuring them.

a 30°	**b** 80°	**c** 65°	**d** 160°	**e** 124°
f 92°	**g** 200°	**h** 245°	**i** 257°	**j** 300°
k 310°	**l** 281°			

Review Draw a shape with the angles shown.

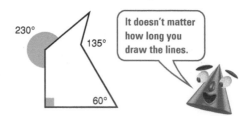

230°

135°

60°

It doesn't matter how long you draw the lines.

Practical

1 **Work in pairs**. Measure your angle of vision. Your partner should stand behind you holding a pencil at your eye height. This pencil should be gradually moved forward until you can see it.

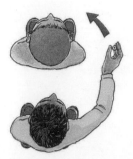

Think carefully about how you can *measure* the angle.

2 Draw a circle of radius 4 cm. Mark the centre.
Mark a point on the circle and label it 1.
Use your protractor to mark the points 2, 3, ... 12 which are evenly spaced around the circle.
Join each point to the point that is three spaces clockwise around the circle i.e. join 1 to 4, 2 to 5, 3 to 6, ... 12 to 3.
Colour the design you have made.

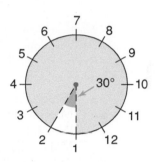

Variation Mark off 24 equally spaced points around the circle.

Variation Join the points using a different spacing e.g. join each point to the point that is four spaces further around the circle.

3 Choose a topic. Make a poster about the angles in this topic.

Suggestions

Sport. *Example* Gymnastics

Farms. Angles between fencelines, the angles through which gates swing, the angles of a loading ramp, the angle between the sides and the bottom of a trough etc.

Cars. Angles between windscreens and bonnets, angles of seat backs, angles through which the doors and bonnet and boot move as they are opened etc.

Clothing. Angles between seams, angles of collar peaks, angles in patterns for a garment etc.

Leisure, for example snooker. You could investigate the angle at which a ball must be hit if it is to be sunk. You would have to consider the balls at different positions on the table.

Parallel and perpendicular lines

Practical

Which lines in these pictures are parallel?
Make a list of them.
Which lines are perpendicular?
Make a list of them.

- Walk around your classroom, school buildings or grounds. Make a list of the parallel and perpendicular lines that you see.

- Choose an activity such as playing football, going shopping, dancing, taking a photograph, playing a card game etc. **Discuss** what parallel and intersecting lines you would be likely to find.

Remember
Parallel lines are always the same distance apart (equidistant).
Perpendicular lines intersect at right angles.

Parallel lines never meet.
We write AB//CD. We read // as 'is parallel to'.

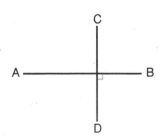

We write AB⊥CD.
The symbol ⊥ is read
as 'is perpendicular to'.

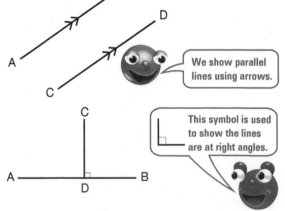

We show parallel lines using arrows.

This symbol is used to show the lines are at right angles.

Exercise 5

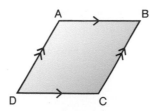

1 Which line is parallel to
   a AB   **b** AD?

2 **a** Which line is perpendicular to SP?
   b Which lines are perpendicular to PQ?
   c Which line is parallel to QR?

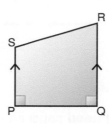

3 Which of // or ⊥ goes in the gaps?
   a DE __ HI   **b** DE __ EF   **c** GF __ DE
   d HG __ GF   **e** EF __ DE

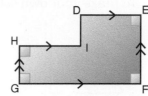

4 This is a drawing of a cuboid.
   a Which edges are parallel to AB?
   b Which edges are perpendicular to DH?

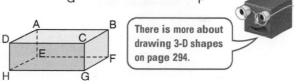

There is more about drawing 3-D shapes on page 294.

Review
a Name three pairs of parallel lines in this shape.
b Name the perpendicular lines.

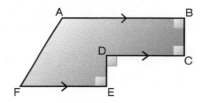

 Practical

You will need a dynamic geometry software package.

1 Draw a line AB as shown in the diagram.
2 Draw a point C and through it construct a line parallel to AB.
3 Construct a line through A perpendicular to AB.
4 Construct a line through B perpendicular to AB.

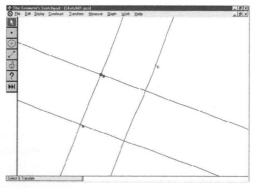

What do you notice about the parallel lines?
If you draw a line perpendicular to one parallel line is it perpendicular to the other parallel line?
Are the two perpendicular lines parallel to each other?

Parallel and perpendicular lines can be drawn with a **set square**.
One angle of a set square is a right angle.

If you do not have a set square, you can make a paper one as shown in the following practical.

 Practical

You will need paper and a stapler.
To make your own set square, follow these steps.

Fold one corner of a piece of paper over, to make a straight edge.

Fold the paper again so that half of the straight edge is *exactly* on top of the other half.

Tidy up the edges. Staple the paper through all thicknesses. Make sure the edges don't move as you staple.

Discussion

● How can you use your set square to find which of these lines are perpendicular? **Discuss**.

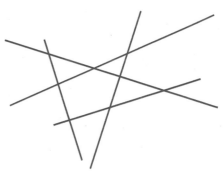

●

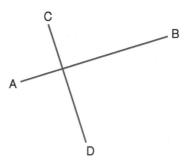

Draw a line segment AB.
How could you use your set square to draw the line segment CD which is perpendicular to AB? **Discuss**.

These diagrams show how to draw two parallel lines using a ruler and set square.

Set
Square
Ruler

Ruler

Ruler

Ruler

Put the set square on the ruler.

Draw a line.

Slide the set square along the ruler.

Draw another line.

Discussion

The diagrams below show how to draw a line through a point C that is parallel to the line AB.

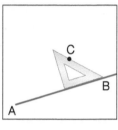
Step 1

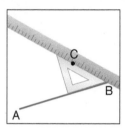
Step 2

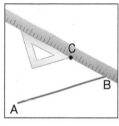
Step 3

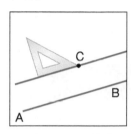
Step 4

Discuss these steps. (**Hint**: in Step 3, the set square slides along the ruler.)

Exercise 6

1 a Draw a line segment, AB.
  b Mark a point, D, anywhere on the line segment.
  c Use your set square to draw a line through D that is perpendicular to AB.

2 a Draw a line segment, CD.
  b Mark a point, E, above the line.
  c Use your set square to draw a line through E that is perpendicular to CD.

3 Repeat question **2** but mark the point, E, below the line segment CD.

4 Draw a line segment GH.
  Using your set square and ruler, draw a line that is parallel to GH.

5 a Draw a line segment PQ.
  b Mark a point, S, above the line segment.
  c Use your set square and ruler to draw a line through S that is parallel to PQ.

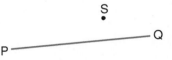

Review 1

a Draw a line segment, MN.
b Mark a point, P, anywhere on the line segment.
c Use your set square and ruler to draw a line through P that is perpendicular to MN.

Review 2 Repeat **Review 1** but mark P above the line segment.

Review 3

a Draw a line segment, KL.
b Use your set square and ruler to draw a line that is parallel to KL.

 Practical

1 Make a logo for your class or school using parallel and/or perpendicular lines and shading.

2 A new village is to be built.
Design a street plan for the village using parallel and perpendicular lines. Give your village a name.

Angles

 Practical

You will need acetate sheets.
This practical can also be done using a dynamic geometry software package.

Draw a line segment AB.
Draw another line segment CD on a separate sheet of acetate.
Lay it on top of the first piece of acetate so the lines intersect.

Move the line segment CD making sure that lines still intersect.

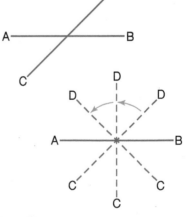

What can you find out about the size of the marked angles in each diagram?
Which ones are equal?
Which ones add to 180°?
Which ones add to 360°?

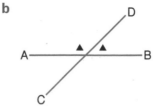

a and *b* are called **vertically opposite angles**.
They are opposite each other when two lines intersect.
Vertically opposite angles are equal.

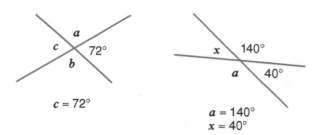

$a = b$
$c = d$

Examples

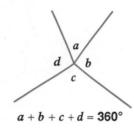

$c = 72°$

$a = 140°$
$x = 40°$

a, *b*, *c* and *d* are called **angles at point**.
Together, *a*, *b*, *c* and *d* make a complete turn.
Angles at a point add to 360°.

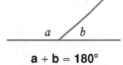

$a + b + c + d = \textbf{360°}$

Worked Example
Find the size of angle *x*.

a

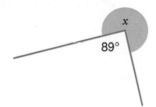

b

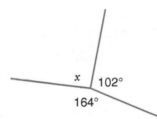

Answer

a $x = 360° - 89°$
   $= \textbf{271°}$

b $x = 360° - 102° - 164°$
   $= \textbf{94°}$

Angles on a straight line add to 180°

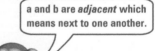

$a + b = \textbf{180°}$

> a and b are *adjacent* which means next to one another.

Worked Example
Find the size of angle *a*

a

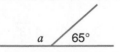

b

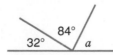

Answer

a $a = 180° - 65°$
   $= \textbf{115°}$

b $a = 180° - 84° - 32°$
   $= \textbf{64°}$

269

Exercise 7

1 Find the value of the unknown.

a

a ⤫ 97°

b

130° a

c

62° d

d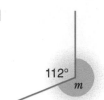

112° m

e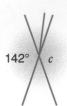

142° c

f

61° e 52°

g

41° f

h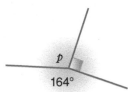

p
164°

i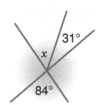

31°
x
84°

j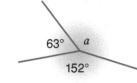

63° a
152°

k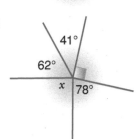

41°
62°
x 78°

l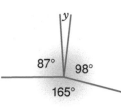

y
87° 98°
165°

2 Find the value of a, b and c.

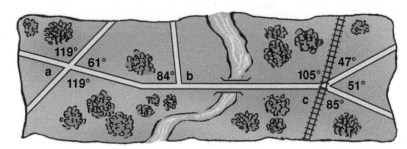

119°
61°
a
119°
84° b
105°
47°
51°
c 85°

T

Review

138°	64°	151°	35°	70°	57°	67°	151°	113°	125°	138°		110°	70°	113°	70°
	N			**N**											
47°	**65°**	122°	70°	**65°**	35°	70°	125°		47°	**65°**	**1963**				

Use a copy of this box. Calculate the shaded angles on the next page.
Put the letter that is beside each diagram above its answer.

 **N** 115° 180° − 115° = 65°

 W 158° 92°

 S 138°

 A 29°

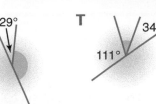 **T** 34° 111°

 K 26°

 V 148°

 E 72° 218°

 I 41° 31° 61°

R 116° 41°

B 68° 125°

O 52° 119°

D 135° 100°

Angles in triangles

Discussion

1
a, c, b

Take an A4 sheet of paper.
Draw a large triangle on it, similar to the one shown, but larger.

2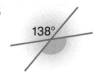
a, c, b

Tear off each of the corners.

3
b, a, c

Put the torn off corners side by side.

You could use a dynamic geometry package to show this.

What can you say about the sum $a + b + c$? **Discuss**.
Is this the same for other triangles? Test some. You could measure to check.

The three angles of a triangle add to 180°

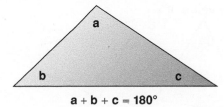
a, b, c

$a + b + c = 180°$

Remember

An isosceles triangle has two equal sides and two equal angles.

The unequal side is called the base.
The equal angles are at each end of the base.
They are called the **base angles**.

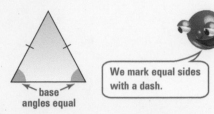

base
angles equal

We mark equal sides
with a dash.

Worked Example

Find the value of the angle marked with a letter.

a

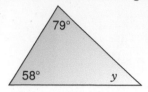

79°

58°   y

b

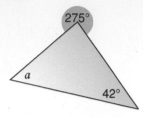

275°

a

42°

Answer

a $y = 180° - 79° - 58°$
      $= \textbf{43°}$

b Before we can find a, we must find the value of the unmarked angle in the triangle
        unmarked angle $= 360° - 275°$ (angles at a point add to 360°)
                    $= 85°$

  $a = 180° - 85° - 42°$
    $= \textbf{53°}$

Worked Example

Find the value of n.

n

68°

Answer

Before we can find n, we must find the value of the unmarked angle in the triangle.
The unmarked angle is one of the base angles of the isosceles triangle.
unmarked angle $= 68°$
$n = 180° - 68° - 68°$
  $= \textbf{44°}$

Exercise 8

1 Find the value of x.

  a

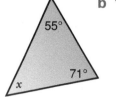

55°

71°

x

  b

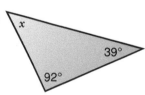

x

92°

39°

  c

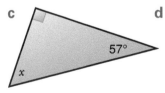

57°

x

  d

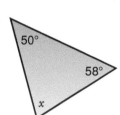

50°

58°

x

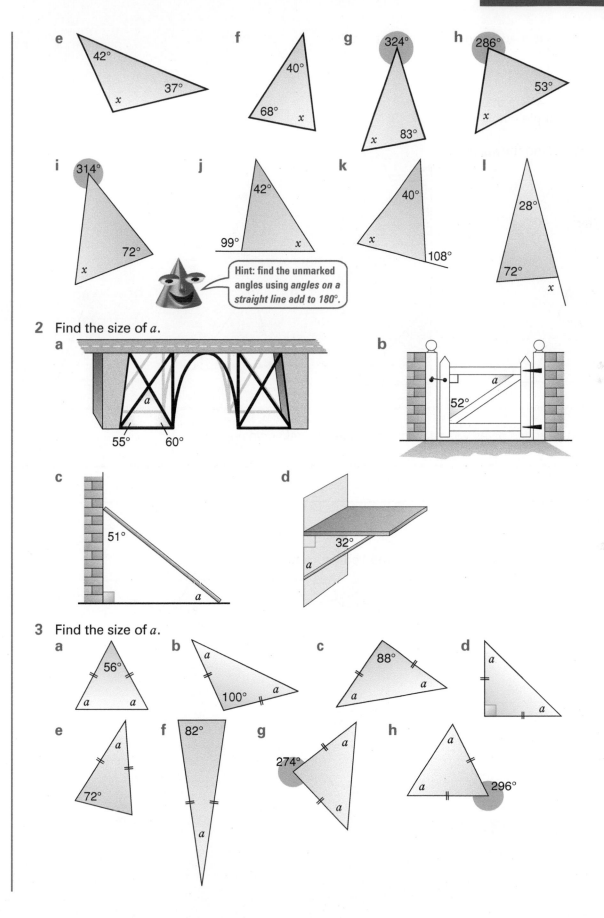

e 42° 37° x

f 40° 68° x

g 324° x 83°

h 286° 53° x

i 314° 72° x

j 42° 99° x

Hint: find the unmarked angles using *angles on a straight line add to 180°.*

k 40° x 108°

l 28° 72° x

2 Find the size of *a*.

a a 55° 60°

b a 52°

c 51° a

d 32° a

3 Find the size of *a*.

a 56° a a

b a 100° a

c 88° a a

d a a

e a 72°

f 82° a

g 274° a a

h a a 296°

4 This tile is made from six identical triangles.
   a Find the size of angle x.
   b Find the size of angle y.

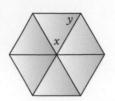

*5 Explain why a triangle can never have a reflex angle within it.

Review   Find the value of x.

a

68°   51°

b

46°

c

39°
288°

d
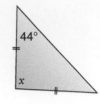
44°

e
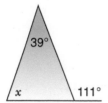
39°
111°

f
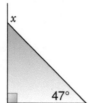
47°

g

76°

h

318°

The next exercise gives you practice at calculating unknown angles.

Exercise 9

1 Calculate the value of the angles marked with letters.

a

115° x

b

139°

c

81°

d

42°

e

88°
156°

f
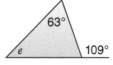
51°
47°

g

53°
319°

h
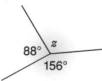
53°
96°

i
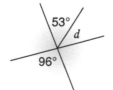
63°
109°

j

k

l

m

n

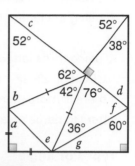

2 Use a copy of this.

  a Find the value of each unknown on the diagram.
  b If the value of the unknown is less than 60°, shade the triangle that contains the unknown.
  c What is special about all the shaded triangles?

Review Use a copy of the crossnumber.
Calculate the value of each unknown.
Using these values, complete the crossnumber.

Across
1 c
2 d
5 g
8 b
9 i

Down
1 h
3 e
4 f
6 a
7 j

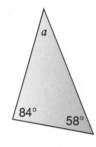

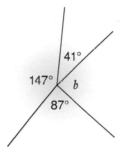

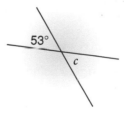

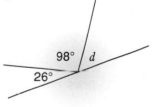

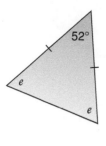

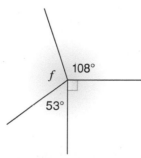

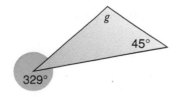

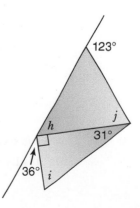

Summary of key points

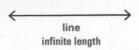

line
infinite length

P   line segment   Q
finite length

Two straight lines on the same **plane** must either **intersect** or be **parallel**.

We **name this angle**

1 using the letter at the vertex, ∠Q

or

2 using three letters, the middle letter being the vertex,

    ∠PQR or ∠RQP   **or**   PQ̂R or RQ̂P

If there is more than one angle at a vertex, always use three letters to name an angle.

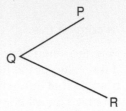

P

Q

R

 Types of angle

one complete
turn or 360°

right angle
or 90°

straight angle
or 180°

acute angle
less than 90°

obtuse angle
between 90° and
180°

reflex angle
between 180° and
360°

 We **measure and draw angles** using a
protractor.

It is best to measure and draw reflex angles
using a 360° protractor.

To measure a reflex angle with a 180°
protractor

1 measure the acute or obtuse angle

2 subtract if from 360°.

Example   Obtuse angle is 138°.

       Reflex angle = 360° − 138°

               = 222°

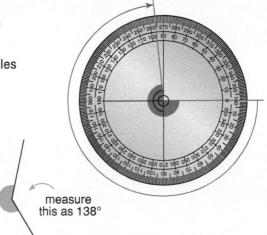

measure
this as 138°

Always **estimate** the size of the angle first.

Do this by comparing the acute or obtuse angle to a right angle.

D **Parallel lines** never meet.

We show parallel lines with arrows.

We show lines are **perpendicular** using
the symbol ─┼─.

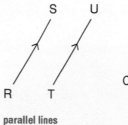

parallel lines

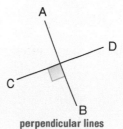

perpendicular lines

Parallel and perpendicular lines can be drawn using a **ruler and set square**.

Example   These diagrams show how to draw parallel lines.

Set
Square

Ruler

Put the set square on
the ruler.

Draw a line.

Slide the set square
along the ruler.

Draw another line.

E

$a = b$
Vertically
opposite angles
are equal.

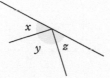

$x + y + z = 180°$
Angles on a straight
line add to 180°.

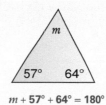

$c + d + e = 360°$
Angles at a point
add to 360°.

F **The interior angles of a triangle add to 180°.**

Example $m = 180° - 57° - 64°$
              $= 59°$

$m + 57° + 64° = 180°$

277

Test yourself

1 Use a single letter to name the marked angles.

a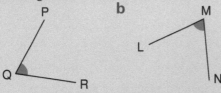
    b

2 Use three letters to name the marked angles in question **1**. Ⓐ

3 a Name each of the marked angles as acute, obtuse, right reflex or straight.
   b Estimate the size of each of the marked angles.
   c Measure each of the marked angles.

Ⓑ Ⓒ

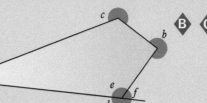

4 Draw these angles.
   a 84°   **b** 152°   **c** 287°

Ⓒ

5 a Name the parallel lines in this shape.
   b Name the perpendicular lines.

Ⓓ

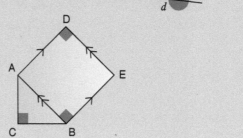

Ⓓ

⊤ **6** Use a copy of this diagram.
Use your ruler and set square to draw
   a a line through C that is perpendicular to AB
   b a line through D that is parallel to AB.
What shape have you drawn?

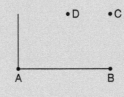

7 Find the angles marked with letters. Ⓔ Ⓕ

a
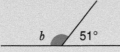

b

c
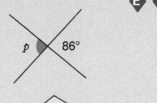

d
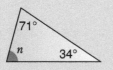

e

f
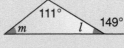

g
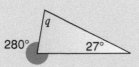

h

i
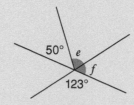

12 Shape, Construction

You need to know

✓ 2-D shapes   — triangles
                 — quadrilaterals
                 — polygons   page 242

✓ symmetry   page 244

✓ 3-D shapes   page 243

✓ faces, edges, vertices   page 243

✓ nets   page 243

Key vocabulary

adjacent, base (of solid), centre, circle, concave, construct, convex, diagonal, edge, face, horizontal, identical, irregular, opposite, measure, net

polygon: hexagon, pentagon, octagon

quadrilateral: arrowhead, delta, kite, parallelogram, rectangle, rhombus, square, trapezium,

regular, shape, side

solid (3-D): cube, cuboid, cylinder, hemisphere, prism, pyramid, sketch, sphere, square-based pyramid, tetrahedron, three-dimensional (3-D)

triangle: equilateral, isosceles, right-angled, scalene,

two-dimensional (2-D), vertex, vertical, vertices

Three of a Kind

Triangles is a game for two students.

You will need a pen each.
Put six dots on a piece of paper so that no three are in a straight line.

Take turns to join two dots with a line.
The first player forced to make a triangle, loses.

Note If the dots are the vertices of a regular hexagon this game is called Sim.

Each of the vertices of your triangle must be on a dot.

Naming triangles

A triangle is named using the capital letters of the vertices. We start with
one of the letters and go round in order clockwise or anticlockwise.
This triangle could be named as △PQR or △QPR or △PRQ.
How else could it be named?

Sides opposite vertices are often named using
the lower-case letter of the opposite vertex

Examples   The side opposite ∠Q is named as q.
The side opposite ∠P is named as p.

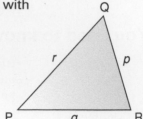

1 Name the red triangle in each.

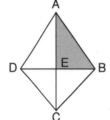 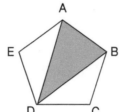 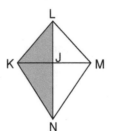 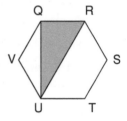

2 Use a single lower-case letter to name the side opposite the marked angle.

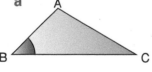 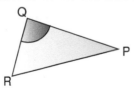  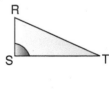

Review 1   Name the purple triangle in each.

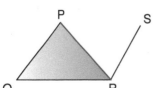 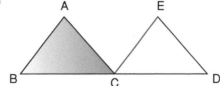

Review 2   Name the pink side of each triangle in Review **1** using a single lower-case
letter.

Describing and sketching 2-D shapes

 Practical

1 Look around your school to find an example of each
   of these shapes.

   square   rectangle   rhombus
   parallelogram   kite   trapezium
   pentagon   hexagon

   You could see
   who can find one
   of each first.

2 **You will need** a partner.
   Without using its name, describe a shape to your
   partner. Can your partner name your shape?
   Swap and get your partner to describe a shape to you.
   What sorts of things did your partner say about the shapes that made it easy
   for you to name them?

Discussion

How might Max and Simone have got their shapes?

Are any other shapes possible? **Discuss**.

Exercise 2

1 Imagine a square with both diagonals drawn on. Remove one of the small triangles. What
   shape is left?

2 Imagine a regular hexagon. Join adjacent mid-points of
   all of the sides of the hexagon. What shape is formed by
   the new lines?

   Adjacent means next
   to one another.

3 Imagine a rectangle with one corner cut off.
   Describe the shape that you have left.

4 Imagine an equilateral triangle. Fold it along one of its lines of symmetry. What angles are
   in the folded shape? Explain your answer.

5 Imagine a square. Cut off all the corners. What shape have you got left? Is this the only
   possible shape?

*6 Imagine folding a piece of square paper in half and then half again to get a smaller square. One of the corners of this smaller square is the centre of the larger square of paper. Cut a small triangle off this corner. If the paper was opened out, what shape must the hole be? Give a reason.

Review 1   Imagine a regular pentagon. Cut it into two pieces.
What shapes have you got? Are these the only possible shapes?

*Review 2   Imagine a semicircle cut out of paper. Fold it in half along its line of symmetry.
Fold it in half again and then in half again.
How many degrees is the angle at the vertex of the shape now?

 Puzzle

1 Move two matchsticks to different places to make 5 equilateral triangles.

2 Remove two matchsticks to leave eight triangles.

*3 Move eight matchsticks to different places to make a shape with just three squares.

Properties of triangles, quadrilaterals and polygons

A **polygon** is a 2-dimensional shape made from line segments which enclose a region.
2-D is short for **two-dimensional**.
2-D shapes have length and width but no height.
We sometimes call 2-D shapes **plane** shapes. A plane is a flat surface.

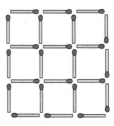
closed region

Discussion

- Jasmine was told that a polygon was a shape with three or more straight sides.
Is this correct? **Discuss**.
Jasmine drew these shapes. Are they all polygons? **Discuss**.

- Jasmine wrote her initials **JT** using polygons.
Can all the letters of the alphabet be written using polygons? **Discuss**.

Investigations

T

A Symmetry

You will need a large copy of each of these shapes.

isosceles triangle

equilateral triangle

right-angled triangle

scalene triangle

regular pentagon

regular hexagon

regular octagon

quadrilateral

square

rectangle

parallelogram

rhombus

trapezium

kite

arrowhead or delta

You may not have seen this one before.

If we fold a rectangle along one of the dotted lines, one half will fit exactly on top of the other.
The dotted lines are lines of symmetry.
How many lines of symmetry do each of the given shapes have?
Test by folding or using a mirror.
Write the number of lines of symmetry on your shape.

Try folding along any diagonals.

B Triangles

You will need a 3 × 3 pinboard and some square dotty paper.
There are eight *different* triangles you can make on a 3 × 3 pinboard.
Make all eight and draw them on square dotty paper.
Write down all the properties you know about each.

is the same triangle as

and

and

Example   Richard made this triangle.
He wrote

*This is an isosceles triangle.
It has two equal sides and
its base angles are equal.
It has one line of symmetry.*

These are properties of an isosceles triangle.

Put the eight triangles into groups.
You could put all the isosceles triangles in one group.
Is there more than one way to group the triangles?

C Triangles and Quadrilaterals

You will need some larger copies of this diagram.
A, B, C, D, E and F are equally spaced around the circumference of the circle.
G is the centre of the circle.
By joining three or four of the points A, B, C, D, E, F and G, how many different triangles and quadrilaterals can you find?
Investigate.
Write down the name of each shape you find.
What can you find out about the angles of these shapes?

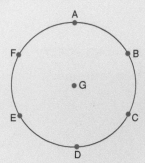

Example   Marcus joined the points A, F, D and B
He wrote.
AFDB is a kite.
It has two equal angles, ∠AFD and ∠ABD.

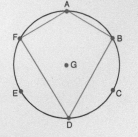

How do you think Marcus found out that these angles were equal?

Properties of Triangles

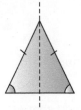

right-angled
one angle is a right angle

isosceles
2 equal sides
base angles equal
1 line of symmetry

equilateral
3 equal sides
3 equal angles
3 lines of symmetry

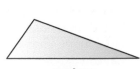

scalene
no 2 sides are equal
no 2 angles are equal

Properties of Quadrilaterals

A quadrilateral has four sides.
These are the **special quadrilaterals**.

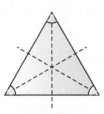 quadrilateral

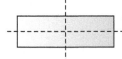

square
4 equal sides
4 right angles
4 lines of symmetry

rectangle
2 pairs of opposite sides equal
4 right angles
2 lines of symmetry

parallelogram
opposite sides equal and parallel

rhombus
a parallelogram with 4 equal sides
2 lines of symmetry
opposite angles equal

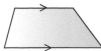

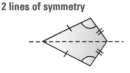

Adjacent sides are next to each other.

trapezium
1 pair of opposite
sides parallel

kite
2 pairs of adjacent sides equal
1 line of symmetry
1 pair of equal angles

arrowhead or **delta**
2 pairs of adjacent sides equal
1 reflex angle
1 line of symmetry
1 pair of equal angles

A shape which has no reflex angles is a **convex shape**.
A shape which has one or more reflex angles is a **concave** shape.

convex shape   concave shape

We can use the properties of 2-D shapes to find unknown angles.

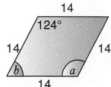

Worked Example
Find a and b.

Answer
This shape is a rhombus.
The diagonals are lines of symmetry.
By symmetry $a = \textbf{124°}$.

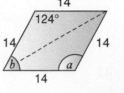

This diagonal cuts angle a and the 124° angle in half.
The blue triangle is isosceles.
$$62° + 62° + b = 180°$$
$$124° + b = 180°$$
$$b = 180° - 124°$$
$$= \textbf{56°}$$

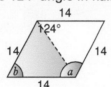

Half of 124° = 62°

Exercise 3

1 Write true or false for each of these.
  a A parallelogram has opposite sides equal and parallel.
  b A kite has two pairs of opposite sides parallel.
  c A trapezium has two pairs of parallel sides.
  d An equilateral triangle has three equal sides and three equal angles.
  e A rhombus has no equal angles.
  f An arrowhead has two pairs of adjacent sides equal.
  g An arrowhead is a concave shape.
  h A rectangle and a rhombus both have two lines of symmetry.
  i A square has only two lines of symmetry.

2 What shape am I?
  a I am a quadrilateral.
     I have only two lines of symmetry.
     I have four right angles.

  b I have three sides.
     I have three lines of symmetry.

  c I am a quadrilateral.
     I have four equal sides.
     I have only two lines of symmetry.

  d I am a quadrilateral.
     My opposite sides are equal and parallel.
     I have no lines of symmetry

  e I am a triangle.
     I have only one line of symmetry.

  f I am a quadrilateral.
     I have two pairs of adjacent sides equal.
     I am a concave shape and I have
     one line of symmetry.

3 Find the angles named with letters.
Use the properties of the shapes to help.

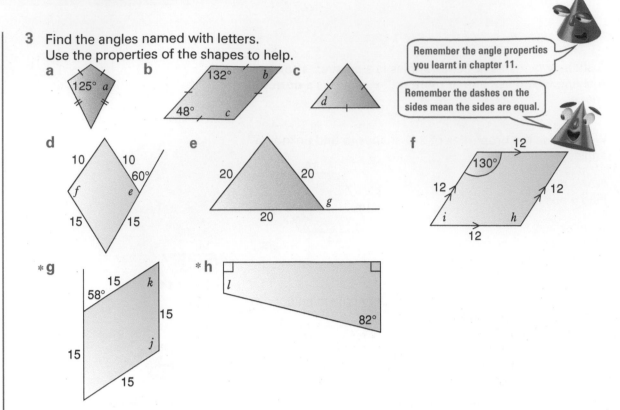

Remember the angle properties you learnt in chapter 11.

Remember the dashes on the sides mean the sides are equal.

*4 Explain why a triangle can never have a reflex angle but a quadrilateral can.

*5 Explain why it is always possible to make an isosceles triangle from two identical right-angled triangles.

*6 Morgan made this shape with three identical yellow and three identical blue tiles.
Each tile is in the shape of a rhombus.
Find the sizes of the angles in the yellow and blue tiles.

65°

T

*7 Use a larger copy of this table.
This trapezium is symmetrical.
It has two pairs of equal angles and one pair of parallel sides.
It is filled in on the table as shown.
   i Fill these in on the table.
   a square   **b** rectangle   **c** kite
   d rhombus   **e** arrowhead (delta)
   f non-symmetrical trapezium
   ii Some of the boxes are still empty. Is it possible to draw quadrilaterals for these? If not, explain why. Use the properties you know about quadrilaterals in your explanation.

		Number of pairs of equal angles	
	0	1	2
Number of 0			
pairs of 1			symmetrical trapezium
parallel sides 2			

A box could have more than one shape in it.

Review 1   Choose the best name from the box for each of the shapes on the next page.

A trapezium	**B** parallelogram	**C** rhombus	**D** arrowhead (delta)
E kite	**F** equilateral triangle	**G** scalene triangle	
H isosceles triangle	**I** rectangle	**J** right-angled triangle	

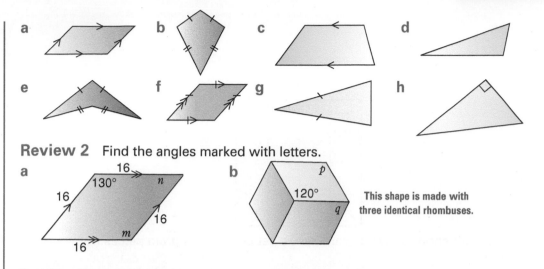

Review 2 Find the angles marked with letters.

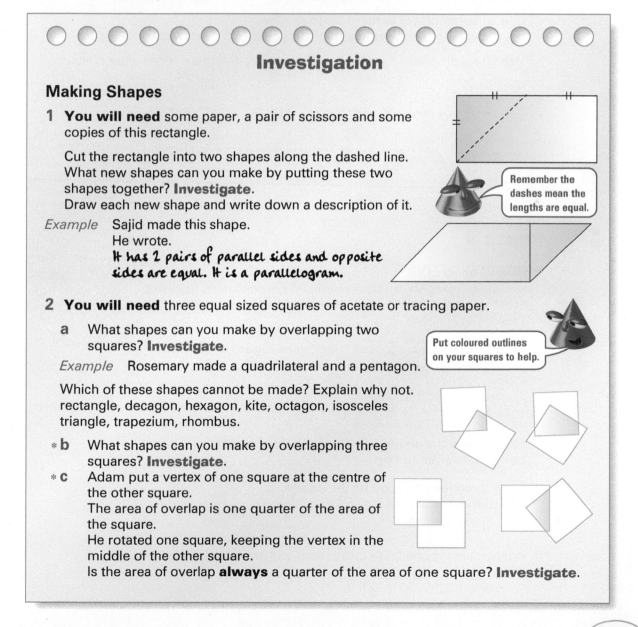

a

b

This shape is made with three identical rhombuses.

*Review 3 Is it possible to draw a quadrilateral with two pairs of parallel sides and one pair of equal sides? If so, draw it. If not, explain why not.

Investigation

Making Shapes

1 **You will need** some paper, a pair of scissors and some copies of this rectangle.

Cut the rectangle into two shapes along the dashed line. What new shapes can you make by putting these two shapes together? **Investigate**.
Draw each new shape and write down a description of it.

Remember the dashes mean the lengths are equal.

Example Sajid made this shape.
He wrote.
It has 2 pairs of parallel sides and opposite sides are equal. It is a parallelogram.

2 **You will need** three equal sized squares of acetate or tracing paper.

a What shapes can you make by overlapping two squares? **Investigate**.

Put coloured outlines on your squares to help.

Example Rosemary made a quadrilateral and a pentagon.

Which of these shapes cannot be made? Explain why not.
rectangle, decagon, hexagon, kite, octagon, isosceles triangle, trapezium, rhombus.

*b What shapes can you make by overlapping three squares? **Investigate**.

*c Adam put a vertex of one square at the centre of the other square.
The area of overlap is one quarter of the area of the square.
He rotated one square, keeping the vertex in the middle of the other square.
Is the area of overlap **always** a quarter of the area of one square? **Investigate**.

✳ Puzzle

You will need squared paper or a set of tangram pieces.

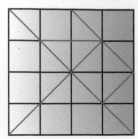

This diagram shows how you can make a set of seven tangram pieces from squared paper.

1 Make a triangle using   **a** two pieces   **b** three pieces   **c** four pieces
   d five pieces   **e** all seven pieces.

2 Use all seven pieces to make these shapes.
   a rectangle   a parallelogram   a trapezium   an irregular hexagon

3 Use all seven pieces to make these digits. Try and make some other digits.

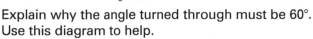
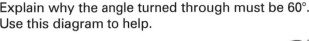

 T

⭐ Practical

You will need LOGO.

1 A regular hexagon can be divided into six equilateral triangles. From this we can work out that each red shaded angle is 120°. Explain how.

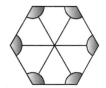

2 These instructions will draw a regular hexagon using LOGO.

　　　　forward　　　　60°
　　　　　↓　　　　　　↓
　　repeat 6 [fd 100 rt 60]
　　　　　　　　↑
　　　　　right turn

Explain why the angle turned through must be 60°.
Use this diagram to help.

3 Draw a regular pentagon on the screen.
   Part of the instructions are

　　　repeat 5 [fd 100 rt ____]

 Fill in the angle.

4 Predict what shape these instructions will draw, then check to see if you are right.
　　　repeat 3 [fd 100 rt 120]

5 Draw a regular octagon and a regular decagon on the screen.

6 Liam McBride's initials are LM.
Draw these initials on the screen.
Draw some other letters on the screen.

*7 Draw a parallelogram.

*8 Draw these shapes.

In all of these diagrams the equilateral triangles are equal distances apart.

*9 These instructions draw five octagons arranged in a circle.

repeat 5 [rt 72 repeat 8 [fd 50 rt 45]]

Draw some other designs.

○ ○ ○ ○ ○ ○ ○ ○ ○ ○ ○ ○ ○ ○ ○ ○ ○ ○ ○

Investigation

Equilateral Triangles

You will need some triangle dotty paper.

1 An equilateral triangle can be drawn using triangle dotty paper.
We can make other shapes by joining **identical** equilateral triangles together along one side.
Here are two shapes drawn by joining four equilateral triangles.
How many **different** shapes can you make by joining four **identical** equilateral triangles? Draw them on dotty paper.

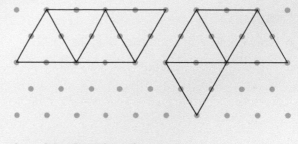

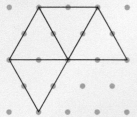

is the same
shape as

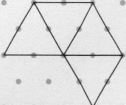

2 How many *different* shapes can you make by joining five identical equilateral triangles? **Investigate**.

*3 There are 12 different ways of joining six identical equilateral triangles together.
Draw them all on spotty paper. Cut them out.

These 12 different shapes can be put together to make a parallelogram.
Investigate how.

289

Constructing triangles

We can **construct triangles** using a **ruler** and **protractor**.

The diagram shows a triangle with the lengths of two sides and the included angle given. To construct this, follow these steps.

Included angle means the angle between the two sides.

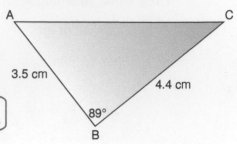

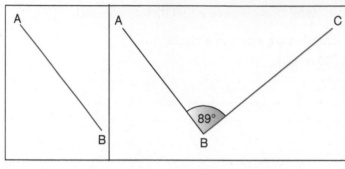

Draw AB, 3·5 cm long.   Draw an angle of 89° at B. Use your protractor.   Draw BC, 4·4 cm long. Join A to C.

To get an **accurate construction**, you must measure and draw very carefully.

Discussion

● How could you construct triangle XYZ? **Discuss**.

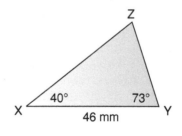

In this exercise, measure and draw very carefully.

1 Use your ruler and protractor to construct the sketched triangles.
   Check your triangle by measuring the size of the third angle.

a

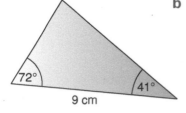

b

c

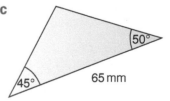

2 Construct the triangle ABC. Measure the size of the third angle in your triangle.
   a CB = 74 mm, angle C = 60°, angle B = 50°
   b AB = 68 mm, angle A = 72°, angle B = 39°
   c BC = 9·2 cm, angle B = 62°, angle C = 42°

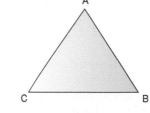

3 Construct the triangle PQR. On your triangle measure the length of the third side.
   a PQ = 5 cm, PR = 8 cm, ∠P = 54°
   b QR = 63 mm, PQ = 55 mm, ∠Q = 70°
   c PR = 8·1 cm, PQ = 6·4 cm, ∠P = 62°

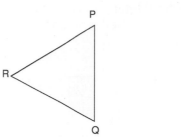

4 Here is a plan of a pathway between three places.
   Construct the triangle. Draw the lines so that 1 cm represents 1 m.
   On your triangle, measure the length of the third side.
   What is the distance between the gate and the bus shelter?

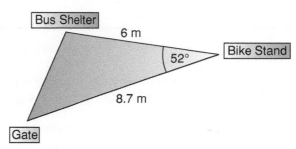

*****5** Construct a rhombus with side 5 cm and one angle 44°.

Review 1 Use your ruler and protractor to construct the sketched triangles.
On your triangle measure the size of angle B.

a
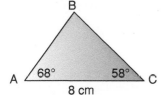

b
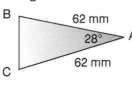

c
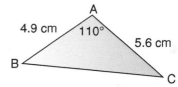

Review 2 Construct the triangle FGH.
Measure the size of angle H.
a FG = 82 mm, FH = 70 mm, ∠F = 64°
b FG = 5·1 cm, FH = 7 cm, ∠F = 39°
c FG = 9 cm, ∠F = 54°, ∠G = 75°

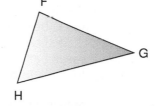

Describing and sketching 3-D shapes

Remember
3-D stands for **three-dimensional**.
3-D shapes have length, width and height.

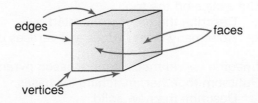

Practical

You will need some Multilink or centimetre cubes and a partner.

1 Sit back to back with your partner.
Without letting your partner see, make a model using six cubes at most.
Example Beth and Tony made these.

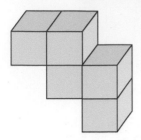 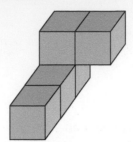

Take turns to tell your partner how to make the model.

2 **You will need** a bag or box that you can put your hands inside and a group to work with.
One pupil makes a model from cubes and puts it in the bag or box.
Take turns to put your hands inside the bag or box and describe the shape to everyone else.
Everyone else tries to make the shape.

Exercise 5

1 Imagine painting a dot on each face of three cubes.
   a Imagine gluing the three cubes together in a row. How many dots would be showing altogether?
   b Imagine gluing the three cubes together in an L-shape. How many dots would be showing altogether?

You are allowed to turn the shape around.

2 Imagine you have two identical cubes.
Put them together, matching face to face.
   a Name and describe the new solid.
   b How many faces, edges and vertices does it have?

3 Imagine a pyramid with a square base.
Cut a slice off the top.
   a Describe the new face created.
   b How many faces, edges and vertices does the new solid have?

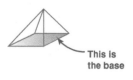
This is the base

4 Imagine two identical square-based pyramids.
Put them together, matching the bases.
   a Describe the new solid.
   b How many faces, edges, vertices does the new solid have?

5 i Imagine you are standing in front of a large square-based pyramid.
A genie gives you a magic potion and you shrink to the size of a mouse.
  a If you walk around the pyramid, what is the maximum number of faces you can see at one time?
  b The genie now gives you wings to fly. You fly above the pyramid. What is the maximum number of faces you can see at one time?
* ii Imagine you are standing in front of a large cube that comes up to your waist.
Is it possible for you to stand so that you can see just
  a one face   **b** two faces   **c** three faces   **d** four faces?
Sketch what you would see each time.

Review 1 Imagine a cube. Cut off one corner as shown.
a Describe the new face created.
b How many faces, edges and vertices does the new solid have?

Review 2 Imagine you are standing in front of a large tetrahedron as tall as a house.
i Is it possible to stand so you can see **a** just 1 face   **b** 2 faces   **c** 3 faces?
ii If you flew above the tetrahedron, what is the maximum number of faces you can see at one time?

> Remember a tetrahedron is a pyramid with a triangular base.

* Review 3 How could you cut a cube to create an hexagonal face?

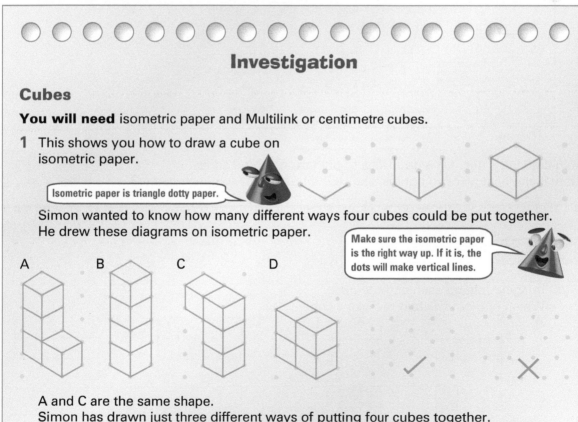

Investigation

Cubes

You will need isometric paper and Multilink or centimetre cubes.

1 This shows you how to draw a cube on isometric paper.

> Isometric paper is triangle dotty paper.

Simon wanted to know how many different ways four cubes could be put together.
He drew these diagrams on isometric paper.

> Make sure the isometric paper is the right way up. If it is, the dots will make vertical lines.

A   B   C   D

A and C are the same shape.
Simon has drawn just three different ways of putting four cubes together.
How many different ways are there of putting four cubes together? **Investigate.**
Draw them on isometric paper.
* How many different ways are there of fitting five cubes together? **Investigate.**
Draw them on isometric paper.

Continued …

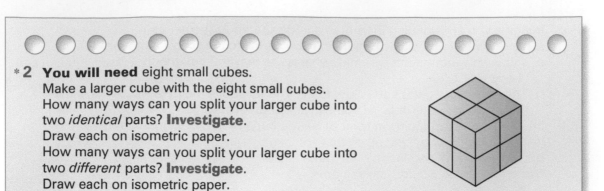

*2 **You will need** eight small cubes.
   Make a larger cube with the eight small cubes.
   How many ways can you split your larger cube into
   two *identical* parts? **Investigate**.
   Draw each on isometric paper.
   How many ways can you split your larger cube into
   two *different* parts? **Investigate**.
   Draw each on isometric paper.

Nets

A 2-D shape that can be folded to make a 3-D shape is called a **net**.
This net folds to make a cube.
The fold lines are dashed.

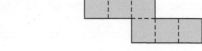

Example   This net folds to make a cuboid.

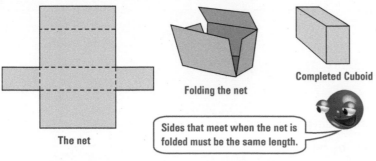

The net

Folding the net

Completed Cuboid

Sides that meet when the net is
folded must be the same length.

When drawing a net, it is a good idea to put tabs on some edges.
Where is the best place to put the tabs so the box is easy to glue together?

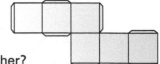

Exercise 6

[T]

1 a Imagine folding each of these nets.
      Which of them will fold to make a cube?

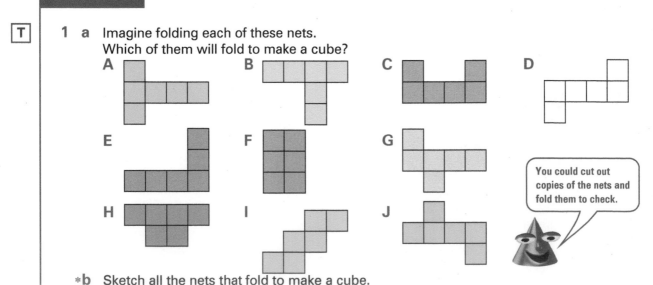

You could cut out
copies of the nets and
fold them to check.

   *b Sketch all the nets that fold to make a cube.

2 Which of the nets below will make this box?

A

B

1cm
2cm
3cm

1cm
1cm

1cm
1cm

3 a Jasmine drew this net for a cube.
She wanted her cube to have a red and a blue strip along the top edges like this.

Use a copy of the net.
Colour a red strip on the other edge that should be red.

b Jasmine drew a different net for a cube.
When the net is folded, which colour will be opposite the green face?

c Copy the net.
Put a dot on the two corners that will meet the corner with the dot on it.

4 Imagine folding these nets.
Which ones will fold to make cuboids?

A   **B**   **C**   **D**   **E**

F

You could cut out copies
and fold them to check.

5 You will need squared paper, a ruler and set square.

a Bobby wanted to make this box to display his cycling medal.
He started the net below.
Use a copy of this.
Use your ruler and set square to finish it.
Put tabs on some of the edges.
Remember you can cut unneeded tabs off but you can't glue them back on.

3cm

4cm

6cm

3cm

6cm   4cm

You might like to
decorate your net.

b Fold the net to make the box.

295

6 Use a ruler and set square to draw nets for these cuboids.
  a length 5 cm, width 4 cm, height 2 cm   **b** length 4 cm, width 2 cm, height 3 cm
  c length 6 cm, width 5 cm, height 2 cm   **d** length 4 cm, width 3 cm, height 1 cm
  Fold your nets to make the cuboids.

7 Sketch the nets for these shapes.
  a   **b**

Remember to put tabs on.

[T]

*8 Joanne started the net below for a regular tetrahedron.
Use two copies of this.
Use a ruler and protractor to finish the net in two different ways.

3.5 cm

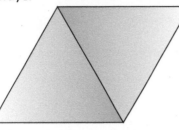

[T]

*9 Robyn made a pyramid-shaped box for an easter egg.
Part of her net is given below.
Use a copy of this.
Finish the diagram for the net accurately.

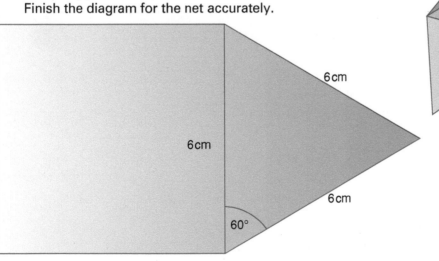

6 cm

6 cm

6 cm

6 cm

6 cm

60°

60°

*10 David made some chocolates for his Aunt's birthday.
He designed a box with triangular ends.
This sketch shows the triangular end.

Construct this triangle.
Using your triangle as part of the net for the box, finish the net accurately.
Fold it to make the box.

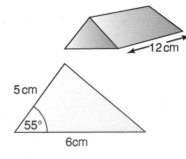

12 cm

5 cm

55°

6 cm

This shape is a *prism*.
It has two identical ends.

*11 Draw a net for each of these.

a

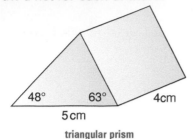

48°   63°   4cm
5cm

triangular prism

b

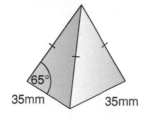

65°
35mm   35mm

square-based pyramid

Review 1   Will this net fold to make a cube?

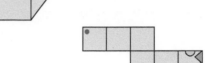

Review 2   Cameron wanted to make this cube.
He drew the net shown.

Use a copy of the net.
a   Draw the other half of the diamonds and the
    circle on your net.
b   Put a dot on the other two corners that
    will meet the corner with the red dot in it.

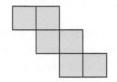

Review 3   Use a ruler and set square to draw nets for these cuboids.

a   length 8 cm, width 3 cm, height 2 cm
b   length 50 mm, width 40 mm, height 25 mm

Fold your nets to make the cuboids.

Review 4   Asha made a box for her little sister's front tooth.
It has a triangular end like the one shown.
Draw a net for this box.

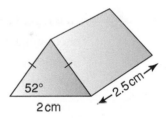

52°   2.5cm
2cm

Practical

1 **You will need** a ruler, set square, protractor and some sheets of paper.
  The number of dots on opposite faces of a dice always add to 7.
  This net can be used to make a dice.

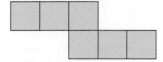

Remember to put
tabs on your net.

Construct this net.
Draw the dots on the net, then make the dice.
Make other nets which fold to make a dice. Put the dots on the faces before
you fold your nets.

T

2 **You will need** card or thick paper.
How products are packaged is important.
Things like chocolates, easter eggs and perfume sell better in nice boxes.

Design a nice box.
Draw the net for the box and then decorate it.
Fold the net to make the box.
Your box does not need to be a cube or cuboid.

3 **You will need** an A4 piece of card or paper.
Using just one piece of A4 card or paper, draw a net for
a basket to hold sweets.
Decorate your net.
Fold the net to make the basket.
Remember to keep a strip of paper for the handle.

Summary of key points

 A **triangle is named** using the capital letters at the vertices.
This triangle could be named as △STR or △SRT or △RST or ...
The side opposite each vertex is named with the lower-case
letter of the vertex.

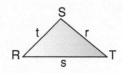

Start with one letter and go
round in order ⤵ or ⤶.

 A polygon is a 2-D shape made from line segments enclosing a region.
2-D is short for two-dimensional.
Properties of triangles
A triangle is a 3-sided polygon.

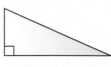

right-angled
one angle is a right angle

isosceles
2 equal sides
base angles equal
1 line of symmetry

equilateral
3 equal sides
3 equal angles
3 lines of symmetry

scalene
no 2 sides are equal
no 2 angles are equal

Properties of Quadrilaterals

A quadrilateral is a 4-sided polygon.

These are the **special quadrilaterals**.

  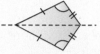 

square
4 equal sides
4 right angles
4 lines of symmetry

rectangle
2 pairs of opposite sides equal
4 right angles
2 lines of symmetry

parallelogram
opposite sides equal
and parallel

rhombus
a parallelogram with
4 equal sides
2 lines of symmetry
Opposite angles equal

trapezium
1 pair of opposite
sides parallel

kite
2 pairs of adjacent
sides equal
1 line of symmetry
1 pair of equal angles

arrowhead or **delta**
2 pairs of adjacent
sides equal
1 reflex angle
1 line of symmetry
1 pair of equal angles

Adjacent sides are
next to each other.

A **concave** shape has at least one reflex angle.

A **convex** shape has no reflex angles.

convex   concave

 We can **construct triangles** using a ruler and protractor.

Example   To construct this triangle
draw PR 2·6 cm long
draw an angle of 85° at R
draw RQ 2·8 cm long
join P to Q.

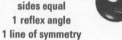

 3-D shapes

3-D stands for three-dimensional. 3-D shapes have length, width and height.

3-D shapes can be drawn on triangle dotty paper.

 A 2-D shape that can be folded to make a 3-D shape is called a **net**.

Example   The net below folds to make this cuboid.

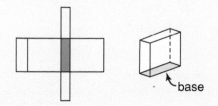
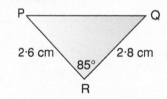
base

Shape, Space and Measures

Test yourself

1 Name these triangles.
a   Q   b   B   c   L

   P   R   A   C   M   N

2 Using a lower-case letter, name the red side of each triangle in question **1**.

3 Write true or false for these.
   a   A kite has two equal angles and one line of symmetry.
   b   A rhombus has two pair of equal angles and two lines of symmetry.
   c   A parallelogram has two lines of symmetry.

4 How many lines of symmetry do these have?
   a   an equilateral triangle   b   a square   c   a rhombus
   d   a parallelogram   e   a kite   f   an arrowhead

5 What shape am I?
   a   I am a triangle.   b   I am a convex quadrilateral.
      I have three lines of symmetry.   I have two lines of symmetry.
                              I have no right angles.

6 Find the value of the angles marked with letters.
a   50°   P   b   40°   8   r   8   q   8   c   126°   s   t

7 Each blue tile is a rhombus.
Each purple tile is an isosceles triangle.
Find the size of angles x, y and z.   50°   x   y   z

8 Is it possible to draw a triangle with only two lines of symmetry?
Explain your answer.

9 Imagine joining adjacent mid-points of the sides of a square.
What shape is made by the new lines?

10 Imagine an equilateral triangle. Fold along a line of symmetry.
What angles will be in the folded shape? Explain why.

300

11 Construct these triangles using a ruler and protractor. Measure the size of angle B on your triangles.

a

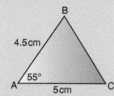

b

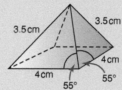

12 Imagine you are standing in front of a square-based pyramid which has been tipped onto its side.
If you could fly above this pyramid, what is the maximum number of faces you could see at one time?

13 Perry wanted to make this cube. He drew the net shown.
a Draw the other part of the pattern on the net.
b Put a dot on the other two corners that will meet the corner with the red dot on it.

14 Use a ruler and set square to draw a net for these.

a

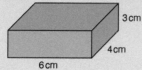

b

c

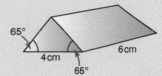

You need to know

✓ coordinates

✓ symmetry

✓ transformations

Key vocabulary

axes, axis, axis of symmetry, centre of rotation, congruent, coordinates, direction, grid, image, intersect, line of symmetry, line symmetry, mirror line, object, origin, order of rotation symmetry, position, quadrant, reflect, reflection, reflection symmetry, rotate, rotation, rotation symmetry, symmetrical, transformation, translate, translation, x-axis, x-coordinate, y-axis, y-coordinate

 ## That Sinking Feeling

Sink the Ships — a game for 2 players

You will need a grid like this for each player.

To Play

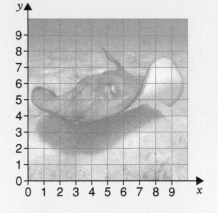

- Each player puts these ships on the grid.
    - 1 submarine (1 dot)
    - 2 destroyers (2 dots each)
    - 1 cruiser (3 dots)
    - 1 battleship (4 dots).
    The ships must not cross.

- Players take turns to name a coordinate.
  If one of the other player's dots is at this
  coordinate this player must say 'hit' and cross out
  that dot.

Example

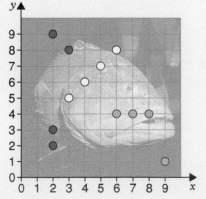

- Once all the dots of a ship have been 'hit' call out
  'ship sunk'.

- The winner is the first person to sink all of the
  other players ships.

302

Coordinates

We use **coordinates** to give the **position** of a point or place on a grid. You may have used coordinates in geography.

The coordinates of the lighthouse are ($^-$1, $^-$2).
$^-$1 is the x-coordinate.
$^-$2 is the y-coordinate.
We always give the x coordinate first. (x, y) is in alphabetical order.
The coordinates of the origin are (0, 0).
The origin is where the x-axis and y-axis **intersect**.

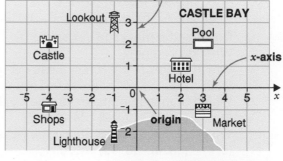

The x-axis and the y-axis make four **quadrants** as shown.

Examples　The point P(3, 1) is in the first quadrant.
　　　　　The point Q($^-$3, 2) is in the second quadrant.
　　　　　The point R($^-$2, $^-$1) is in the third quadrant.
　　　　　The point S(4, $^-$3) is in the fourth quadrant.

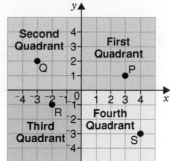

We name the quadrants in an anticlockwise order.

Exercise 1

1　This map shows Pleasure Island.
　a　Write down the coordinates of
　　i　Halfway Bush
　　ii　Lakeland
　　iii　Tyson Track
　　iv　Shopping Centre
　　v　Marshland
　　vi　Waterfall
　b　Anne's favourite place on the island has coordinates (4, $^-$1).
　　What is Anne's favourite place?
　c　Dave often goes to the place with coordinates ($^-$2, 1).
　　Where is this?
　d　What is at each of these coordinates?
　　i　($^-$4, 2)　　　ii　(3, 4)　　　iii　(4, 3)
　　iv　($^-$2, 4)　　　v　(0, 2)　　　*vi　(5, 1·5)

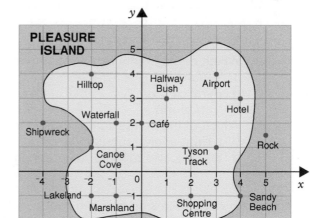

T

2 a Write down the letters at each of the following
coordinates. You will get 4 words.

1. (⁻4, 0) (1, 1) (⁻4, ⁻6) (⁻2, ⁻2)
2. (4, 5) (0, ⁻6) (2, ⁻6)
3. (⁻3, ⁻1) (2, ⁻6) (4, 5) (4, 5) (2, ⁻6) (3, 3) (1, 5)
4. (7, 5) (⁻3, ⁻1) (2, 2) (0, ⁻6) (7, 5) (7, ⁻1) (2, ⁻6)
   (4, 5) (⁻4, ⁻6) (5, 1) (7, 5) (⁻3, ⁻1) (⁻3, ⁻1) (6, 5)

b Do what the message in **a** says.
What do you get?

c Which quadrant is each of these in?
(4, 3)   (⁻3, ⁻4)   (⁻3, 2)   (2, ⁻1)

d Which of these points are on the *x*-axis?
(0, 4)   (⁻3, 0)   (0, ⁻2)   (4, 0)

e Which of these points are on the *y*-axis?
(0, 4)   (⁻3, 0)   (0, ⁻2)   (5, 0)

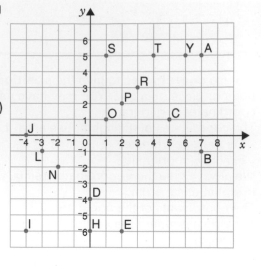

3 Draw *x* and *y* axes on squared paper.
Number the *x*-axis from ⁻8 to 8 and the *y*-axis from ⁻6 to 10.

a These are the coordinates of the vertices of a shape.
Plot them on your grid.
Join them in this order.
(⁻1, 2), (5, 2), (6, 0), (2, ⁻1), (⁻2, 0), (⁻1, 2)

b How many lines of symmetry does this shape have?

c The points (⁻4, 0), (1, 0), (1, ⁻3) are three of the vertices of a rectangle.
Write down the coordinates of the fourth vertex.

d The points (⁻2, 2) and (3, 2) are two of the four vertices of a rectangle.
What might the coordinates of the other two points be?

*4 Draw and label axes on squared paper.
Choose a suitable scale so that you can plot these points.

A (⁻4, ⁻1)   B (4, 2)   C (0, ⁻2)   D (1, 1)
E (4, ⁻1)   F (1, ⁻1)   G (1, 2)   H (⁻2, 1)

a Which of the points could you join to make a
   i square
   ii parallelogram
   iii trapezium?

b Which of the points could you join to make
   i a right-angled triangle
   ii an isosceles triangle without a right angle?

*5 Draw *x* and *y* axes on squared paper.
Number both axes from ⁻4 to 10.

a Plot the points (⁻2, 1), (1, 2), (⁻1, 3).

b What fourth point could you plot to make
   i a parallelogram
   ii a kite
   iii an arrow head?

c Is it possible to make a rectangle that is not a square?
Explain your answer.

T

Review 1 The places shown on this graph are the places that 5-year-old Jake often visits.

a In which quadrant is the sweet shop?

b Write down the coordinates of these
school   post office   bus station   granny

c One Sunday morning, while his parents were still asleep, Jake decided to go out. From his home he walked to (⁻1, 2) then (1, 0) then (2, 1) then (2, ⁻2) then (⁻1, ⁻2) and then (0, ⁻1) and then back home. Write down the names of all the places, in the order that he visited them.

d Use a copy of the graph. Beginning at his home, join each point on Jake's walk with a straight line.

e The diagram for **d** should tell you which place Jake likes best. Which place is this?

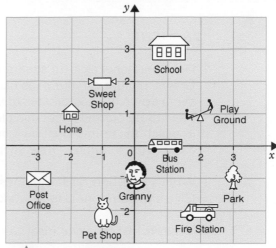

What does the arrow point to?

Review 2 Draw and label axes on squared paper. Have the x-axis from ⁻7 to 5 and the y-axis from ⁻10 to 5.

a Plot the points (⁻2, 3), (⁻1, ⁻2), (⁻6, ⁻3).
What fourth point could you plot to make a square?
Is it possible to make a rectangle that is not a square? Explain why or why not.

b Plot the points (⁻4, ⁻3), (2, ⁻1), (⁻2, 1).
What fourth point will make   **i** a kite   **ii** a parallelogram?

T

Four in a Line – a game for 2 players

You will need a copy of this grid,
a blue pen for one player,
a red pen for the other.

To play

- Players take it in turns to give the coordinates of a point and then plot the point.
- The winner is the first player to draw a straight line joining four of his or her points.
- None of the other players' points may lie on this line.
- If a player plots a wrong point for the coordinates they have named, the other player may plot two points at the next turn.

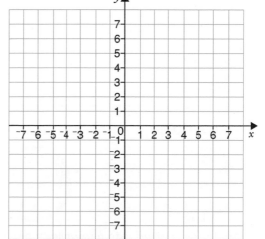

Alternative The winner is the first person to draw a square by joining four of his or her points.

⭐ Practical

1 Draw and label axes on squared paper.
Draw a picture on your axes using only straight lines.
Write down the coordinates of each point that would
need to be plotted to get this picture. Give the order
in which the points must be joined.
Swap with someone. Draw each others picture.

2 Write a question similar to question **2a** of Exercise 1.
Give it to someone else to work out your message.

3 **You will need** a graphical calculator.

a Use the `plot` and `line` on your graphical calculator to draw these.

 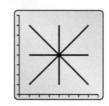 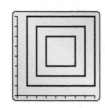

b Fabian Peters put his initials on the screen.
Draw your initials on the screen.

c Plot the points (4, 2) and (7, 2).
These two points are the ends of one side of a trapezium.
Plot two other points to finish the trapezium.
Plot two different trapezia with the points (4, 2) and (7, 2)
as the ends of one side.
Make other shapes which have the points (4, 2) and (7, 2)
as the ends of one side.
What if the points are (4, 2) and (5, 6) instead of (4, 2) and (7, 2)?

Trapezia is the
plural of trapezium.

d Draw this outline of a cube on your screen.

Reflection

Discussion

As Jenny walked into a jeweller's she saw these reflections of three clocks in a mirror.
What time was each of these clocks actually showing?

• The word AMBULANCE is to be written across the front of an ambulance. The driver
of a car in front of the ambulance can see AMBULANCE in the rear mirror.
How should the word AMBULANCE be written on the front of the ambulance? **Discuss**.

When we **reflect** an **object** in an **axis of symmetry** or **mirror line** we get an **image**.
ABCDE has been reflected in the mirror line to get A'B'C'D'E'.
We say ABCDE maps onto A'B'C'D'E'.

To map A'B'C'D'E' onto ABCDE we use the **inverse** reflection.
The inverse reflection is the reflection in the same mirror line.

The image is **congruent** (same shape
and size) to the original.

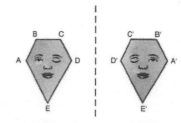

> We name image points
> with a dashed letter.

⭐ Practical

1 You will need a sheet of paper.
Draw a mirror line on your paper.
Draw a triangle on the paper. Label it ABC.
Fold the paper along the mirror line.
Put a pin through the paper at A.
Open the paper and label the new point A'.
Join AA'. Is AA' at right angles to the mirror line?
Repeat for points B and C.

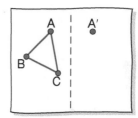

2 You will need a geometry software package like Geometer's Sketchpad.
Draw a line to be a mirror line on your screen.
Draw a shape on one side of the line.
From one vertex of your shape, draw a
line perpendicular to the mirror line.
Extend this line to a point an equal
distance on the other side of the mirror
line. This will be the image point.
Repeat this for the other vertices.
Join the points to give the image.

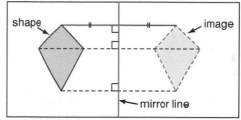

Use the cursor to drag the vertices of the original shape.
What happens to the image?
What happens if one of the vertices of the original shape is on the mirror line?
What happens if the original shape crosses the mirror line?

The line joining any point to its image is perpendicular to the
mirror line.

Example   AA' and CC' are perpendicular to the mirror line.
The image is the same distance behind the mirror line as the
object is in front.
Points **on** the mirror line do not change their position when
reflected.

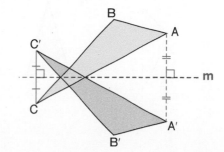

Discussion

Zoe was asked to reflect this shape in the mirror line.
She used tracing paper to find the image.
How might she have done this? **Discuss**.

Exercise 2

T

1 Use a copy of these shapes.
Reflect them in the mirror line, m.

a   m

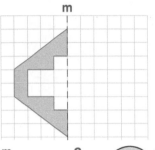

b

c   m

d m   **e**

_m_

You could check your answers using a mirror.

2 ABC is reflected in the dashed line.
    a Which points will not change their position? Why?
    b Draw the reflection of ABC.
       Label it A'B'C'.
    c A'B'C' is reflected in the dashed line.
       What happens?

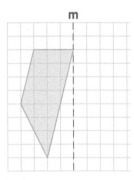

3 Copy these shapes.
Reflect these shapes in the dashed line.
Write down the name of the quadrilateral you get.

a   **b**   **c**   **d**

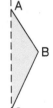

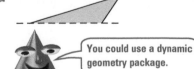

You could use a dynamic geometry package.

4

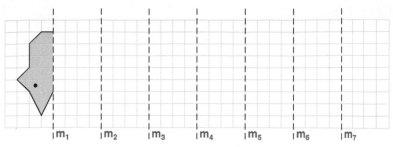

Use a copy of this.
a Reflect the shape in m_1.
b Reflect the image you get in **a** in m_2.
c Continue reflecting the images in m_3, m_4, m_5, m_6 and m_7.
Describe what happens.
d Use another copy of the parallel mirror lines.
Begin with a different shape of your choice.
Does the same thing happen?

5 Use a copy of this.
Reflect these shapes in m.

a 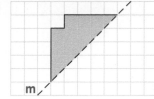 **b** 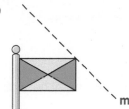 **c**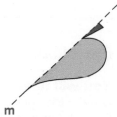

6 Use a copy of this.
Reflect these words in the mirror line.

a  **b**  **c** **d**

7 Explain why A'B'C'D' is not the reflection of ABCD in the mirror line m.

a 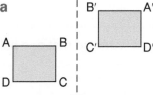 **b** 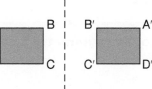 **c**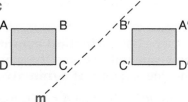

*8

M M
A A
H H

E B C
E B C

M A H
H A M

E Ǝ
B B
C Ɔ

a Some letters look the same after reflection in a vertical mirror line.
What do these letters have in common?
b Some letters look the same after reflection in a horizontal mirror line.
What do these letters have in common?

Shape, Space and Measures

Review 1 Use a copy of these shapes.
Reflect them in the mirror line, m.

a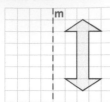

b

c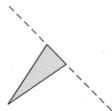

Review 2 Write down the name of the quadrilateral that you got in Review **1c**.

Review 3 Use a copy of these.
Reflect each in the dotted line.

a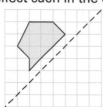

b LANDMINE

c

⭐ Practical

You will need a dynamic geometry software package.
Draw a triangle.
Reflect it in one of it sides.
Name the shape that the image and object together make.
What if you reflect the triangle you drew in one of its
other sides?
Try and make these shapes by reflecting a triangle in
one of its sides. Which ones can you make?

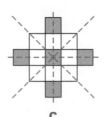
mirror
line

*The image and object
together make a kite.*

rectangle	square	kite	parallelogram
arrowhead	rhombus	trapezium	

Reflection symmetry

A shape has **reflection symmetry** or **line symmetry** if a line can be found so that one half of
the shape reflects onto the other half.
The line is a **line of symmetry**.

Examples Shapes A and C have reflection symmetry.

A
I line of symmetry

B
0 lines of symmetry

C
4 lines of symmetry

Which letters of your name have reflection symmetry?

Exercise 3

1 Use a copy of these.
Which of these patterns have reflection symmetry?
Draw the lines of symmetry on these patterns.

a    b 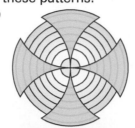   c

d    e 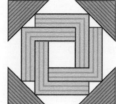   f

2 Which of these shapes have reflection symmetry?
  a equilateral triangle   b isosceles triangle   c rhombus
  d parallelogram   e regular pentagon

3 Imagine this is how a piece of paper looks after it has been folded twice,
each time along a line of symmetry.
What shape might the original piece of paper have been?
Is there more than one answer?

4 Show how you can put these shapes together to make a single shape with reflection
symmetry.
There is more than one way for each. Find all the ways.

a 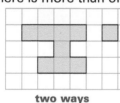   b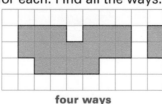
             two ways   **four ways**

*5 Draw a pattern on squared paper that has
  a two lines of symmetry   b four lines of symmetry   c three lines of symmetry.

Review 1   Use a copy of these.
Which of these patterns have reflection symmetry?
Draw the lines of symmetry on these patterns.

a 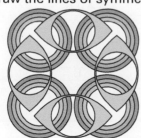   b    c

T

T

Review 2 Use a copy of these. Show how you can put these shapes together to make a single shape with reflection symmetry. Find all the ways.

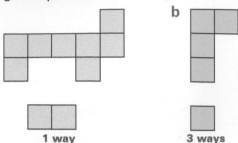

a

b

1 way

3 ways

Rotation

Jordan went to the fair.
He went on two rides where he was **rotated**.
A **rotation** is a movement round a centre point.

$\frac{1}{4}$ turn = 90° $\frac{1}{2}$ turn = 180° $\frac{3}{4}$ turn = 270°

In both of the diagrams the purple shape has been **rotated** 90° anticlockwise about O.
A maps to its image point A'.
O is called the **centre of rotation**.
The centre of rotation can be inside the shape or outside the shape.
The image is congruent to the original.

The centre of rotation stays in the same place as you rotate a shape.

To rotate a shape you need to know **the centre of rotation** and the **angle of rotation**.
If the direction is not given, the angle of rotation is always in an **anticlockwise** direction.

Worked Example
Rotate ABC 90°
a about (0, 0)
b about (1, 1).

Answer
a We can using tracing paper.
  1 Trace ABC onto tracing paper.
  2 Put a pin or sharp pencil through the tracing paper at (0, 0).
  3 Turn the tracing paper 90° anticlockwise.
  4 Press hard on the tracing paper to give the outline of A'B'C'.
  5 Remove the tracing paper. Draw A'B'C'.
b Use (1, 1) instead of (0, 0) as the centre of rotation.
  Follow the same steps as in **a**.

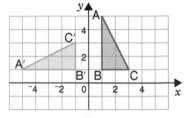

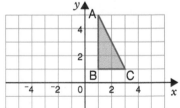

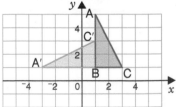

Exercise 4

1 Use a copy of these.
   Rotate these shapes about (0, 0). The angle of the rotation is given.

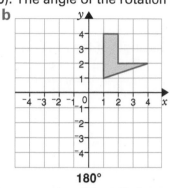

a

90°

b

180°

c

270°

2 Use another copy of question **1**.
   a Rotate the shape in **a** through 180° about the point (2, 0).
   b Rotate the shape in **b** through 90° about the point (1, 1).
   c Rotate the shape in **c** through 270° about the point (0, ⁻1).

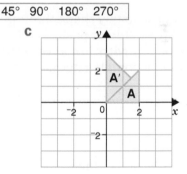

Remember if no direction is given, rotate anticlockwise

3 Use another copy of question **1**
   a Rotate the shape in **a** through 270° using (⁻1, ⁻1) as the centre of rotation.
   b Rotate the shape in **b** through 180° using (1, 0) as the centre of rotation.
   * **c** Rotate the shape in **c** through 90° using (2, 2) as the centre of rotation.

4 Shape **A** has been rotated about the origin to shape **A**′.
   Write down the angle of rotation.
   Choose from the box.

   45° 90° 180° 270°

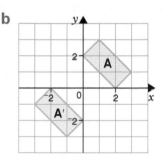

a

b

c

5 A rotation through 90° about (0, 0) maps **A** onto **A**′.
   Which of these rotations maps **A**′ onto **A**? There
   are two correct answers.
   a rotation through 180° about (0, 0)
   b rotation through 90° about (0, 0)
   c rotation through 270° about (0, 0)
   d rotation through 90° clockwise about (0, 0)

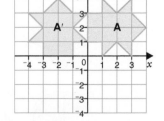

6 A rotation through 270° about (1, 2) maps B onto B′.
   Which of these maps B′ onto B?
   There are two correct answers.
   a rotation through 90° about (0, 0)   **c** rotation through 270° clockwise about (1, 2)
   b rotation through 90° about (1, 2)   **d** rotation through 180° about (1, 2)

*7 Copy this using just one of the words in red in each gap.
The inverse of a rotation is either an **equal**/**unequal** _____ rotation about **the same/
a different** _____ point in the **same/opposite** _____ direction,
or a rotation about **the same/a different** _____ point in the **same/opposite** _____
direction so that the angle of the original rotation and the angle of the inverse rotation
add to **180/360°** _____ .

[T] **Review 1** Use a copy of these.
Rotate these shapes about (0, 0). The angle of the rotation is given.

a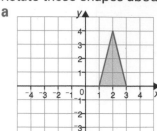

b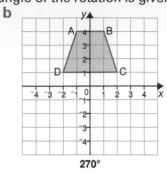

180°   270°

[T] **Review 2** Use another copy of Review **1**.
a Rotate the shape in **a** through 90° about the point (1, ⁻1).
*b Rotate the shape in **b** through 180° about the point (0, 3).

Rotation symmetry

If we rotate this shape through a complete turn, it will fit on to itself
in three different positions. This shape has **rotation symmetry**. A
shape has **rotation symmetry** if it fits onto itself more than once
during a complete turn.

Discussion

- Which of these shapes have rotation symmetry?
  Give reasons. **Discuss**.

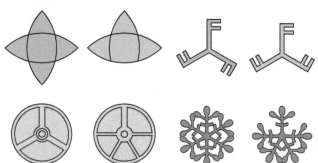

- Flowers and wheels often have rotation symmetry. **Discuss** some examples of these.
  What else has rotation symmetry? **Discuss**.

The **order of rotation symmetry** is the number of times a shape fits exactly onto itself during a rotation of 360°.

All shapes fit onto themselves at least once when they are rotated through 360°. So all shapes have order of rotation symmetry of at least one. Only shapes with rotation symmetry of order 2 or more are said to have rotation symmetry.

Examples   The order of rotation symmetry of these road signs is given.

order 3   order 2   order 4

Exercise 5

1

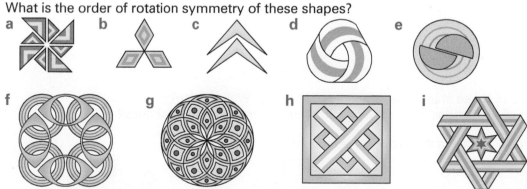

Shape 1   Shape 2   Shape 3   Shape 4

   a   Which of these shapes does *not* have rotation symmetry?
   b   What is the order of rotation symmetry of the other shapes?

2   What is the order of rotation symmetry of these shapes?

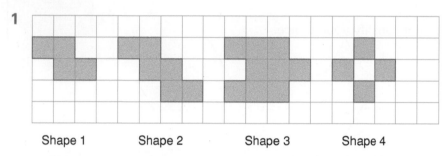

  a square   **b** regular pentagon   **c** kite   **d** rhombus   **e** parallelogram   **f** regular octagon

3   What is the order of rotation symmetry of these shapes?

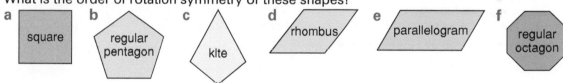

4   In question **3** what is the smallest angle each shape must be turned through to map onto itself?

T

5 a Rochelle put these two shapes together, edge to edge. Her new
shape had just one line of symmetry. Show two ways Rochelle
might have done this.

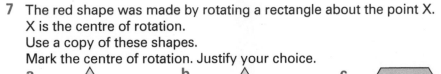

b Show a way of putting the two shapes together edge to edge to
get a new shape with
  i rotation symmetry of order 2 but no lines of symmetry (two ways)
  * **ii** two lines of symmetry *and* rotation symmetry of order 2.

6 Show how you can put these shapes together to make a single shape with
rotation symmetry of the order given.

a

order 2

b

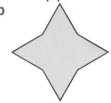

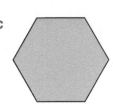

order 2

There are more than
eight ways for b.

T

7 The red shape was made by rotating a rectangle about the point X.
X is the centre of rotation.
Use a copy of these shapes.
Mark the centre of rotation. Justify your choice.

a   **b**   **c**

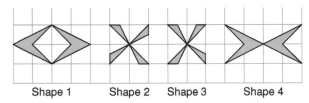

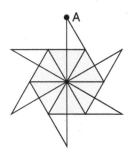

*8 This shape has rotation symmetry of order 6.
As it is rotated through 60° angles about the centre point, it
maps onto itself in six different positions in a full turn.
a Explain how you know that the yellow shape in the middle is
a regular hexagon.
b If a pen was attached to point A, what would it draw as the
whole shape is rotated 360° about its centre?

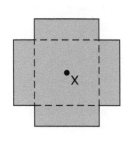

Review 1

Shape 1   Shape 2   Shape 3   Shape 4

a Three of these shapes have rotation symmetry. Which three?
b Two of the shapes have rotation symmetry of order 2. Which shapes?

Review 2 What is the order of rotation symmetry of these?

a   **b**   **c**

*Review 3   This shape has rotation symmetry of order 8.
As it is rotated through 45° angles about the centre point, it maps onto itself in
eight different positions in a full turn.
a   Explain how you know that the blue shape in the middle is a regular octagon.
b   If a pen is attached to point A, what would it draw as the whole shape is
rotated 360° about its centre?

★ **Practical**

You will need Logo or a similar ICT package or a dynamic geometry package.
Make a shape on your screen with rotation symmetry of order

a 2   b 4   c 5   d 6   e 8   f 9

Example   Melanie made this shape by rotating a right-angled
triangle through 45° angles.
Her shape has rotation symmetry of order 8.

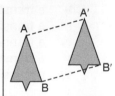

Translation

Sliding movements are called **translations**.
The pieces on a chess board are translated.
Jim moved his knight 2 squares across to the right and
1 square up.
The ringed shape is called the image.

When we **translate** a shape, every point moves the
same distance in the same direction.

Example   A has moved the same distance and in the same
direction to A', as B has moved to B'.
The image is congruent to the original.
The original shape is not turned during translation
(it has the same orientation).

Examples   ABCDE has been translated to A' B' C' D' E'.
The translations are described under
each diagram.

> The inverse of any translation is an
> equal move in the opposite direction.

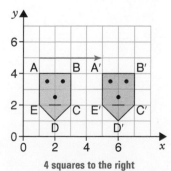

4 squares to the right

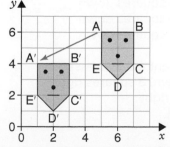

4 squares to the left and 2 squares down

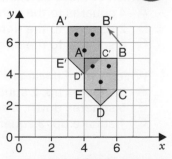

1 square to the left and 2 squares up

Exercise 6

T

1 Use a copy of these.
   Draw the image after these translations.

a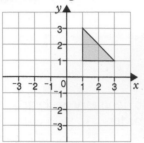

3 units left

b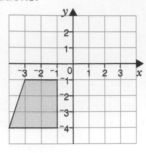

2 units right and 3 units up

c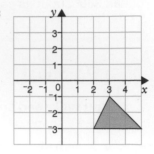

2 units left and 3 units up

d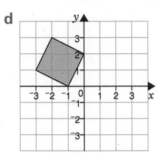

4 units right and 3 units down

e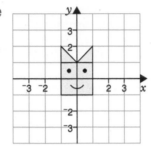

1 unit right and 2 units down

2 ABCD has been translated 5 units to the right and 3 units
   down to A′ B′ C′ D′.
   a What translation would map A′ B′ C′ D′ to ABCD?
   b What translation is the inverse of these?
      i   a translation 5 units to the left
      ii   a translation 4 units to the right
      iii a translation 3 units up
      iv a translation 7 units down
      v   a translation 4 units to the right and 6 units down
      vi a translation 3 units to the left and 4 units up

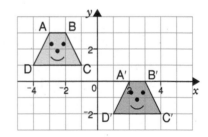

The *inverse translation* undoes the translation.

3 A shape has been translated 3 squares to the right and 5 squares down.
   The image is then translated 1 square to the left.
   Which of these translations would have given the same result?
   A 4 squares to the right and 5 squares down
   B 2 squares to the right and 4 squares down
   C 2 squares to the right and 5 squares down

4 Which two of these are equivalent translations?
   A 3 squares up, 2 squares down, 3 squares right
   B 5 squares down, 3 squares right
   C 1 square up, 3 squares right

Review 1 Use a copy of these.

Draw the image after these translations.

a

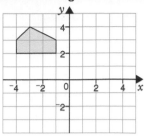

4 units down

b

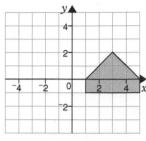

4 units left and 3 units up

c

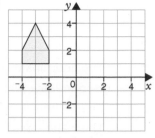

5 units right and 2 units down

Review 2 Which two of these are equivalent translations?

A 3 squares right, 4 squares down and 2 squares left
B 1 square right and 3 squares down
C 1 square right and 4 squares down

⭐ **Practical**

1 **You will need** a pinboard and square dotty paper.

Find all the possible translations of this triangle
on a 3 × 3 pinboard?
Start with the triangle in the position shown.
Draw each on square dotty paper.

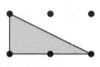

What about a 4 × 4 pinboard?
What about a 5 × 5 pinboard?

2 **You will need** some square dotty paper.

Some Islamic art is made by translating a symmetrical pattern.
The pattern on the left is translated to make the design on the right.

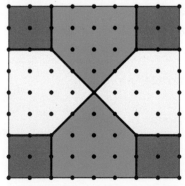

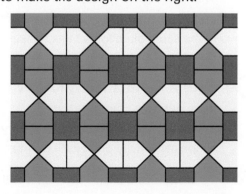

Draw a symmetrical pattern, like the one on the left, on square dotty paper.
Translate your pattern to make a design.

319

Summary of key points

 A We use **coordinates** to give the position of a point on a grid.

The coordinates of A are (⁻2, 1). We always give the *x*-coordinate first.

The coordinates of the **origin** are (0, 0).

The *x* and *y* axes make four quadrants as shown.

 B **Reflection**

ABC has been reflected in the mirror line to get the image A′B′C′.

The line joining a point to its image is perpendicular to the mirror line.

A′B′C′ is the same distance behind the mirror line as ABC is in the front.

The **inverse of a reflection** is a reflection in the same mirror line. If a reflection maps A onto A′ then it also maps A′ onto A, using the same mirror line.

 C A shape has **reflection symmetry** if one half of the shape can be reflected in a line to the other half. The line is a **line of symmetry**.

4 lines of symmetry

 D To **rotate** a shape we need to know the **centre of rotation** and the **angle of rotation**. The centre of rotation is a fixed point. It can be inside or outside the shape being rotated.

Example ABCD has been rotated 270° using (0, 1) as the centre of rotation.

An angle of rotation given as 90° means 90° in an **anticlockwise** direction.

The inverse of a rotation is either an equal rotation about the same point in the opposite direction **or** a rotation about the same point in the same direction so that the sum of the angles of the rotation and the inverse add to 360°.

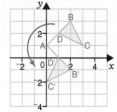

E A shape has **rotation symmetry** if it fits onto itself more than once during a complete turn.

The **order of rotational symmetry** is the number of times a shape fits exactly on to itself during one complete turn.

order 2

 F To **translate** a shape we slide it without turning.

Every point on the shape moves the same distance in the same direction.

Example This shape has been translated 4 units right and 3 units up.

The **inverse of a translation** is an equal move in the opposite direction.

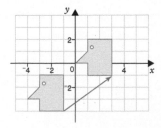

Test yourself

T

1 Use a copy of this grid.

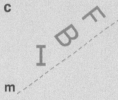

a These are the coordinates of the vertices of a shape. Plot them on your grid and join them in the order given.

(⁻1, 3), (1, 1), (⁻1, ⁻1), (⁻3, 1)

b How many lines of symmetry does the shape have?

c The points (1, 4) and (4, 2) are two of the four vertices of a rectangle.
What might the coordinates of the other two vertices be?
Is it possible to make a square? Explain why or why not.

d Which quadrant does the point (⁻3, 4) lie in?

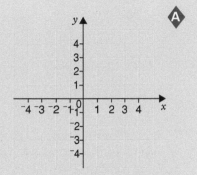

T

2 Use a copy of these shapes.
Reflect them in the mirror line, m.

a

b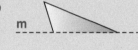

c

3 What quadrilateral do the shape and its image make in question **2b**.

4 Explain why P′ is not the reflection of P in the line S.

T

5 Use a copy of these shapes.

a

b

Draw on the lines of symmetry.

6 Join these shapes together to make a single shape with reflection symmetry.

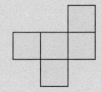

Find all five ways of doing this.

T 7 Use a copy of these. The centre of rotation and angle of rotation are given.
Draw the image shapes.

a (0, 0)
90°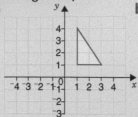

b (1, 1)
180°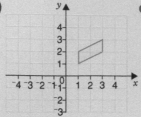

c (2, 2)
270°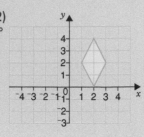

8 A rotation through 90° about (2, 1) maps P onto P'.
Which of these maps P' onto P? There are two correct answers.
A a rotation through 90° about (2, 1)
B a rotation through 270° about (2, 1)
C a rotation through 90° about (1, 2)
D a rotation 90° clockwise about (2, 1)

9 What is the order of rotation symmetry of these?

a b c  d

10 Join the shapes in question **6** together to make a single shape with rotation symmetry.

T 11 This shape has rotation symmetry order 4.
Use a copy of it.
Mark the centre of rotation.

T 12 Use a copy of these.
Draw the image after these translations.
a 3 units down
b 2 units left and 3 units up
c 4 units right and 3 units down

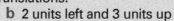 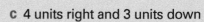
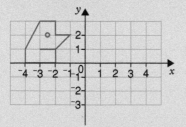 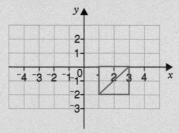 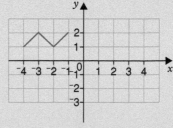

13 What translation is the inverse of these?
a 5 units right and 3 units down
b 4 units left and 6 units up

14 Which two of these are equivalent translations?
A 3 units left and 2 units down, 4 units right
B 1 unit left, 2 units down
C 5 units left and 2 units down, 4 units right

14 Measures

Key vocabulary

**area: square centimetre, square kilometre, square metre,
   square millimetre
capacity: centilitre, gallon, litre, millilitre, pint
length: centimetre, kilometre, metre, mile, millimetre
mass: gram, kilogram, pound
temperature: degrees Celsius (°C), degrees Fahrenheit (°F)
time: century, day, decade, hour, millennium, minute, month,
   second, week, year
depth, distance, height, high, perimeter, surface area, width**

I Shrunk the Kids!

- In the following 'story' most of the measurements are in the wrong places.

  Rewrite the story so the measurements are in the correct places.

  *On the way to the football I saw a dog that weighed 5 kg, a
  car that was 3 cm long, a cat with a 20 cm tail, a baby
  that weighed 40 kg and a house that was 2 m tall. At the
  football I bought a pie that was 3 m thick from a man
  who was 6 m tall.*

  Write another 'story' in which you mix all, or most,
  of the measurements up.

  Ask someone else to rewrite your story.

- Two systems for measuring lengths, weights and capacities are the **metric** system
  and the **imperial** system.

  Think of some metric units.

  Think of some imperial units which are still used in Britain.

  Which system is easier to use? Why?

323

Metric measurements

Fiona needs 2·5 centimetre long nails to make a bookcase.
At the store, all the measurements are in millimetres.
She needs to change 2·5 centimetres to millimetres.
We often need to **change one metric unit into another**.

Remember

Length
1 kilometre (km) = 1000 metres (m)
1 metre (m) = 100 centimetres (cm)
1 metre (m) = 1000 millimetres (mm)
1 centimetre (cm) = 10 millimetres (mm)

Mass
1 kilogram (kg) = 1000 grams (g)

Capacity
1 litre (ℓ) = 1000 millilitres (mℓ)
1 centilitre (cℓ) = 10 millilitres (mℓ)
1 litre (ℓ) = 100 centilitres (cℓ)

kilo: means 1000 times
centi: means one hundredth
milli: means one thousandth.
These prefixes go in front of the
base unit. The base unit for
   length is metre
   mass is gram
   capacity is litre.

Discussion

Think of lots of times when one metric unit needs to be changed into another. **Discuss**.

Think of other words that have the prefix kilo, centi or milli. **Discuss**.

We can use this chart to help us convert between one metric unit and another.

← Changing from larger to smaller, we multiply.

	×10	×10	×10	×10	×10	×10
1000	**100**	**10**	**1**	**0.1**	**0.01**	**0.001**

	1000	100	10	1	0.1	0.01	0.001
length	km	–	–	m	–	cm	mm
mass	kg	–	–	g	–	–	–
capacity		–	–	ℓ	–	cℓ	mℓ

÷10 ÷10 ÷10 ÷10 ÷10 ÷10

← Changing from smaller to larger, we divide.

Changing larger to smaller units

Examples

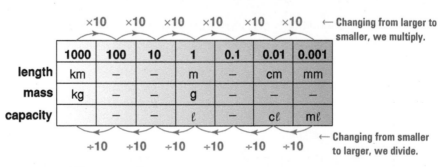

5 km = 5000 m
7 m = 700 cm = 7000 mm
42 cm = 420 mm

5000 m and 5 km are the same distance. m are smaller than km so there are more of them.

Worked Example

Jade put a 2·5 cm border around a poster.
She was asked how wide it was in millimetres.
What should Jade answer?

Answer

2·5 cm = 2·5 × 10 mm
       = **25 mm**

> If you need to remind yourself
> how to multiply by 10, 100 and
> 1000 go to page 19.

Worked Example

Change   **a** 0·42 kg to g   **b** 14·2 ℓ to mℓ.

Answer

a 0·42 kg = (0·42 × 1000) g   **b** 14·2 ℓ = (14·2 × 1000) mℓ
         = **420 g**   = **14200 mℓ**

Exercise 1

> When changing larger units to smaller units,
> your answer will be a bigger number.

1 Change
    a 7·2 cm to mm   **b** 16·4 cm to mm   **c** 3·2 km to m   **d** 7·56 cm to mm
    e 3·125 km to m   **f** 0·64 m to mm   **g** 0·8 cm to mm   **h** 0·03 m to mm.

2 Change
    a 3·4 kg to g   **b** 5·7 kg to g   **c** 8·7 kg to g   **d** 18·6 kg to g
    e 15·4 kg to g   **f** 7·25 kg to g   **g** 0·52 kg to g   **h** 0·05 kg to g

3 Change
    a 3·2 ℓ to mℓ **b** 8·642 ℓ to mℓ **c** 0·68 ℓ to mℓ **d** 0·6 ℓ to cℓ
    e 0·5 ℓ to cℓ **f** 1·4 ℓ to cℓ **g** 0·06 ℓ to cℓ **h** 6 cℓ to mℓ
    i 25 cℓ to mℓ **j** 4·6 cℓ to mℓ **k** 18·3 cℓ to mℓ

4 John took some measurements, in cm, for a model he was making. He then realised he
needed the measurements in mm. Change these to mm for him.
    a 10 cm   **b** 8·4 cm   **c** 23·6 cm   **d** 8·75 cm   **e** 0·3 cm

5 Find the missing numbers.
    a 3·6 km = ___ m   **b** 0·89 m = ___ cm   **c** 1·57 m = ___ mm
    d 2·6 cm = ___ mm   **e** 0·6 cm = ___ mm   **f** 14·8 m = ___ cm

6 Maryanne wanted to divide up some bulk food she had bought. To do this she needed to
convert the kg into grams. Convert these to g.
    a 2·6 kg   **b** 3·5 kg   **c** 0·8 kg
    d 0·05 kg   **e** 2·534 kg

7 Jugs made by 'Useful Plastics' had the capacity, written in ℓ and in mℓ, on the side. How
many mℓ did each of these jugs have written on them?
    a 3·54 ℓ **b** 6·75 ℓ **c** 5·5 ℓ
    d 0·6 ℓ **e** 0·95 ℓ **f** 0·09 ℓ

8 What is your height in metres?
Change it to   **a** cm   **b** mm.

T Review

| $\overline{6\cdot8}$ | $\overline{7\cdot2}$ | $\overline{6800}$ | $\overline{720}$ | $\overline{820}$ | $\overline{72}$ | $\overline{6\cdot8}$ | | $\overline{7\cdot2}$ | $\overline{680}$ | $\overline{7200}$ | $\overline{6800}$ | $\overline{112}$ | $\overline{6\cdot8}$ |

| $\overline{7200}$ | $\overline{6800}$ | $\overline{680}$ | | $\overline{720}$ | $\overline{12}$ | | $\overline{112}$ | $\overline{7\cdot2}$ | $\overline{680}$ | $\overline{720}$ | $\overline{6800}$ |

| $\overline{7\cdot2}$ | $\overline{680}$ | $\overline{7200}$ | $\overset{\textbf{D}}{\overline{82}}$ | $\overline{6\cdot8}$ |

Use a copy of this box.
What goes in the gap? Write the letter beside each question above its answer.

D 8·2 cm = **82** mm **E** 0·68 km = ___ m **A** 7·2 ℓ = ___ mℓ **I** 0·72 kg = ___ g
N 0·12 ℓ = ___ cℓ **T** 11·2 cℓ = ___ mℓ **R** 6·8 kg = ___ g **S** 0·68 cm = ___ mm
P 0·072 km = ___ m **M** 8·2 m = ___ cm **H** 0·072 m = ___ cm

Changing smaller to larger units

Worked Example
Michael's bed is 190 cm long.
How long is this in metres?

Answer
190 cm = (190 ÷ 10 ÷ 10) m
      = (190 ÷ 100) m
      = **1.9 m**

If you need to remind yourself how to divide by 10, 100 and 1000 go to page 19.

Worked Example
Change **a** 530 mm to m **b** 1526 g to kg **c** 3 cℓ to ℓ

Answer
a 530 mm = (530 ÷ 1000) m **b** 1526 g = (1526 ÷ 1000) kg **c** 3 cℓ = (3 ÷ 100) ℓ
        = **0·53 m**   = **1·526 kg**   = **0·03 ℓ**

Exercise 2

1 Change
  a 5000 mm to m   **b** 800 cm to m   **c** 3900 m to km   **d** 30 mm to cm
  e 560 cm to m   **f** 480 mm to m   **g** 89 cm to m   **h** 360 m to km
  i 9 cm to m   **j** 92 mm to m   **k** 8 mm to cm   **l** 4 mm to m
  m 68 m to km   **n** 4 m to km   **o** 1386 cm to m.

2 Change
  a 4200 mℓ to ℓ **b** 800 mℓ to ℓ **c** 400 mℓ to ℓ **d** 2850 mℓ to ℓ
  e 1396 mℓ to ℓ **f** 47 mℓ to ℓ   **g** 54 mℓ to ℓ   **h** 400 cℓ to ℓ
  i 600 cℓ to ℓ   **j** 750 cℓ to ℓ **k** 855 cℓ to ℓ.

3 Change
  a 8600 g to kg **b** 860 g to kg **c** 4700 g to kg **d** 5230 g to kg
  e 864 g to kg   **f** 75 g to kg   **g** 9 g to kg.

4 A science experiment asked for all masses to be in kg. Courtney had weighed everything in g. Change these to kg for her.
   a 3520 g   **b** 5140 g   **c** 700 g   **d** 38 g   **e** 427 g

5 Yuri's baby sister was very small. She weighed just 855 g. How many kg did this baby weigh?

6 Find the missing numbers.
   a ___ m = 50 mm   **b** 386 m = ___ km   **c** 23 mm = ___ cm   **d** ___ kg = 450 g
   e 47 cm = ___ m   **f** ___ ℓ = 385 mℓ   **g** ___ ℓ = 420 cℓ

7 What is your handspan in mm?
   Change this to   **a** cm   **b** m.

T **Review**

							E				
0·04	8·6	0·352	8·6	0·086	8·6	4	0·4	8·6	8	0·04

	E		E										
3·52	0·4	0·04	0·4	0·086	0·075	0·86	8	8·6	0·086	8·6	0·75	0·86	0·8

Use a copy of this box.
What goes in the gaps?

E 400 mm = 0.4 m   **M** 4000 g = ___ kg   **N** 80 mm = ___ cm   **I** 86 cm = ___ m
T 75 g = ___ kg   **D** 3520 mℓ = ___ ℓ   **B** 750 m = ___ km   **C** 80 cℓ = ___ ℓ
R 86 g = ___ kg   **H** 352 mℓ = ___ ℓ   **S** 40 m = ___ km   **A** 8600 mℓ = ___ ℓ

Mixed conversions

Exercise 3

T **1** Use a copy of this cross number. If decimal points are needed give them a full space.

Across
Write
   1 1450 cm in m
   4 37 mm in cm
   6 2·8 cm in mm
   7 0·048 ℓ in mℓ
   10 0·707 kg in g
   12 564 cm in m
   14 7·35 m in cm
   15 31·42 ℓ in cℓ
   16 9090 g in kg
   18 0·026 km in m
   19 4100 mℓ in ℓ
   20 3·4 cm in mm
   22 0·523 kg in g
   23 7070 mm in m

Down
Write
   1 0·12 ℓ in mℓ
   2 0·048 m in mm
   3 5·2 cm in mm
   4 0·39 km in m
   5 0·074 kg in g
   8 8·05 kg in g
   9 3460 m in km
   10 7400 cℓ in ℓ
   11 7·7 ℓ in mℓ
   12 510 mm in cm
   13 6·2 cm in mm
   15 3·425 kg in g
   16 9870 mℓ in ℓ
   17 0·932 kg in g
   21 4·3 m in cm
   22 5·6 cm in mm

*2 Choose the best answer.

a	5340 g + 6·2 kg =	**A** 115·40 g	**B** 11·54 kg	**C** 5420 g	**D** 5960 g
b	14·6 m − 127 cm =	**A** 19 cm	**B** 1·9 m	**C** 13·33 m	**D** 133·3 cm
c	7·3 km − 576 m =	**A** 15·4 m	**B** 154 m	**C** 67·24 m	**D** 6·724 km
d	3 ℓ − 250 mℓ =	**A** 2·5 ℓ	**B** 2·75 ℓ	**C** 50 mℓ	**D** 27·5 ℓ
e	2650 mℓ − 2·4 ℓ =	**A** 2647·6 mℓ	**B** 24·1 ℓ	**C** 0·25 mℓ	**D** 250 mℓ

Review Write

a 17·82 m in cm   **b** 5·04 kg in g   **c** 9461 mℓ in ℓ   **d** 804 mm in cm

e 1842 cℓ in ℓ   **f** 56 mℓ in ℓ   **g** 542 m in km   **h** 0·31 ℓ in cℓ

Puzzle

Use a copy of this.

Cut out the nine squares.

Fit the nine squares together so that the edges which touch show measures which are equal to one another.

4cm / 3kg / 4.56km / 5600mℓ 400mm	504cm / 4.56m / 18ℓ / 40mm	2.4ℓ / 18000mℓ 1.8km 18m / 5.04m / 2400mℓ
1800m / 8.64kg / 40cm	8640g / 5.6ℓ / 456cm	3000g / 1800cm / 4560m

Solving measures problems

Remember

1 minute = 60 seconds	1 day = 24 hours	1 year = 12 months or 52 weeks or 365 days
1 hour = 60 minutes	1 week = 7 days	1 millennium = 1000 years
1 decade = 10 years	1 leap year = 366 days	

Mohammed trained for 2 hours and 20 minutes each week.
He trained for the same time each day.
His trainer asked him how long he trained each day.
Mohammed needs to divide 2 hours and 20 minutes by the number of days in a week, 7.
It will be easier if he converts 2 hours and 20 minutes to minutes.
2 hours and 20 minutes = 120 + 20 minutes
$$= 140 \text{ minutes}$$
$$\frac{140}{7} = 20$$

Mohammed trained for 20 minutes each day.

When solving measures problems;
1 Write down what you know.
2 Work out what you have to do to find the answer.
3 Make sure the measures are in the same units, or write units you want for the answer.

Worked Example

An apple weighs 55 g. How many kilograms would 200 of these apples weigh?

Answer

55 × 200 = 11 000 g
11 000 g = 11 000 ÷ 1000 kg
     = **11 kg**

> Change the grams to kilograms because the question asked for kilograms.

Exercise 4   **Only use your calculator if you need to.**

1   A swimming pool is 50 m long.
    How many lengths must be swum in a 2·5 km race?

2   How many 5 mℓ spoons of medicine can be taken from a 0·2 ℓ bottle?

3   In gymnastics the length of the beam is 50 times its width. The length of the beam is 5 m.
    How many cm wide is the beam?

4   A recipe needed 1·2 kg of tomatoes.
    Tina had 560 g in one bag and 575 g in another. Did Tina have enough?

5   Patty typed eight pages in 1 hour and 40 minutes. How many minutes on average did it
    take to type each page?

6   Jamie made 2·2 ℓ of chutney. He poured this into four jars of the same size. If each of these
    jars was filled to the top, find the capacity in mℓ, of each jar.

7   An art class made a mural using three sheets of paper with lengths 42 cm, 45 cm
    and 852 mm.
    How long was the mural? (Answer in m.)

8   Tyrone is a tour guide for 'Lakes Tours'. Each tour takes four days.
    Last year he guided 20 tours. How many weeks and days did this take?

9   A speed skating track is 0·4 km long.
    a   How many metres is this?
    b   How many times does Joel need to skate round the
        track to travel 20 km?

10 Tony walks 450 m to get to school each morning. He
    walks the same distance to get home in the afternoon.
    How many kilometres does Tony walk to and from
    school each week? (Tony goes to school five days a week.)

11 Ken bought 500 g of tomatoes at £1·89 per kilogram.
    How much did the tomatoes cost to the nearest penny?

12 An ice cream tub is 12 cm deep. It is full to within 13 mm of the top.
    What depth of ice cream is in the tub?

13 Kate wanted to lighten her hiking pack. Her food weighed 3·2 kg.
    She took out a tin of soup which weighed 432 g.
    How much did her food weigh, in kg, then?

14 On a map 1 cm represents 5 km.
    A lane on the map is 1·2 cm long.
    What is the length of the lane in kilometres?

Link to ratio and
proportion page 138.

15 A 1·2 kg box of cereal is enough for 25 servings.
    How many grams is each of these servings?

16 Sweets cost £1·55 for 250 g.
Selina buys 1·5 kg of these sweets for her party.
How much change does she get from £10?

17 Khalid made 450 cℓ of curry sauce.
How many 75 mℓ servings will he get from this?

18 Ash Wednesday is the first day of Lent, coming $6\frac{1}{2}$ weeks before Easter Sunday. In the year 2000, Easter Sunday was on the 23rd April.
What date was Ash Wednesday in the year 2000?

19 Paul and Asmat are stacking boxes in a wardrobe which is 1·88 m high.
Each box is 24 cm high.
How many layers will fit in the wardrobe?

Round sensibly.

20 A box weighs 800 g. The box is filled with packets of crisps.
Each packet weighs 85 g. The full box weighs a total of 4·88 kg.
How many packets of crisps are in the box?

21 How many decades are there in a millennium?

*__22__ Tangerines cost £1·80 per kilogram.
One Tangerine weighs 150 g.
Find the cost of one Tangerine.

*__23__ It takes 23 hr 56 min 4 sec for the Earth to make one complete rotation on its axis.
It takes 24 hr 37 min 23 sec for Mars to make one complete rotation on its axis.
How much longer does it take for Mars to make one rotation?

*__24__ Plums cost £1·36 per kilogram.
A bag of plums weighs 750 g.
How much does this cost?

*__25__ Each day Manisha bikes to school. She leaves at 8·15 a.m. and arrives at about 8·35 a.m.
Each Tuesday, she posts a letter on the way to school and doesn't arrive until 8·45 a.m.
How long does Manisha spend coming to school
a each week   **b** each term of 16 weeks   **c** each year of 39 weeks?

Give the times in sensible units.

*__26__ A box of fruit punch holds 1250 mℓ.
Zeta likes her fruit punch strong. She adds between 3 and 3·5 parts lemonade to the fruit punch.
About how many litres of drink can she make with one box of fruit punch?

1250ml
FRUIT PUNCH
Add 4 parts lemonade

*__27__ Ribbons are 40 cm long.
Adam joins some ribbons together to make a streamer.
They overlap each other by 4 cm.

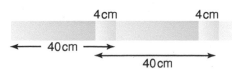

4 cm   4 cm

← 40 cm →

40 cm

i How long is a streamer made from
    a 2 ribbons   **b** 5 ribbons   **c** 10 ribbons?
ii Nasreen made a streamer that was 220 cm long.
    How many ribbons did she use?

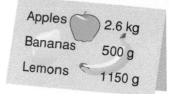

Apples 2.6 kg
Bananas 500 g
Lemons 1150 g

Review 1   What is the total weight, in kg, of this fruit?

Review 2   Max bought 2 m of wrapping paper to wrap gifts. From this he cut two lengths, one of 34 cm and the other of 85 cm. How much paper did he have left?

Review 3   To keep in shape Charlotte swims 1 km twice a week. If the pool is 40 m long, how many lengths does Charlotte swim each week?

Review 4   A large tin of tomato sauce contains 500 cℓ.
How many 250 mℓ bottles can be filled from this tin?

Review 5   How many books each weighing 760 g can be sent in this service lift?

**Maximum
Load
60 kg**

Review 6   The first woman to orbit Earth was Valentina Tereshkova in the Vostok 6.
She was launched on June 16th, 1963 at 9·30 a.m. (Greenwich time). She returned after 2 days, 22 hours and 46 minutes. When did Valentina land back on Earth?

*Review 7   Susan has made 8 litres of grape juice. To store it she has bottles of two different sizes: 1·25 ℓ and 750 mℓ.
To store the juice, bottles should be filled right to the top. If as many large as small bottles are used, how many bottles are filled?

Puzzle

1   A pile of ten £5 notes is 1 mm high. How much money is in a pile of £5 notes that is 1 m high?

2   Each link of this chain is 24 mm from outside edge to outside edge.
Each link is made from brass 5 mm thick.
What is the length of the 6-link chain?

24mm

3   100 kg of potatoes are given to 100 people. Each adult gets 3 kg, each teenager gets 2 kg and each small child gets 0·5 kg. How many adults and teenagers and small children were there? (There are many answers to this problem. Find as many as you can.)

4   Rose needed to measure out exactly 2 ℓ of water. She had two containers, one that held 8 ℓ and one that held 5 ℓ. She began by filling the 5 ℓ container. How did she continue?

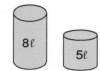

8ℓ 5ℓ

5   Jimmy has four containers which, when full, will hold 2 ℓ, 3 ℓ, 4 ℓ and 5 ℓ of wine. The 3 ℓ and 5 ℓ containers are full. By pouring wine from one container to another, show how to get 2 ℓ of wine into each of the containers in as few moves as possible.

6 Three winemakers had between them 21 barrels, seven were empty (E), seven were half-full of wine (H) and seven were full of wine (F). These barrels were divided between the winemakers so that each had the same number of barrels and the same amount of wine. There are two ways in which this could be done. Find them.

*7 Josh, Becky and Ryan have to cross a river in a rowing boat which can carry at most 140 kg. Josh weighs 100 kg, Becky weighs 50 kg and Ryan weighs 75 kg. They all have rucksacks. Josh's weighs 40 kg, Becky's weighs 15 kg and Ryan's weighs 25 kg. How do they cross the river if none of them will leave their rucksacks with any of the others?

*8 Shalome is given nine £1 coins, one of which is fake. This fake coin looks the same as the others but it is lighter. Shalome is allowed to keep all the coins if she can find the fake one by just two weighings on a balance like this. She is not allowed to use any weights during her weighings. How could she find the fake one in two weighings?

Metric and imperial equivalents

Discussion

Hien is helping on her family farm.
Elizabeth is helping at a holiday camp.
Alex is helping at his football club.
Megan is helping to care for her baby brother.
Barry is helping to plan a party.
Bonny is helping to care for the animals at the RSPCA.
Tim is helping out at his mother's car sales business.

What imperial measures will Hien, Elizabeth, Alex, Megan, Barry, Bonny and Tim be likely to use? **Discuss**.

Learn these approximate **metric and imperial equivalents**.

1 gallon ≈ 4·5 litres	8 km ≈ 5 miles
1 pint is just over half a litre	1 kg ≈ 2·2 pound (lb)

Worked Example
A litre of orange drink costs 59p.
About how much would one gallon cost?

Answer
One gallon is about 4·5 litres.
4·5 litres would cost 4·5 × 59p = 265.5p
$$= £2·66 \text{ to the nearest penny.}$$
One gallon would cost about **£2·66**.

Exercise 5

1 About how many litres are these?
   a 2 gallons   b 10 gallons   c 4 pints   d 8 pints
   e 15 pints   f 3 gallons   g 12 pints

2 About how many gallons are these?
   a 9 litres   b 18 litres

Try and use a mental method first.

3 About how many lbs are these?
   a 3 kg   b 4 kg   c 12 kg   d 4·5 kg

4 About how many kg are these?
   a 2·2 lb   b 6·6 lb   c 11 lb   d 22 lb

5 About how many kilometres are these?
   a 5 miles   b 10 miles   c 25 miles

6 About how many miles are these?
   a 16 km   b 32 km   c 64 km

7 Rachel weighs 50 kg. About how many lbs does she weigh?

8 Suzannah's dog weighed 44 kg. The vet said it had to lose at least 5 kg.
   Suzannah's scales only weighed in pounds. About how many pounds does the dog have to lose?

9 Mr Taylor's rucksack sprayer holds 15 ℓ of spray.
   He makes up 30 pints of spray.
   Will he be able to fill his rucksack sprayer with this? Explain.

10 Tricia's aunt lives 30 miles south of Manchester.
   About how many km from Manchester does she live?

11 A tank holds 40 gallons of water. About how many litres is this?

12 A litre of petrol costs 93·6p. About how much would a gallon cost?

Review 1 About how many lbs are in a 5 kg bag of potatoes?

Review 2 Rachel went to Germany. A sign told her it was 56 km to Berlin. About how many miles is this?

Review 3 Holly's sink holds about 9 ℓ. About how many gallons is this?

Review 4 Two litres of fruit juice cost £1·20. About how much would 2 pints cost?

Reading scales

There are ten divisions between 1 and 2.
Each division on this scale is 1 m ÷ 10 = $\frac{1}{10}$ m = 0·1 m.

The pointer is at 1·4 m.

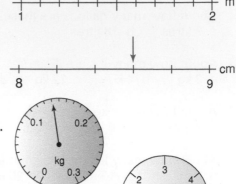

There are five small divisions between 8 and 9.
Each division is 1 cm ÷ 5 = $\frac{1}{5}$ cm = 0·2 cm.
The pointer is at 8·6 cm.

There are ten small divisions between 0·1 and 0·2.
Each division on this scale is 0·1 kg ÷ 100 = 0·01 or $\frac{1}{100}$ kg.

The pointer is at 0·14 kg.

Sometimes we have to estimate the reading on a scale.
The pointer is about halfway between 5 ℓ and 6 ℓ or at about $5\frac{1}{2}$ ℓ.

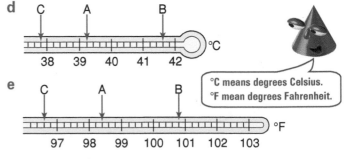

Exercise 6

1 Find the measurements given by the pointers A, B and C.

a

b

c

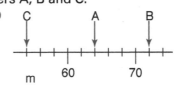

d

°C means degrees Celsius.
°F mean degrees Fahrenheit.

e

f

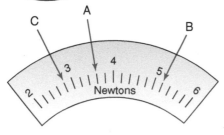

g

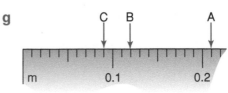

h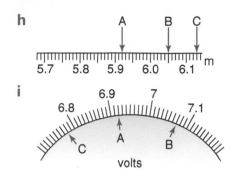

i

j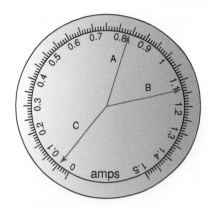

2 Use a copy of this.

Use arrows to show these. Label each arrow.

a 1·7 cm   **b** 4·1 cm   **c** 5·0 cm   **d** 96·5 cm   **e** 95·8 cm   **f** 99·3 cm

3 Use a copy of this.

Use arrows to show these. Label each arrow.

a 2·3 m   **b** 2·35 m   **c** 2·42 m   **d** 2·17 m   **e** 2·08 m   **f** 2·26 m

4 What reading, in grams, is given by the pointer?

5 Maria only had broken rulers to measure with.
Find the length of these.

a

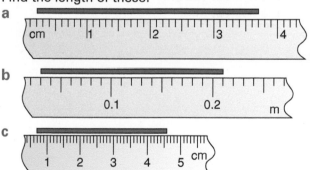

b

c

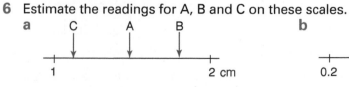

6 Estimate the readings for A, B and C on these scales.

a 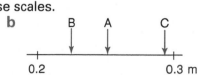   **b**

7 Use a copy of these scales.
Use an arrow to show the number given beside the scale.

a 2.9   **b** 4.7

8 What is the volume, to the nearest 10 mℓ?

  a   **b**   **c**

9 What is the mass to the nearest 100 g?
   Estimate the mass to the nearest 10 g.

  a   **b**   **c**

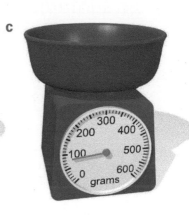

10 How long is the string to
   a the nearest 10 cm   **b** the nearest 100 cm
   c the nearest cm   **d** the nearest mm?

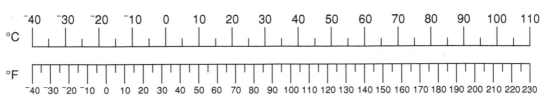

11 Use this scale to convert these to °F.
   a 40 °C   **b** 100 °C   **c** 0 °C

°C
```
⁻40  ⁻30  ⁻20  ⁻10   0   10   20   30   40   50   60   70   80   90   100  110
```

°F
```
⁻40 ⁻30 ⁻20 ⁻10  0  10 20 30 40 50 60 70 80 90 100 110 120 130 140 150 160 170 180 190 200 210 220 230
```

12 Use the scale in question **11** to convert these to °C.
   a 50 °F   **b** 200 °F   **c** 140 °F

Review 1 What measurement is given?

a

b

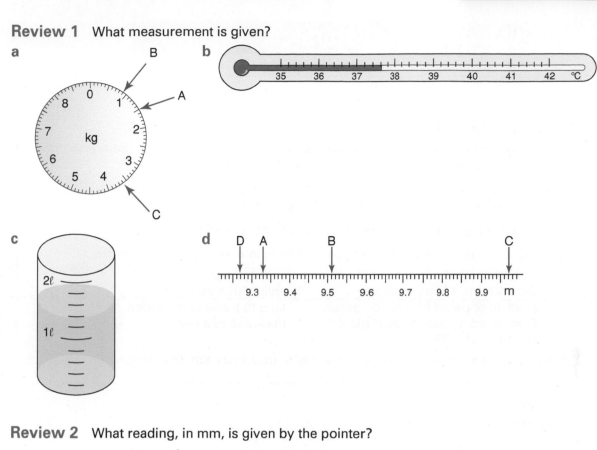

c

d

Review 2 What reading, in mm, is given by the pointer?

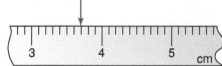

Review 3 Use a copy of the scale in Review **2**.
Use an arrow to show 4·6 cm.

Review 4 Estimate the readings for A, B and C.

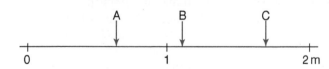

Review 5 Use a copy of the scale in Review **4**.
Use an arrow to show 1·6 m.

Review 6 Use the scale opposite to convert these to kilograms.
a 84 lb   **b** 130 lb   **c** 74 lb

Review 7 Use the scale opposite to convert these to pounds.
a 65 kg   **b** 30 kg   **c** 25 kg

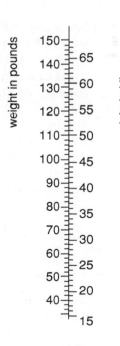

Choosing units and measuring instruments

Discussion

- What units of measurement might you see in these places? **Discuss**.

    supermarket; chemist; garage; bus or train station; on TV; garden centre; weather station; bakery; hospital; sports ground; farm

- Think of as many different measuring instruments as you can. What sort of things might each be used to measure? **Discuss**.

- What measuring instruments are used to measure time? **Discuss**.

- Suggest an imperial or metric unit you could use to measure these. **Discuss**.

    weight of a nail   width of a pencil
    distance from Mars to Jupiter   weight of a car
    amount of petrol in a petrol tanker   time to travel to the moon
    time to grow a water cress plant   thickness of a hair
    diameter of a 20p coin

    Suggest a measuring instrument and a way to measure each. **Discuss**.

Exercise 7

| kg | g | km | m | cm | mm | ℓ | mℓ | hours | min | sec |

1 Which of the units in the box would you use to measure each of the following?
   a width of a netball court   b length of a motorway
   c mass of a packet of biscuits   d mass of an apple
   e mass of a bag of potatoes   f capacity of a bucket
   g capacity of a cup   h time to count to ten
   i time to fly from London to New York j mass of a person
   k length of an ant   l height of a cathedral
   m thickness of a coin   n length of the Nile river
   o distance to the moon   p capacity of a fridge
   q capacity of a tablespoon   r mass of a steak
   s time to eat an ice cream

Review

| kg | g | mg | km | m | cm | mm | ℓ | mℓ | min | sec |

Which of the units in the box would you use to measure each of the following?

a length of a building   b time to eat dinner
c mass of a school bag   d amount of tea in a tea bag
e thickness of a ruler   f diameter of an aspirin tablet
g length of a Boeing 747   h time to run the length of a cricket pitch
i mass of a calculator   j distance around the Equator
k mass of a suitcase   l amount of milk added to a cup of coffee
m time for eight heartbeats   n amount of water used in a shower

Practical

Think about how you could measure these as accurately as possible. Measure them.

the thickness of one sheet of paper
the weight of one paper clip
the amount of liquid in one drop from an eye dropper

＊ Write a short report on how you measured them.

Estimating measurements

Estimation – a game for a group

You will need a tape–measure and ruler.

To play

● Choose someone to start.
● This player chooses a length or a height and a unit of length for this measurement.

Examples   height of my desk in centimetres
or   length of Dale's hair in millimetres
or   width of the room in metres

● Each player writes down an estimate for this.
● The length or height is measured.
● The player with the closest estimate gets 5 points.
● Take turns to name the length or height and unit for this.
● The winner is the person with the most points at the end of the game.

Note: You could play this game again but choose a mass and a unit of mass or choose a capacity and a unit of capacity.

When we **estimate a measurement** it is a good idea to give a **range** for it.

Jordan was asked to estimate the width of the teacher's desk to see if it would fit through the door. He said 'It's bigger than 0·8 m but less than 1 m'.

He wrote this as   0·8 m < width of desk < 1 m.
This gives the range for the estimate.

> Remember < means is less than.

Discussion

Kay wanted to paint one wall of her bedroom. She estimated the length of the wall so she could work out how much paint she needed.

Think of as many situations as you can when you might need to estimate a length, mass or capacity. **Discuss.**
What unit would you use for each?

Practical

1 **You will need** a tape measure and chalk.
   Mark a point on the ground with chalk.
   Estimate a distance of 1 m from this mark and put another mark.
   Measure the actual distance.
   How good was your estimate?

   Repeat for other distances.

   Estimate the length of some things such as the length of a netball court, width of a window, length of a gymnasium, length of a paper clip, width of a book, ...
   Give a range for your estimate.
   Measure the actual lengths.
   You could draw a table like this one.

Estimate	Range	Actual
	_ < _ < _	

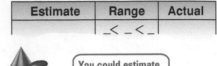

You could estimate by pacing.

2 **You will need** a trundle wheel or long measuring tape.
   Choose two objects in the school grounds. Estimate the distance between them. Give a range for your estimate.
   Use a measuring tape or trundle wheel to measure the actual distance.
   Was it within your range?

3 **You will need** scales.
   Estimate the masses of some objects such as a calculator, a school bag, a brick, a pencil case, a golf ball, ten 20p coins, ... Give a range for each estimate.
   Check your estimates by weighing.

4 **You will need** a measuring jug.
   Estimate the capacities of some containers such as a yoghurt carton, a cup, a bowl, a bucket. Give a range for each estimate.
   Fill the containers with water from the measuring jug to check your estimates.

5 Bring a collection of tins, packets etc. to school.
   Estimate various measurements such as height, capacity or mass. (Don't look at the labels!)

6 **You will need** a stopwatch.
   Work in pairs to estimate intervals of time such as 1 minute, 5 minutes etc.
   Before you begin, decide how you are going to do this.

   Estimate the time to run 200 m. Give a range for your estimate.
   Check by timing someone to run 200 m.

7 'Invent' a person who is about your age. Give your person a name.
   Estimate sensible measurements for height, weight, arm length, foot length, waist measurement, ... Make a poster showing your 'person'.

Exercise 8

1 a Which is most likely to be about 5 m?
   A the width of a coach
   B the length of a tennis racquet
   C the height of a tree
   D the height of a rose bush

b Which is most likely to be about 2 mm?
   A the length of a pencil
   B the thickness of this page
   C the diameter of a cricket ball
   D the thickness of a coin

2 a The length of an eyelash could be
   A 3 mm   **B** 20 mm   **C** 30 mm   **D** 10 cm.

b The distance between two cities could be
   A 200 cm   **B** 200 km   **C** 20 m   **D** 200 mm.

c An apple has a mass of about
   A 1 g   **B** 10 g   **C** 200 g.

d A man could have a mass of about
   A 80 kg   **B** 8 kg   **C** 800 kg.

e A bottle of suntan cream could hold
   A 10 mℓ   **B** 200 mℓ   **C** 1 ℓ.

f A pot full of soup could hold
   A 100 mℓ   **B** 1 ℓ   **C** 100 ℓ.

3 Which of km, m, cm, mm is missing?
   a A house is about 8 ___ high.   **b** John's thumb is about 50 ___ long.
   c Brian can run about 250 ___ in one minute.
   d London is about 350 ___ from Liverpool.
   e Laura is 150 ___ tall.   **f** Louise walks 200 ___ to school.
   g On her holiday abroad Rasha flew 920 ___.

4 Estimate the length of each of these lines.
   Measure each line to the nearest 0·1 cm (nearest mm).
   a ————————   **b** ————————

   c ————————   **d** ————————

   e ————————

5 Choose the best range.
   a **A** 1 m < width of classroom door < 2 m   **b** **A** 100 mℓ < capacity of a cup < 300 mℓ
       B 0·5 m < width of classroom door < 1 m   **B** 0 < capacity of a cup < 500 mℓ
       C 1·0 m < width of classroom door < 1·1 m   **C** 500 mℓ < capacity of a cup < 1 litre
   c **A** 500 g < mass of small apple < 1 kg   **d** **A** 1 m < length of bed < 2 m
       B 50 g < mass of small apple < 100 g   **B** 1 m < length of bed < 5 m
       C 1 kg < mass of small apple < 2 kg   **C** 1·5 m < length of bed < 2·6 m

6 Write A, B or C for each item.

i

Item	Weight between		
	A 10 g and 500 g	**B** 0·5 kg and 1·5 kg	**C** 1·5 kg and 5 kg
a a pen			
b a cup			
c a large cat			
d a ruler			
e litre of milk			
f a pumpkin			

ii

Item	Capacity between		
	A 0 mℓ and 100 mℓ	**B** 100 mℓ and 1 ℓ	**C** 1 ℓ and 10 ℓ
a a basin			
b a bucket			
c a shampoo bottle			
d a can of cola			
e a teaspoon			
f a perfume bottle			

Review 1 What should go in the gaps?
Choose from the numbers given in the box
below.
Use each number just once.

30	4000	
1500		
12		
45	150	9
	4	
	500	

a	the temperature on a cold day	___ °C
b	weight of an elephant	___ kg
c	time to run 100 metres	___ sec
d	length of a bus	___ m
e	amount of coke in a large bottle	___ mℓ
f	length of a cat's tail	___ cm
g	weight of a boy	___ kg
h	length of a calculator	___ mm
i	mass of a box of chocolates	___ g

Review 2 Choose one of km, m, cm, mm, ℓ, mℓ, kg, g, mg to complete these.

a A small parcel weighs about 200 ___.
b The length of a hospital corridor is about 50 ___.
c The depth of a bath is about 30 ___.
d A van could weigh about 1500 ___.
e The width of a wedding ring could be 5 ___.
f The capacity of a wheelbarrow could be 100 ___.
g Joanna sailed her yacht 50 ___.
h Joanna weighs about 52 ___.

Review 3 Write A, B, or C for each of the following to show which box would give your estimate for its length.

Item	Length between		
	A 0 m and 1 m	B 1 m and 2·5 m	C 2·5 m and 5 m
a a book			
b a telephone			
c a tissue box			
d a car			
e a room			
f a mountain bike			

Area of a triangle

Area is the amount of space covered by a shape.
Area is measured in square kilometres (km^2), square metres (m^2), square centimetres (cm^2) or square millimetres (mm^2).

Discussion

James made a triangular flag by cutting a piece of rectangular material in half like this.

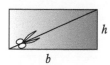

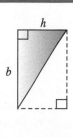

James said that the area of his flag was one half of the area of the rectangle or $\frac{1}{2} b \times h$.
Is he correct? **Discuss**.

Do you always get a right-angled triangle when a rectangle is cut in half like this?
Can all right-angled triangles be made by cutting a rectangle in half? **Discuss**.

What is the formula for the area of a right-angled triangle of base, b, and height, h? **Discuss**.

Check by cutting rectangles yourself.

Area of rectangle = base × height = bh
Area of triangle = $\frac{1}{2}$ area of rectangle
$= \frac{1}{2} bh$

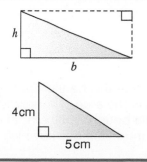

Example   Area of a $\triangle = \frac{1}{2}$ bh
$= \frac{1}{2} \times (5 \times 4)$
$= \frac{1}{2} \times 20$
$= 10$ cm^2

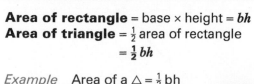

Remember to put the units in your answer.

Exercise 9 **Except questions 3 and 4**

1 Calculate the area of these triangles.

a
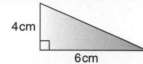
4cm
6cm

b
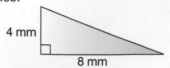
4 mm
8 mm

c
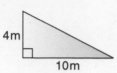
4m
10m

d
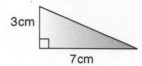
3cm
7cm

e
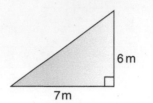
6 m
7m

f
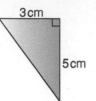
3cm
5cm

g
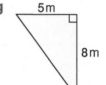
5m
8m

h

7m
10m

i
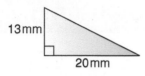
13mm
20mm

j
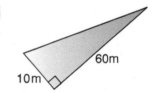
60m
10m

k
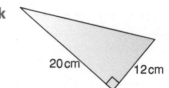
20cm
12cm

2 What is the area of this sail?

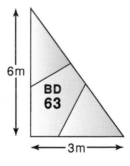
6m
BD
63
3m

3 What are the areas of the tops of these spa pools?

a
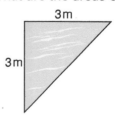
3m
3m

b
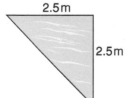
2.5m
2.5m

c
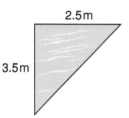
2.5m
3.5m

4 Amelia made this banner for her school netball team.
   a What is the area of Amelia's banner?
   b Amelia bought 2500 cm^2 of material to make the banner.
      How much did she have left over?

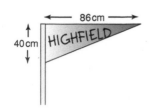
86cm
40cm
HIGHFIELD

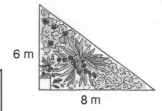

5 a What is the area of this flower bed?
   b Mary wants to fertilise the flower bed.
   One box of fertiliser is needed for each 10 m².
   How many boxes does she need?
   ∗c What fraction of a box will she have left?

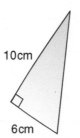

∗6 The area of this triangular glass window hanging is 24 cm².
   What might the lengths of **a** and **b** be?

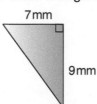

Review 1   Calculate the area of these triangles.

a

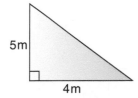

b   7 mm

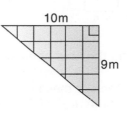

   9 mm

c

10 cm

6 cm

Review 2

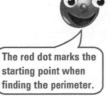

10 m

9 m

a What is the area of this courtyard?
b Mr Nathan sprayed the courtyard with moss killer. One bottle is
   needed for each 30 m². How many bottles does he need?
∗c What fraction of a bottle will he have left?

Perimeter and area of shapes made from rectangles

Remember
Perimeter is the distance around the outside of a shape.
Perimeter is measured in km, m, cm, mm.

Example   To find the **perimeter** of this shape we need to
   calculate the missing lengths.

   blue side = 14 − 8 = 6 cm
   red side = 12 − 5 = 7 cm
   Perimeter = 14 + 5 + 8 + 7 + 6 + 12
      = **52 cm**

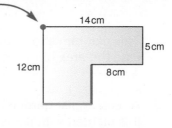

14 cm
5 cm
12 cm
8 cm

The red dot marks the
starting point when
finding the perimeter.

Example   To find the **area** of this shape divide it into
   rectangles. One way is shown.

   Area of A = 12 × 6
      = 72 cm²
   Area of B = 8 × 5
      = 40 cm²
   Total area = **112 cm²**

Add 72 and 40.

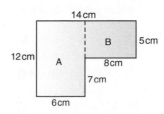

14 cm
B   5 cm
12 cm
A   8 cm
7 cm
6 cm

Exercise 10 **except question 3**

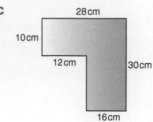

Remember to note your starting point

1 Find the perimeter of these shapes.

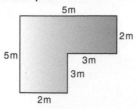

a — 5m, 2m, 5m, 3m, 3m, 2m

b — 8mm, 4mm, 2mm, 9mm, 5mm, 6mm

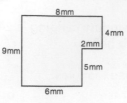

c — 28cm, 10cm, 12cm, 30cm, 16cm

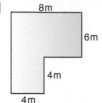

d — 8m, 6m, 4m, 4m

e — 12cm, 6cm, 4cm, 11cm

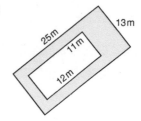

f — 4m, 2m, 5m, 1m, 3m, 1m

2 Find the area of each shape in question **1**.

3 In each of these rectangular lawns a flower bed has been cut out. Find the area of the remaining lawn.

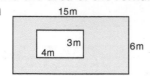

a — 15m, 3m, 4m, 6m

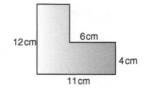

b — 24m, 4m, 12m, 16m

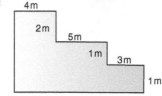

c — 25m, 11m, 12m, 13m

4 Sarah's lawn was 10 m by 12 m. She cut a rectangular flower bed 2 m by 5 m, into the lawn. Find the area of lawn left.

5 This diagram shows a path through a courtyard. Find the area of the path (shaded grey).

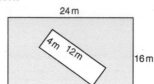

9m, 7m, 1m, 2m

6 A rectangular garden is 4 m by 6 m.
It is planted with flowers which cost £5·85 per square metre to plant.
How much does it cost to plant the garden?

*7 A floor 4·5 m by 6 m is covered with square tiles of side 25 cm.
   a Find the number of tiles needed.
   b If each tile cost £2·25, find the cost of tiling the floor.

*8 Johnny uses this shape to make a pattern.

   What is the perimeter of the pattern?

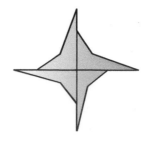

1.5cm, 2cm, 1cm, 1cm

346

Review 1   Find the perimeter and area of the green shapes.

a

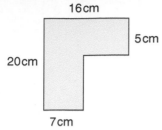

16 cm

20 cm

5 cm

7 cm

b

15 m

5 m

4 m

5 m

3 m

15 m

c

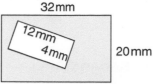

32 mm

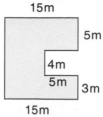

12 mm

4 mm

20 mm

Review 2   Hayden made a wall hanging from rectangles and triangles. Each rectangle is 12 cm².

a   What is the area of the whole wall hanging?
b   The perimeter of the wall hanging is 38 cm.
     What is the longest side of each triangle?

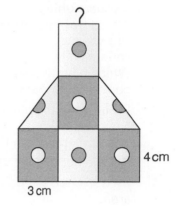

4 cm

3 cm

★ **Practical**

You will need LOGO or a dynamic geometry package.

Draw a square.
Now draw a square which is one quarter of the area of the first square.

How do you know that the area of the second square is one quarter of the area of the first square? Explain.

Discussion

Pierre drew this red quadrilateral on square dotty paper.
The dots are 1 cm apart.
He found the area of the quadrilateral like this.

Area of blue square = 9 cm²
Sum of areas of triangles *outside* the quadrilateral
= 2 cm² + 1 cm² + 1 cm² + $\frac{1}{2}$ cm² = 4$\frac{1}{2}$ cm²

Area of quadrilateral = area of blue square – sum of areas of
                    triangles outside the quadrilateral
                    = 9 – 4$\frac{1}{2}$
                    = 4$\frac{1}{2}$ cm².

Discuss Pierre's method.
How else could Pierre have found the area of the quadrilateral?

How could you find the area of this green quadrilateral? **Discuss**.

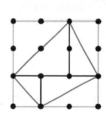

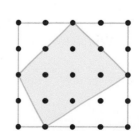

Investigation

Perimeter and Area

1 All of these shapes have an area of 3 square units.
Which two have the same perimeter?
Draw some more shapes with area 3 square units.
Which of your shapes has the greatest perimeter? **Investigate**.

Do any two of your shapes have the same perimeter? **Investigate**.

2 These shapes both have a perimeter of 12 cm.

Draw other shapes with a perimeter of 12 cm.

Investigate to find the shape with the greatest area.

3 This shape has an area of 16 cm².
Draw other shapes with an area of 16 cm².
Investigate to find the shape with the greatest perimeter.

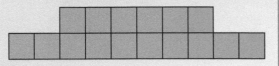

4 Both of these rectangles have a perimeter of 16 m.

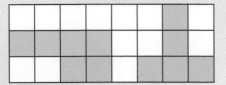

Draw more rectangles which have a perimeter of 16 m.

Copy this table.
Fill it in for the other rectangles you have drawn.
Which rectangle gives you the greatest area?

Rectangles with perimeters of 16m		
Length (m)	Width (m)	Area (m²)
7	1	7
6	2	12

Draw all the rectangles you can with a perimeter of 36 m. Draw a table like the one above. Fill it in for rectangles you have drawn. Which one has the greatest area?

5 A rectangle has an area of 24 cm².

Investigate to find possible lengths and widths of this rectangle.
Which one has the greatest perimeter?

Continued . . .

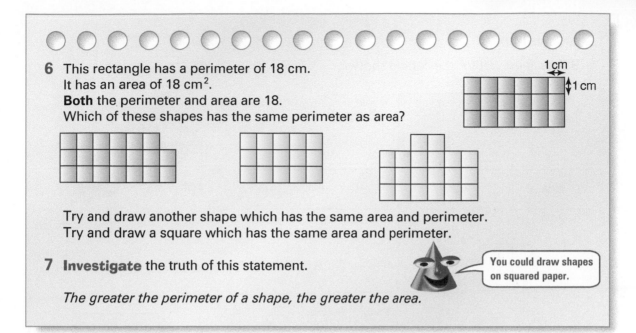

6 This rectangle has a perimeter of 18 cm.
It has an area of 18 cm².
Both the perimeter and area are 18.
Which of these shapes has the same perimeter as area?

Try and draw another shape which has the same area and perimeter.
Try and draw a square which has the same area and perimeter.

7 Investigate the truth of this statement.

The greater the perimeter of a shape, the greater the area.

> You could draw shapes on squared paper.

Surface area

The **surface area** of a 3-D shape is the total area of all the faces.

 Practical

You will need some unit cubes or centimetre cubes.

1 Janet made this shape from centicubes.
Each face has an area of 1 cm².
She counted 18 faces so the surface area of her shape is 18 cm².

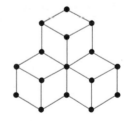

Make some other shapes using cubes.
Find the surface area of each.

∗2 Make as many different cuboids as you can with 20 cubes.
Do they all have the same surface area?

349

Discussion

- Below is the net for the cuboid shown.

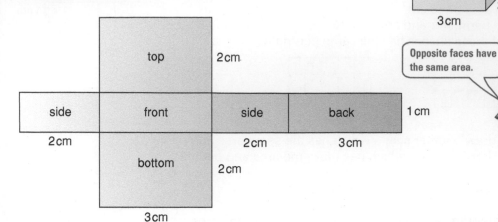

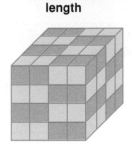

Opposite faces have the same area.

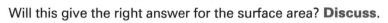

Brendon wrote

top and bottom ↓   front and back ↓   sides ↓

Surface area of cuboid = 2(3 × 2) + 2(3 × 1) + 2(2 × 1)

Will this give the right answer for the surface area? **Discuss.**

How could you use this to write a formula for the surface area of a cuboid? **Discuss.**

- What is the surface area of this cube? **Discuss.**
  If each face is painted in the same pattern as the faces shown, how much of the surface area is red? **Discuss.**

Practical

1 **You will need** some empty cuboid-shaped packets.

Unfold each packet until you have a net.
How could you use the net to find the surface area of the packet?

2 **You will need** some cuboids such as a cereal packet, matchbox, tile, box, book.

Estimate the surface area of each cuboid.
Check by measuring the dimensions of each face and then calculating the surface area.

Surface area of a cuboid = 2(length × width) + 2(length × height) + 2(height × width)
= **2 ℓw + 2 ℓh + 2 hw**

Worked Example
Brad wanted to glue green cardboard onto the surface of this wooden cuboid.
What area of cardboard will he need?

Answer
Total surface area = 2 ℓw + 2 ℓh + 2 hw
= 2(12 × 10) + 2(12 × 5) + 2(5 × 10)
= 2 × 120 + 2 × 60 + 2 × 50
= 240 + 120 + 100
= **460 cm²**

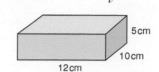

Exercise 11

1 These shapes are made from unit cubes.
Find the surface area of each shape.

a **b** **c** **d**

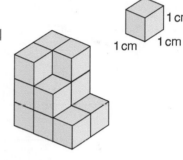

2 Calculate the surface area of each of these cubes and cuboids.

a 2 cm cube **b** 4 m cube **c** 8 mm cube **d** 14 cm cube

  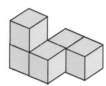

e 2 cm 5 cm 1 cm **f** 2 m 8 m 2 m **g** 16 cm 12 cm 4 cm **h** 14 mm 24 mm 16 mm

  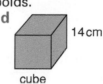

3 This small box is made from cardboard.
 a If the box has a lid, how much cardboard is needed?
 b How much cardboard is needed if the box has no lid?

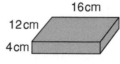 85 mm 66 mm 100 mm

*4 Sally places three cubes as shown to make a stand. The top cube is in the middle of the other two.
She covers the stand with sticky paper.
Sketch the shapes of the sticky paper she could use and say how many of each are needed.

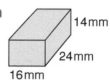 2 cm 2 cm 2 cm

*5 Calculate the surface area of these shapes.

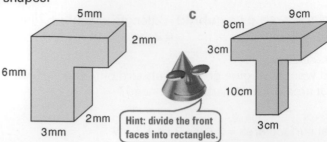

a   4cm
1cm
3cm
2cm
1cm

b   5mm
2mm
6mm
2mm
3mm

c   9cm
8cm
3cm
10cm
3cm

Hint: divide the front faces into rectangles.

*6 **a** Dylan makes 12 coloured cubes.
Each cube has 1 cm edges.
Each is covered with sticky paper.
How much paper is needed?

b Amber makes this shape from 12 cubes.
She covers it with sticky paper.
What area of paper did she use?

c What other shapes could Amber have made with 12 cubes?
Which shape uses the least paper?

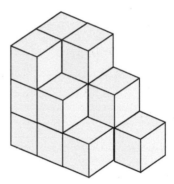

Review 1 Find the surface area of these shapes.

Each is made from unit cubes of length 1 cm.

a   b   c

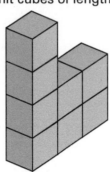

Review 2 Calculate the surface area of these cubes.

a   5cm   b   11mm

Review 3 Bon Bon the clown stands on this platform for his performance. He wants to paint it.
a What is the surface area of the platform?
b Bon Bon buys 1 ℓ of red paint. It will cover 24000 cm².
Will he have enough paint?

20cm
100cm
80cm

*Review 4 Mohammad built this platform to display the cups he had won.
Find the surface area of the platform.

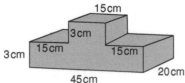

15cm
3cm
15cm
3cm   15cm
45cm   20cm

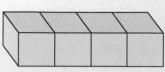

Investigation

* Surface Area

You will need some unit cubes or centimetre cubes.

This shape is made with four centicubes.
The area of all of the surfaces (top, bottom, front, back and both sides) is 18 cm².

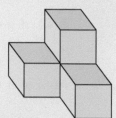

What is the surface area of these?
Can four centicubes be arranged in other ways?

Investigate to find the arrangement which has the smallest surface area.

Investigate to find the arrangement which has the greatest surface area.

What if we began with three cubes?

What if we began with five cubes?

Summary of key points

 A You need to know these **metric conversions**.

length	mass	capacity (volume)
1 km = 1000 m	1 kg = 1000 g	1 ℓ = 1000 mℓ
1 m = 100 cm		100 cℓ = 1 ℓ
1 m = 1000 mm		1 cℓ = 10 mℓ
1 cm = 10 mm		

Examples 0·62 kg = 0·62 × 1000 g   520 mℓ = 520 ÷ 1000 ℓ   42 cm = 42 ÷ 100 m
                    = 620 g   = 0·52 ℓ   = 0·42 m

 B You need to know these **metric and imperial equivalents**.

length	mass	capacity
8 km ≈ 5 miles	1 kg ≈ 2·2 lb	1 gallon = 4·5 ℓ
		1 pint is just over ½ ℓ

 C When **reading scales** you need to work out the value of each small division.

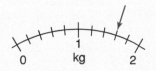

Example There are five small divisions between 0 and 1.
Each small division is 1 ÷ 5 = ⅕ or 0·2 kg.
The pointer is at 1·6 kg.

 D When we **estimate** a measurement it is a good idea to give a **range** for the estimate.
Example 200 g < mass of calculator < 500 g

When measuring we must choose the best **unit** to use and a suitable **measuring instrument**.

Example   When measuring the length of a pin we could measure in mm using a ruler.

 Area is the amount of space covered by a shape.
Area is measured in km^2, m^2, cm^2, mm^2.

Area of a rectangle = length × width
$$= \ell w$$

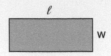

Area of a triangle = $\frac{1}{2}$ area of rectangle
$$= \frac{1}{2} \times \text{base length} \times \text{height}$$
$$= \frac{1}{2} bh$$

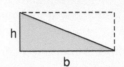

 We can find the **perimeter** and **area** of a **shape made from rectangles**.

Examples   To find the perimeter, we find the missing lengths.

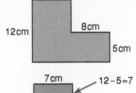

Perimeter = 7 + 7 + 8 + 5 + 15 + 12
$$= 54 \text{ cm}$$

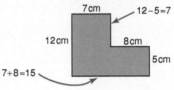

To find the area we divide the shape into two rectangles.
Total area = area of A + area of B

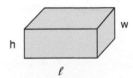

Area of A = 12 × 7   Area of B = 8 × 5
$$= 84 \text{ cm}^2 \qquad\qquad = 40 \text{ cm}^2$$

Total area = 84 cm² + 40 cm²
$$= 124 \text{ cm}^2$$

 The **surface area of a cuboid**

= 2(length × width) + 2(length × height) + 2(height × width)
= $2\ell w + 2\ell h + 2hw$

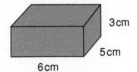

Example   Surface Area = $2\ell w + 2\ell h + 2hw$
$$= 2(6 \times 5) + 2(6 \times 3) + 2(3 \times 5)$$
$$= 60 + 36 + 30$$
$$= 126 \text{ cm}^2$$

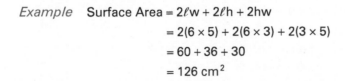

Test yourself

1 Find the missing numbers.
   a 3·2 cm = ___ mm   b 2·8 km = ___ m   c 3·6 m = ___ cm
   d 0·68 m = ___ mm   e 7·6 ℓ = ___ mℓ   f 470 g = ___ kg
   g 860 m = ___ km   h 6852 mℓ = ___ ℓ   i 752 cm = ___ m
   j 52 mm = ___ cm   k 864 mm = ___ m   l 0·06 kg = ___ g
   m 5·85 cℓ = ___ ℓ   n 386 cℓ = ___ mℓ   o 6·2 ℓ = ___ cℓ

2 Grant is training for cycling.
   a He trains on a 2500 m track. Yesterday he travelled 40 km on the track.
      How many laps did he do?
   b One drink bottle holds 750 mℓ. In a 50 km race he
      needs to drink 300 cℓ. How many bottles will he need?
   c A road race is 56 km. About how many miles is this?
   d His bike weighs 10 kg. About how many lb is this?
   e This is a map for a road race.
      How long is the race?
      About how many kilometres is this?

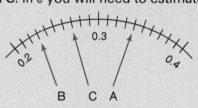

3 Find the measurements given by pointers A, B and C. In **c** you will need to estimate.
   a   b

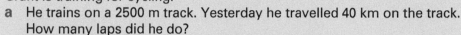

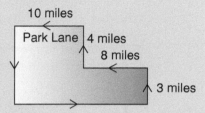

   c

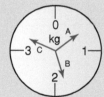

4 What was the measurement you wrote down for question **3a** pointer A?
   About how many kilograms is this?

5 Choose the best range.
   a A 0·5 m < length of woman's arm < 1 m
      B 1 m < length of woman's arm < 1·5 m
      C 1·5 m < length of woman's arm < 2 m
   b A 10 g < mass of pen < 1 kg
      B 10 g < mass of pen < 20 g
      C 10 g < mass of pen < 200 g

6 Suggest a metric unit, a measuring instrument and a way to measure each of these.
   a the mass of an egg
   b the length of your foot
   c the capacity of an egg cup

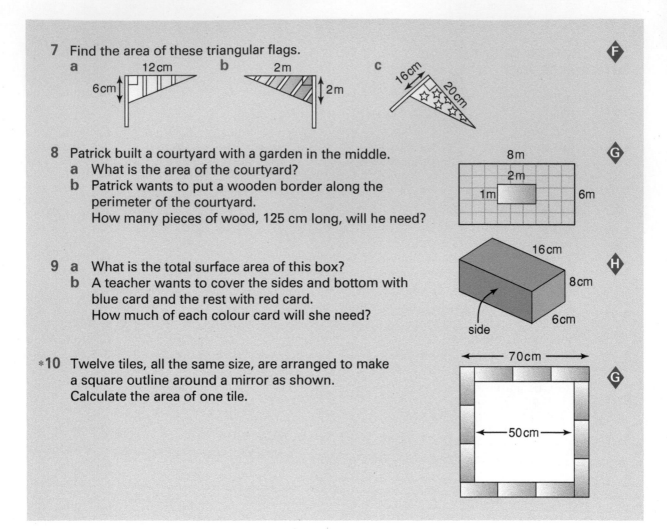

7 Find the area of these triangular flags.

a 12cm 6cm

b 2m 2m

c 16cm 20cm

8 Patrick built a courtyard with a garden in the middle.
  a What is the area of the courtyard?
  b Patrick wants to put a wooden border along the
     perimeter of the courtyard.
     How many pieces of wood, 125 cm long, will he need?

8m
2m
1m
6m

9 a What is the total surface area of this box?
  b A teacher wants to cover the sides and bottom with
     blue card and the rest with red card.
     How much of each colour card will she need?

16cm
8cm
6cm
side

*10 Twelve tiles, all the same size, are arranged to make
    a square outline around a mirror as shown.
    Calculate the area of one tile.

70cm
50cm

Handling Data Support

Collecting data

We often collect data using a **tally chart**.
The tally marks are put in groups of five. **JHT** is 5.

Example   Liz wrote down what her friends bought at the tuck shop.

What was bought	Tally	Frequency
crisps	JHT JHT	10
bun	JHT I	6
baked potato	IIII	4

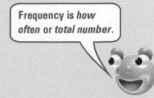

Frequency is *how often* or *total number*.

This table is sometimes called a **frequency table**.

Practice Questions 1, 2

Displaying data

This is a **pictogram**.

It shows the number of bikes in the stands at Grangehill School.

There were 30 bikes in Stand A and 25 in Stand B.

Stand A	🚲	🚲	🚲		
Stand B	🚲	🚲	🚲		
Stand C	🚲	🚲	🚲		
Stand D	🚲	🚲	🚲	🚲	🚲

🚲 represents 10 bikes

This is a **bar chart**.

It shows the number of Year 9 students at Brakehill School who biked to school each day last week.

One day last week it snowed.
Which day do you think it was?

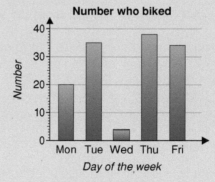

In hospital Jane's temperature was taken every four hours.

At 8 a.m. it was 37 °C, at noon it was 38 °C, at 4 p.m. it was 37·5 °C and at 8 p.m. it was 37·8 °C.

This **line graph** shows these temperatures. It was drawn by plotting the temperatures at 8 a.m., noon, 4 p.m., 8 p.m. and joining the points with straight lines.

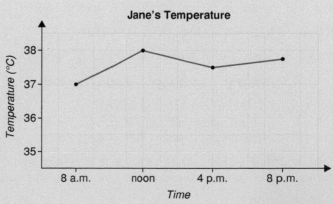

Practice Questions 5, 6, 8, 10, 15, 19

Databases

A **database** is a file of information. It can be written on paper or stored on a computer.

Example   Maths marks for Year 7 students.

Name	Age	Class	Test 1	Test 2	Test 3
D. Asch	11	7B	8	9	7
B. Aitken	11	7T	7	8	7
P. Archer	11	7A	6	4	5

A **computer database** organises large amounts of information.

Examples of databases that your school might use are:
> a database that stores details of each student
> a database that stores details of each teacher
> a database that stores details of the timetables of each student

Practice Question 18

Sorting diagrams

A class was surveyed about board games they played one wet lunch time.
This **Venn diagram** shows that

> 4 played both Scrabble and Monopoly
> 6 played Scrabble but not Monopoly
> 9 played Monopoly but not Scrabble
> 2 played neither Scrabble nor Monopoly.

This **Carroll Diagram** tells us about the weather in April.
It was cold and raining on 8 days, cold but not raining on 2 days, mild and raining on 5 days, mild and not raining on 15 days.

Practice Questions 3, 11

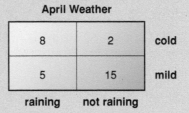

April Weather

8	2	cold
5	15	mild

raining   not raining

Mode and range

The **range** is one way of measuring the spread of data.
We find the range by subtracting the lowest value from the highest value.

Range = highest value – lowest value

Example   Eric wrote down the weights of some parcels he was posting for Christmas.

> **500 g   750 g   860 g   940 g   625 g   910 g**

The highest value is 940 g.
The lowest value is 500 g.
The range is 940 – 500 = 440 g.

The **mode** of a set of data is the value that happens most often.
Some sets of data have more than one mode and some have no mode.

Example   Julie wrote down what size tins of cat food people bought one hour.

| 450 g | 750 g | 250 g | 750 g | 750 g | 750 g |
| 450 g | 750 g | 450 g | 250 g | 750 g | 750 g |

The mode of this data is 750 g.
It is the size that was bought most.

Example   Julie wrote down the sizes bought another hour.

| 450 g | 750 g | 250 g | 450 g | 450 g | 750 g |
| 250 g | 750 g | 450 g | 750 g | 450 g | 750 g |

This data has two modes, 450 g and 750 g.

Practice Questions 12, 14, 17

Probability

The **probability** of something happening can be either **impossible**, **unlikely**, **likely** or **certain**.

Examples   When you throw a dice
it is impossible you will get a 7
it is unlikely you will get a 6
it is likely you will get a number greater than 1
it is certain you will get a 1, 2, 3, 4, 5 or 6.

We also use the following words to say **how likely** an event is to happen.
certain
good chance
even chance or fifty-fifty chance
poor chance
no chance

Examples   There is a poor chance I will fly to the moon before
I die.
There is a good chance I will use the phone next
year.
There is an even chance of getting a head when
I toss a coin.

Sometimes we show probability on a scale.

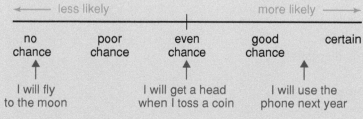

Practice Questions 4, 7, 9, 13, 16, 20

Handling Data

Practice Questions

T

1 Mardi thought that sparrows were the most common bird in her garden. She collected this data one morning to check.

  a Use a copy of the chart. Fill in the frequency column.

  b Which bird was most common that morning?

Bird	Tally	Frequency
Robin	ⵏⵏ III	8
Thrush	IIII	
Sparrow	ⵏⵏ ⵏⵏ III	
Starling	ⵏⵏ ⵏⵏ	
Blackbird	III	

T

2 Paul sells dental supplies.
Paul wrote down how many dental drills each dentist he visited had.

March	1	4	3	3	2	2	4	3	3	1
April	1	1.	4	3	1	2	3	3	1	1
May	1	3	1	3	2	4	3	1	3	1
June	4	1	2	3	1	4	1	3	3	3

Paul started a tally chart to show how many dentists had 1, 2, 3 and 4 drills.
Use a copy of Paul's chart.
Finish it.

Number of drills	Tally	Frequency
1		
2		
3		
4		

3 This diagram shows the people seen, on one day, by a doctor.

16	15	**Adult**
4	6	**Child**

Female   **Male**

  a How many were female adults?
  b How many were adults?
  c What else can you tell from this diagram?

4 Put these events in order of likelihood.
Put the most likely first.
  A A lion will eat you tomorrow.
  B The sun will rise tomorrow.
  C The Prime Minister will visit you this year.
  D You will fly to Hollywood some day.
  E Someone in your class will be sick tomorrow.
  F You will eat breakfast tomorrow.

5 Beth's soccer team played against the 'Raiders' once each week.
This bar chart shows the number of goals scored by her team.

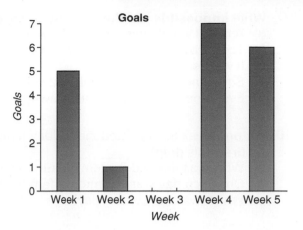

  a How many goals did Beth's team score in week 1?
  b Which week did they score no goals?
  c Which week did they score the most goals?
  d How many more goals did they score in week 1 than in week 2?

6 This pictogram shows the number of goals scored by hockey teams at a tournament.

  a How many goals were scored by
    i the Gaydene team
    ii the Merrihill team?
  b Which team scored the most goals?

Team	Goals
Harrowend	
Gaydene	
Merrihill	
Cavely	
Duneleigh	
Key: represents 2 goals	

7 Masi and Sue are playing a game.

Masi has these cards in her hand. Sue takes one of Masi's cards without looking.
Write **impossible, poor chance, even chance, good chance, certain,** for each of these.

  a Sue's card will have a number less than 10 on it.
  b Sue's card will have an even number on it.
  c Sue's card will have the number 12 on it.
  d Sue's card will have a number less than 8 on it.

8 Use a copy of this.
The pictogram shows the number of animals in a safari park in Kenya.
Each animal in this pictogram represents 10 animals in real life.

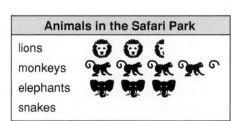

  a How many elephants are there in the park?
  b How many lions are there in the park?
  c Lyn looked at the pictogram.
    She thought there were 45 monkeys in the park.
    Explain why Lyn was wrong.
  d 22 snakes are brought into the park.
    Show what you would draw on the pictogram to show this.

9 Write **impossible**, **unlikely**, **likely** or **certain** for each of these.
   a A rabbit will fly past your window today.
   b Tomorrow someone in the world will cry.
   c It will become dark tonight.
   d Someone in your class will be away one day next week.
   e Everyone in your class will be away one day next week.

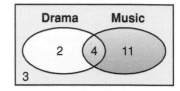

Midday Temperatures

10 For one week Sam plotted the midday temperatures.
   He drew this graph.
   a What was the midday temperature on Saturday?
   b On which day was the midday temperature 8 °C?
   c Which was the coldest day?

11 This diagram shows the number of students who are
   interested in Drama and Music.
   How many are interested in
   a both Drama and Music
   b neither Drama nor Music
   c Drama but not Music?

12 Find the range of these sets of marks.
   a 3, 7, 4, 2, 8, 9
   b 24, 26, 20, 25, 19
   c 36, 58, 96, 21, 72, 89, 32, 18, 73
   d 360, 372, 384, 336, 392, 390, 315
   e 186, 571, 386, 252, 864, 352, 164, 573

13 Which of these have a fifty-fifty chance of happening?
   a The next baby born will be a boy.
   b The next time you walk up stairs you will trip.
   c The next time you roll a dice you will get an even number.
   d When this spinner is spun it will stop on blue.

14

Temperatures in °C							
	Mon	Tue	Wed	Thu	Fri	Sat	Sun
London	15	15	12	13	14	11	8
Paris	18	20	16	13	15	17	14
New York	24	20	19	15	22	23	20

   a Calculate the range of temperatures for each city.
   b Which city had the greatest range?

15 Reece measured a plant at the end of each week.
   He drew this graph.
   a How tall was the plant after 1 week?
   b How tall was the plant after 6 weeks?
   c How long did it take for the plant to grow to
     40 mm?

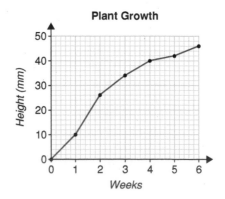

Plant Growth

16 Draw a scale like this one.

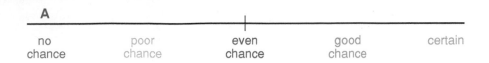

Write the letter which is beside each of these on the line to show how likely it is to happen.
A is done for you.

A A seal will drive a car down your street.
B You will watch TV tonight.
C It will snow next Christmas.
D You will grow taller.
E There will be no clouds in the sky tomorrow morning.
F You will visit a relative this weekend.
G You will see the city or town where you were born this year.

17 Write down the mode or modes of these test results.

Jan 5	Jon 6	Paul 7	Asmat 7	Ben 6	Jen 8
Peter 9	Ikram 7	Todd 6	Marie 5	Maha 9	Fran 8

18 Amanda gathered data from her friends about the board games they played one holiday.
This is what she put into a database.

Name	Monopoly	Pictionary	Draughts	Trivial Pursuit	Chess
Marie	N	N	Y	Y	Y
Beth	Y	N	N	Y	N
Riffet	N	N	Y	Y	Y
Holly	Y	Y	Y	Y	N
Chelsea	N	Y	Y	N	N
Sarah	Y	N	Y	Y	N
Heather	N	Y	N	N	Y
Pam	Y	N	N	N	N
Amanda	Y	N	N	Y	N

Amanda printed these lists of names:

a those who played Trivial Pursuit
b those who did not play Monopoly
c those who played both Pictionary and Monopoly
d those who played either Chess or Draughts
e those who played Monopoly but not Trivial Pursuit.

Write down the lists that were printed.

T

19 Sam's class made things for a sales table at sports day.
Sam started this bar chart of what they sold.
Use a copy of Sam's bar chart.

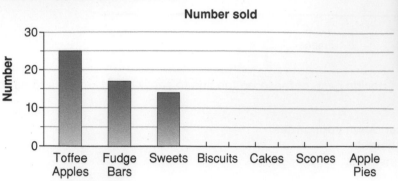

Number sold

They sold 29 biscuits
19 cakes
23 scones
17 apple pies.

a Draw the bars for these on your bar chart.
b How many fudge bars did they sell?
c How many sweets did they sell?
d What did they sell most of?
e Toffee apples cost 80p each.
How much did they make on these altogether?

20 Write some sentences.
Use the following in your sentences.

poor chance   *good chance*   *even chance*
no chance   *probable*   *very unlikely*
likely   *very likely*   *unlikely*
certain   *impossible*   *possible*

15 Planning and Collecting Data

······ **Key vocabulary** ··

**class interval,   data,   database,   data collection sheet,   experiment,
frequency,   frequency chart,   grouped data,   interval,
questionnaire,   survey,   table,   tally**

 Viewing Figures

Television has become a large part of many people's
lives.
There are lots of questions we might like answered
about TV.

>   How much TV do children under 5 watch each
>   day?
>   What percentage of households watch TV while
>   eating dinner?
>   What programmes do 12 year olds like best?

These questions could be answered by doing a
survey.
Think of a survey you might like to do to find out something about TV.
What might you find out in your survey?

Surveys — planning and collecting data

Before you carry out a **survey** you need to **plan it**.

Step 1 **Decide on the purpose of your survey**.
What is the problem or the question you would like answered?

Example How will the volume of traffic on motorways change over the next 20 years compared with traffic on other roads?

Step 2 **Decide what data needs to be collected**.
It is useful to think of possible results you might get from your survey.
This helps you decide what data needs to be collected.

Example For the question 'What computer resources do pupils use and what for?', these might be some of the results.

Most pupils use the Internet to get information for school projects.
Those who use a spreadsheet, use it for school work.
Few pupils use databases.
Some pupils use e-mail to keep in contact with friends.
The software used for the longest time is the Internet.

You might decide to collect the following data from each pupil you survey.

Resource used	Time used each week	Used for
Word processing		
Internet		
Database		
Spreadsheet		
E-mail		
Games		
Other ____		

Step 3 **Decide where to collect the data from and how much data to collect**.
You could collect data from these sources.

a A **survey of people**.
   Sometimes it is not practical to ask *every* possible person to answer the survey. We choose a **sample** of people. The sample size should be as large as it is sensible to make it.

   > You have to decide how many people to ask.

   Example If we want to know what music 12 year olds like we can't ask *every* 12 year old.
   We choose a **sample** to ask. How big the sample is will depend on time, cost, who the survey is for, ...

b An **experiment** where you observe, count or measure something.

   Example If we want to know if boys or girls have longer arms, we have to measure.

c **Other sources** such as reference books, newspapers, Internet websites, historical records, databases, CD-ROMs, ...

   Example If we want to know the birth-rate over the last ten years we have to use historical records.

a and b are called **primary sources** because you collect the data yourself.
c are called **secondary sources** because someone else has collected the data.

Discussion

For each of the questions given below suggest

- some possible results
- what data you would need to collect
- sources you could collect data from
- how much data to collect.

Think how you might use ICT to help you.

Discuss

- Does the school need more rubbish bins? If so where?
- How do pupils travel to school? Why do they travel this way?
- Do different types of magazines have different types of advertisements? If so, why?
- How will the number of cars in more developed countries change over the next 50 years compared with the number of cars in less developed countries?
- Do different types of newspapers use words or sentences of different length? If so why?
- What improvements could be made at the local gymnasium?
- Which is your best catching hand?
- Is the population ageing?

Once you have planned your survey, design a **data collection sheet** or **questionnaire**. The data collection sheet could be a **tally chart** or **frequency table**.

Examples **Tally Chart**
Survey on rubbish

Pieces of Rubbish		
Place	**Tally**	**Frequency**
Outside Block A	ЖТ ЖТ ЖТ II	17
Outside Block B	ЖТ II	7
Outside Library	II	2
In Playing Fields	ЖТ ЖТ ЖТ ЖТ ЖТ ЖТ I	31

Remember ЖТ is 5.

Frequency means *how often* or *how many*.

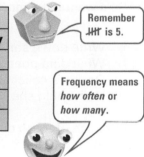

Frequency Chart
Survey on ways of travelling to school

Name	Year	Form of Travel	Distance (m)	Time Taken (minutes)

You need to decide what units to use for any measurements.

Discussion

Discuss how to design a collection sheet to collect the data needed for the questions given in the Discussion on the previous page.
As part of your discussion, design a collection sheet for each.
Remember to decide what units to use for measurements.

1 Reece did a survey to find out which was the most popular place to go for a class trip.
He asked this question.
Which of these places do you want to go to most?

   Swimming Pool   Funfair   Cinema   Picnic

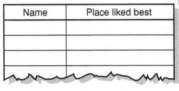

  a Write down two things Reece might find out from his survey.
  b Which of these collection sheets do you think is best?
     Say why.

Place	Tally	Number
Swimming		
Funfair		
Cinema		
Picnic		

Sheet 1

Name	Place liked best

Sheet 2

2 Suki wanted to know which of these items people would buy to help raise funds.

   cakes   cards   sweets   pies   pens   diaries

  a Write down some possible results Suki could get from this survey.
  b What data does she need to collect?
  c Design a collection sheet to collect this data.
  d How could she collect the data?

3 For each of these questions, what would be the best way to collect the data?
   Choose from the box.

   A questionnaire or data collection sheet   **B** experiment   **C** other source

  a What percentage of motor vehicle accidents in Britain occur after 6 p.m.?
  b How often do people go to the cinema? At what times?
  c Do most people in your class have a reaction time of less than $\frac{1}{2}$ second?
  d Which items does the canteen run out of most often at lunch time?
  e What sport do people in your class watch most often?
  f Is the height of most people the same as three times the distance round their head?
  g How has the birth-rate increased or decreased over the last 20 years?

4 Write down what data you would need to collect to answer these questions.
  a Which is the most popular brand of toothpaste?
  b Do right-handed people in your class catch better than left-handed people?
  c Which month of the year has the highest rainfall?
  d Can taller people hold their breath longer than shorter people?

5 Suggest some possible answers for the questions in question **4**.

6 Design a data collection sheet or questionnaire to collect the data for questions **4a–c**.

Review Danya wanted to know how many times each week pupils used the school library, when they used it and for how long.
a Write down some possible results Danya could get from this survey.
b What data does she need to collect?
c Design a collection sheet or questionnaire to collect the data.
d Explain how she could collect the data.

Grouped data

Ben was doing a survey on what people spent at the canteen.
He wrote down the amounts spent one lunchtime.

£4·95	£2·50	£3·90	£1·30	£10	£3·50
£3	£5	£8·50	£2	30p	£6·99
£9	£4·60	£5	£1·50	£9	£5·20
£7·20	£2·30	£4·70	£7·60	50p	£3·49
£4	£15				

It is best if this data is grouped.
The groups must be of **equal width**.

Example   £0·01–£1, £1·01–£2, ... rather than £0·01–£0·50, £0·51–£3
      ↑ equal width ↑   ↑ not equal width ↑

The groups are called **class intervals**.
We can group the data on a tally chart or frequency table.

Example
Ben drew this frequency table.

Amount Spent	Tally	Frequency
£0·01–£2	ЖНГ	5
£2·01–£4	ЖНГ II	7
£4·01–£6	ЖНГI	6
£6·01–£8	III	3
£8·01–£10	IIII	4
over £10	I	1

Sometimes the last class interval is *open*.

Discussion

In the previous example, would it have been more useful if Ben had grouped the data in intervals of £1? **Discuss**.
What about intervals of £5?
What about intervals of 50p?
Which interval is the most useful and why? **Discuss**.

Exercise 2

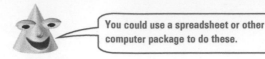

You could use a spreadsheet or other computer package to do these.

1 This data gives the number of CDs owned by Sam's friends.

| 5 | 12 | 16 | 3 | 24 | 0 | 9 | 0 | 27 | 0 | 4 |
| 2 | 17 | 14 | 18 | 24 | 6 | 0 | 4 | 9 | 10 | 8 |

Copy and complete the frequency chart.

Number of CDs	0–4	5–9	10–14	15–19	20–24	25–29
Tally						
Frequency						

2 This data gives the points scored by the people who entered a 'talk non-stop' competition run by a radio station.

| 29 | 19 | 25 | 28 | 26 | 21 | 37 | 25 | 24 | 29 | 42 | 27 |
| 28 | 32 | 35 | 41 | 18 | 20 | 27 | 48 | 22 | 17 | 28 | 29 |

a Put this data into a frequency chart. Use the class intervals 15–19, 20–24, 25–29, etc.
b Regroup the data using the intervals 15–24, 25–34, ... Have the last class interval, >44.
c Which way is more useful?

3 Charles wrote down the ages of the people who came to the class play.

| 28 | 62 | 34 | 29 | 32 | 47 | 68 | 17 | 37 | 48 |
| 37 | 70 | 49 | 43 | 32 | 82 | 61 | 29 | 38 | 64 |

a Why are the intervals 0–9, 10–19, 20–29, etc. more useful than the intervals 0–19, 20–39, 40–59, etc.?
b Put this data onto a frequency chart.
Leave the last class interval open as '80 or over'.

Review Victoria wrote down the number of words in the first 20 sentences of an article in 'Smash Hits' magazine.

| 10 | 21 | 41 | 5 | 48 | 13 | 25 | 38 | 13 | 25 |
| 11 | 53 | 6 | 8 | 32 | 34 | 5 | 8 | 46 | 20 |

a Put this data into a frequency chart. Use the class intervals 1–10, 11–20, 21–30 etc.
b Regroup the data using the intervals 1–5, 6–10, 11–15, etc.
c Which way is more useful?

Collecting data

 Practical

Plan a survey.
You could choose one of the surveys already mentioned in this chapter **or** you could use one of the suggestions below **or** you could make up your own.
Check your choice with your teacher.
Follow the steps given on page 366.
Design a collection sheet or questionnaire. Remember you may need to group the data. Collect the data.

Suggestions
How old are the grandparents of pupils in your class?
What items should the canteen sell that it doesn't at the moment?
How many fast-food meals do people in your class eat each week and when?
What is the life expectancy of people in different countries? Has this changed in the last 20 years?
Do right-handed people spell better than left-handed people?

Summary of key points

 Use these steps to **plan a survey**.

Step 1 **Decide on the purpose of the survey**.

Step 2 **Decide what data needs to be collected**.
Think of possible results you might get from your survey.

Step 3 **Decide where to collect the data from, and how much to collect.**
You could collect data by

a doing a survey using a questionnaire or collection sheet

b doing an experiment where you count or measure something

c gathering data from other sources such as books, newspapers, the Internet, historical records or a database.

 Once a survey has been planned a **collection sheet** or **questionnaire** often has to be designed.

 Some data is best if it is **grouped**.

Examples ages, amount of money, times, ...

The groups are often called **class intervals**.
Class intervals must be of equal width.
A tally chart or frequency chart can be used to collect grouped data.

Example This frequency chart shows the ages of 30 contestants in a quiz show.

Age	20–29	30–39	40–49	50–59	over 59
Tally	IIII	JHT II	JHT JHT II	JHT I	I
Frequency	4	7	12	6	1

This is an open class interval.

The last class interval may be open.

Test yourself

1 Hal did a survey to find out the most popular activity of Year 9 pupils.
   He asked this question.

   Which of these activities do you enjoy doing most?

   playing sport   watching TV or videos   doing homework
   talking with friends   doing an artistic activity (dancing, drama, art, craft)

   a Write down two things Hal might find out from his survey.
   b Which of these collection sheets do you think is best? Say why.

Name	Class	Activity liked best

Sheet 1

Activity	Tally	Number
Sport		
TV / Videos		
Homework		
Friends		
Artistic		

Sheet 2

2 For each of these questions, how would you collect the data?
   Choose from the box.

   > A questionnaire or data collection sheet
   > B experiment
   > C data someone else has already collected

   a What percentage of accidents are caused by drunk drivers?
   b Do boys or girls have longer big toes?
   c Which fast-food is the most popular?

3 Mendel wanted to know how often people eat take-aways and on which days.
   a Write down two things Mendel might find out from his survey.
   b What data does Mendel need to collect?
   c Design a data collection sheet or questionnaire for this data.
   d How could he collect the data?

4 Harry wrote down the number of books some Year 8 pupils had taken out of the
   library in the last six months.

   26   12   18   39   57   89   116   98   36   4
   39   47   89   104 32   20   71   82   23   46

   a Put this data into a frequency chart with class intervals 0–19, 20–29, 30–39, ...
   b Regroup the data using the intervals 0–49, 50–99, 100–150.
   c Regroup the data using the intervals 0–24, 25–49, 50–74, 75–99, over 99.
   d Which class intervals do you think are the most useful?

5 Choose a survey topic.
   Plan your survey using the steps on page 366
   Design a data collection sheet or questionnaire to collect the data.

16 Mode, Range, Median, Mean, Displaying Data

You need to know

✓ mode and range   page 358

✓ displaying data   page 357

·· Key vocabulary ··

average, bar chart, bar-line graph, frequency, interval, label, mean, median, modal class/group, mode, pie chart, range, represent, statistic, title

Averaging it Out

- We often hear the word 'average' used.

  The average house price is ...

  The average length for a 6 month old baby is ...

  The average wage is ...

  The average annual rainfall in York in January is ...

  Think of other examples where average is used.
  Make a collection of labels from packets where the
  word average is used.
  Make a collection of articles from newspapers or
  magazines where the word average is used.

- The 'average' is a way of summarising the data into a single number.
  'The average age of the pupils in 7P is about 11'
  'The average height of the pupils in 7P is about 1.5 m'
  'The number of children in the families of 7P is about 2·4'
  Estimate some 'averages' about your class.

Mode and range

Remember

The **mode** is the most commonly occurring data value

Example   For 8, 10, 12, 12, 12, 14, 15 the mode is 12.
For 3, 6, 4, 7, 6, 8, 2, 3, 6, 3 the modes are 3 and 6.

The range is a number not an interval.

The **range** is the difference between the highest and lowest values.

Example   The range of the above set of data is 15 − 8 = 7.

The mode is the only suitable statistics to find for non-numerical data.

Example   A radio station wanted to know which type of music was most popular. They asked listeners to phone in and tell them.

The modal type of music is **pop**.

This is non numerical data.

Type	Number of people
pop	83
country	21
dance	27
rap	31

If a frequency table has grouped data we find the **modal class** or **modal group**.

Example   This frequency table shows the number of people on a bus each morning in July.

Number of people on bus

Number on bus	0–9	10–19	20–29	30–39
Frequency	8	16	5	2

The modal class is 10–19 because this class interval has the highest frequency.

Discussion

The radio station in the example above wanted to know the modal type of music so it could play more of this.

Think of other instances when it would be useful to know the mode. **Discuss**.

Exercise 1

1   Write down the mode or modes of these test results.
   a   5, 6, 6, 7, 7, 7, 7, 8, 9
   b   10, 11, 11, 11, 12, 12, 12, 13, 15, 18
   c   36, 37, 38, 39, 39, 41, 41, 43, 46, 47

2   Write down the range of each set of test results in question **1**.

3 Mrs Grey wanted to organise an after school programme for children in her area.
She collected this data.

	Number of children in family aged 5–12												
Baker St	1	3	2	2	1	2	2	3	4	3	1	4
Grove St	5	3	2	3	4	3	2	3	4	3	1	3
Heath St	2	1	1	3	2	1	3	2	1	3	2	1

  a What is the modal number of children in each family in
   i Baker St   **ii** Grove St   **iii** Heath St?
  b What is the modal number of children in all three streets?

4 Mr Salt drew this table to show the sizes of women's trainers he sold last week.
  a What is the modal shoe size?
  b What is the range of shoe sizes?
  c Why might Mr Salt want to know the modal shoe size sold?

Shoe Size	Tally	Frequency
4	ⵏⵏⵏ III	8
5	ⵏⵏⵏ ⵏⵏⵏ I	11
6	ⵏⵏⵏ II	7
7	III	3
8	I	1

5 These tables give the results Sally's class got in Maths and Science.
Write down the modal class for each of them.

a   *Maths*

Score	Tally	Frequency
1–10	III	3
11–20	ⵏⵏⵏ I	6
21–30	ⵏⵏⵏ II	7
31–40	ⵏⵏⵏ ⵏⵏⵏ II	12
41–50	III	3

b   *Science*

Percentage	Tally	Frequency
41–50	ⵏⵏⵏ I	6
51–60	ⵏⵏⵏ ⵏⵏⵏ	10
61–70	I	1
71–80	ⵏⵏⵏ ⵏⵏⵏ	10
> 80	III	3

*6 **a** Write down five pieces of data which have mode 6.
  b Write down seven pieces of data which have no mode and range 10.
  c Write down six pieces of data which have modes 8 and 10 and range 8.

Review 1 Olivia wrote down the number of books she borrowed from the library on her last eight visits.   4   3   6   2   7   3   3   3
a What is the modal number of books she borrowed?
b What is the range of the number of books she borrowed?

Review 2 This table gives the time spent on homework by six girls on three nights.
a What is the modal time spent on homework over all three nights?
b What is the modal time spent on Wednesday?
c What is the modal time spent on Thursday?
d What is the range of the times spent on Wednesday?

Time Spent on Homework

Tuesday	Wednesday	Thursday
1 hour	50 min	20 min
50 min	45 min	15 min
20 min	50 min	10 min
45 min	50 min	15 min
30 min	30 min	30 min
55 min	45 min	25 min

Review 3 50 students each tossed a dice 100 times.
This table shows the number of sixes tossed. What is the modal class?

Number of sixes	0–4	5–9	10–14	15–19	20–24	25–29
Frequency	3	8	12	18	7	2

Mean

Olivia 2   Joseph 5   Emily 8   Thomas 3   Rebecca 2

These pictures show the number of Easter eggs given to five children.
By sharing the Easter eggs equally between the children we find the **mean** number of eggs given to each child.
Altogether the children were given 20 Easter eggs.
If these were shared equally between the children each would have four.

$$20 \div 5 = 4$$

The mean number of Easter eggs given to each child is four.

The **mean** of a set of data is the sum of the data values divided by the number of data values.

> The mean is often called the 'average'. In fact, the mean, median and mode are all measures of the average.

$$\text{mean} = \frac{\text{sum of data values}}{\text{number of values}}$$

Worked Example

This data gives the points scored by the horses in a show.

78   59   74   64   75   82   71   77   88   78   67   69

Find the mean number of points scored.

Answer

$$\text{Mean} = \frac{78 + 59 + 74 + 64 + 75 + 82 + 71 + 77 + 88 + 78 + 67 + 69}{12}$$

$$= \mathbf{73 \cdot 5}$$

> You can use the calculator when there are lots of values.

Discussion

- Three meals have an average price of £5.
  What might the price of these meals be? **Discuss**.
- In five tests, Tina's average mark was 66.
  What might Tina's marks be? **Discuss**.
- Ben's marks for his first four maths tests were   72   64   63   69.
  What was Ben's mean mark?
  What mark does Ben need to get in his next test to raise his mean to 70? **Discuss**.

> People often call the mean 'the average'.

Exercise 2

1 Calculate the mean.
   a   2, 2, 3, 4, 4, 4, 5, 5, 7, 7, 8, 9
   b   2, 3, 7, 11, 12, 19, 23, 25, 29, 31, 36
   c   4, 7, 7, 7, 7, 8, 8, 9, 10, 10, 11, 13, 14, 14, 15
   d   22, 19, 14, 16, 27, 11, 32, 41, 17, 43
   e   5·2, 4·6, 2·7, 7·1, 11·4, 5·3, 6·2, 2·3
   f   8·7, 1·4, 3·7, 2·6, 8·9, 5·9, 1·2, 0·9, 4·3, 6·2
   g   24·6, 30, 25·3, 82·4, 63·9, 51·1, 80, 32·7, 48·75

You can use your calculator if you need to.

2 The number of goals scored by a football team in 15 matches were
   3   0   1   0   0   2   1   2   0   3   0   4   0   1   1
   Find the mean number of goals scored.

3 Olivia was studying beetles in Science.
   She wrote down the lengths, in centimetres, of 7 beetles.
   2.2   1.8   2.5   2.1   1.9   2.4   2.5

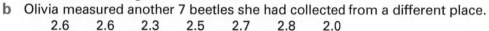

Link to science.

   a   Find the mean length of the beetles.
   b   Olivia measured another 7 beetles she had collected from a different place.
      2.6   2.6   2.3   2.5   2.7   2.8   2.0
      What was the mean length of these beetles?
   c   What was the mean length of all 14 beetles?
      Could you find this by finding the mean of the answers to **a** and **b**?

4 Jan wrote down her results for some of her subjects.

ENGLISH	68	75	71	74		
MATHS	81	62	79	82	86	
SCIENCE	70	75	58	68	72	47
FRENCH	50	46	58	55	56	

   a   What is her mean for Science?
   b   In which subject does Jan have the highest mean?
   c   Jan has to drop one of these subjects next year. Which one do you think she should drop? Why?

5 Matthew weighed himself five times on a set of very accurate scales. The five readings were 61·380 kg, 61·284 kg, 60·982 kg, 61·034 kg, 61·000 kg.
   Find the mean of the five readings.

6 The mean number of hours of training done by a group of ten swimmers last week was 12.
   a   How many hours training did the ten swimmers do in total last week?
   b   Another two swimmers were asked how much training they did last week. One said 14 hours and the other 10 hours.
      Calculate the new mean number of hours training for all 12 swimmers.

7 What three numbers could go in these boxes so that the mean of the three numbers is 7?

8 Jack has four number cards.
Their mean is 5.
Jack picks another card.
The mean of his five cards is 6.
What number is on the new card?

9 Asad got these marks in his first four mental tests.
8   7   9   8
The mean is 8.
After his fifth test, the mean of his marks was still 8.
What mark did he get in his fifth test?

*10 Tamara is playing a game.
To win she has to get a mean score of exactly 80 in three rounds.
In the first two rounds she scored 84 and 75.
What must she score in the third round to win?

Review 1

Maths Test Marks

Callum	68	73	55	74	65
Dylan	93	39	65	73	70

a Who has the highest mean mark? By how much?
b After the next test Callum's mean was 67.
What did he score in this test?

Review 2   The mean hourly rate for a group of ten workers at a factory is £7.
a How much do the ten workers earn in total each hour?
b Another two workers are employed. One has an hourly rate of £8·50 and the other an hourly rate of £6·70.
Calculate the new mean hourly rate for all twelve workers.

Review 3   Joshua wrote down how late the bus was each morning.

	Monday	Tuesday	Wednesday	Thursday	Friday
Minutes Late	1	8	5	2	

The mean for Monday to Thursday is 4 minutes.
After Friday the mean for the five days is 5 minutes.
How many minutes late was the bus on Friday?

Data is sometimes given in a **frequency table**.
We can find the mean using the table.

Example   This frequency table shows how many sick days Pamela's friends had last term.

Number of sick days	Frequency	Total number of sick days
1	4	4
2	4	8
3	2	6
4	1	4
5	1	5

The total number of sick days is found by multiplying the number of sick days by the frequency.

$$\text{mean} = \frac{\text{sum of data values}}{\text{number of values}}$$

We add up the total sick days column.

$$= \frac{\text{total number of sick days}}{\text{total number of students}}$$

Add the frequencies.

$$= \frac{4+8+6+4+5}{4+4+2+1+1}$$

$$= \frac{27}{12}$$

$$= 2\cdot25$$

The mean number of sick days is **2·25**.
If we wrote out the numbers of sick days each person had we would get

$$\underbrace{1,1,1,1}_{4} , \underbrace{2,2,2,2}_{8} , \underbrace{3,3}_{6} , \underbrace{4}_{4} , \underbrace{5}_{5} = 27$$

It is quicker to multiply the number of sick days by the frequency to get the total number of sick days column.

Exercise 3   **Only use the calculator when you need to.**

1 The number of goals scored by a hockey team are given in this table.
   Find
   a   the mode   b   the range
   c   the mean.

Number of goals	0	1	2	3
Frequency	1	3	3	5
Total goals				

2 This frequency table gives the class sizes at Aleisha's school.
   Find
   a   the modal class size
   b   the range in class size
   c   the mean class size.

Class size	28	29	30	31	32
Number of classes	1	0	2	5	2
Total number of pupils	28	0			

3 Joel sold some raffle tickets to his neighbours.
   He drew this table to show how many each bought.
   a   What was the modal number of tickets bought?
   b   What was the range?
   c   What was the mean number of tickets bought?

Number of Tickets per Person	Frequency	Total number sold
0	3	0
1	11	11
2	10	
3	8	
4	5	
5	3	

Handling Data

Review This table shows the marks scored in a 'parents' quick quiz' at a school mathematics evening.
a What was the range of marks?
b What was the modal mark?
c What was the mean mark?

Number of marks	Number of parents	Total marks
0	2	
1	7	
2	5	
3	2	
4	3	
5	1	

Median

These pictures show the number of Easter eggs given to five children.

These pictures show the number of eggs put in order from smallest to largest.

The median number of Easter eggs is the number that is in the middle picture.
The median number of Easter eggs is 3.

The median is the middle value when a set of data is arranged in order of size.

When there is an even number of values the median is the mean of the two middle values.

Worked Example
This list gives the number of squares of pizza eaten by 9 friends.
     2, 5, 6, 6, 7, 10, 11, 12, 15
What is the median?

Answer
The values are already in order of size.   2, 5, 6, 6, ⑦, 10, 11, 12, 15
The middle value is 7. The median is **7**.

Worked Example
Find the median of
a 13, 14, 17, 19, 18, 16, 14, 13, 18, 25, 23   b 21, 10, 72, 15, 43, 20, 56, 83

Answer
a Put the values in order.
   13, 13, 14, 14, 16, ⑰, 18, 18, 19, 23, 25
   The middle value is 17.
   The median is **17**.

b Put the values in order.
   10, 15, 20, ㉑, ㊸, 56, 72, 83
   There are two middle values, 21 and 43.
   The median is the mean of these two values.
   median $= \frac{21+43}{2}$
   $\qquad = \frac{64}{2}$
   $\qquad = 32$
   The median is **32**.

Exercise 4

1 Find the median of
   a 1, 5, 8, 12, 16, 24, 29
   b 3, 9, 5, 12, 16, 12, 7
   c 13, 24, 16, 33, 17, 31, 28
   d 80, 20, 30, 50, 40, 10, 32
   e 4, 9, 5, 9, 7, 12
   f 42, 38, 16, 57, 32, 41, 28, 52

2 In a Mental Test, a group of students got these marks:
   17   16   11   16   18   12   16
   What was the median mark?

3

Wei-Hsin	Rajiv	Victoria	Rebecca	Caleb	Charlotte	Jeremy	Anna	Sandra
1·47 m	1·57 m	1·61 m	1·39 m	1·65 m	1·62 m	1·76 m	1·53 m	1·49 m

   What is the median height of these students?

4 James collected this data from his friends.

	James	Naim	Toby	Casey	Tom	Isla	Sam
Shoe size	6	7	5	6	7	8	8
Maths mark	65	62	78	52	76	49	83
Family size	6	5	5	7	3	4	4
Height (in cm)	158	152	160	154	161	159	163

   a What is the median shoe size?
   b James thinks he has the median maths mark.
      Is he correct?
   c Who has the median family size?
   d What is the median height?

5 This table shows the ages (in years) of the cats in
   Baker Street and Ridge Street.

Baker Street	2	5	0	7	12	5	7	8	9
Ridge Street	1	3	8	7	14	5	3	8	

   a Find the median age of the cats in i Baker Street   ii Ridge Street.
   b Find the median age of all of the cats.

6 This bar chart shows the number of pupils who were
   late to school each day one week.
   What is the median number of pupils late per day?

   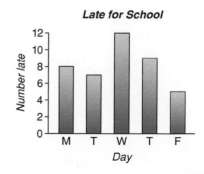

Review 1 Jeremy got these points in five computer games.
98 000, 89 000, 99 000, 95 000, 96 000
What was the median number of points?

Review 2 Tina counted the people at the gym on Wednesday and Friday lunchtimes in the first term.

| Wednesday | 26 | 28 | 21 | 18 | 22 | 16 | 18 | 21 | 12 | |
| Friday | 35 | 37 | 33 | 30 | 28 | 24 | 27 | 19 | 21 | 8 |

a Find the median for **i** Wednesday **ii** Friday.
b Find the median for both days together.

Mean, median, mode

Investigation

∗Basketball Scores

You will need a calculator.
There are four basketball teams in a competition.

Team	Heights (cm)
Green Valley	149, 152, 160, 158, 160, 162, 158, 156
Red Ferns	155, 156, 154, 158, 152, 157, 160, 157
St Peters	143, 147, 150, 152, 155, 149, 150, 154
Albion	150, 146, 159, 152, 144, 158, 157, 158

The heights of the team members are given in this table.
Find the modal height, range, mean and median for each team.

Team	Scores							
Green Valley	35	17	33	35	36	16	27	41
Red Ferns	16	24	22	31	29	15	31	32
St Peters	18	21	31	22	12	39	20	19
Albion	14	16	18	31	31	32	30	22

This table gives the scores for the last eight games of each team.
Find the modal score, range, mean and median for each team.

Does there seem to be any relationship between the heights of the players in a team and the scores? **Investigate**.

Bar charts and bar-line graphs

We often draw graphs to show the results of our data collection. A 'picture' is often worth a thousand words.

This frequency table and **bar-line graph** show the number of glasses of water Kathy drank each day for a month.

Number of glasses	0	1	2	3	4	5
Tally	II	IIII	III	⊞ ⊞ ⊞	IIII	II
Frequency	2	4	3	15	4	2

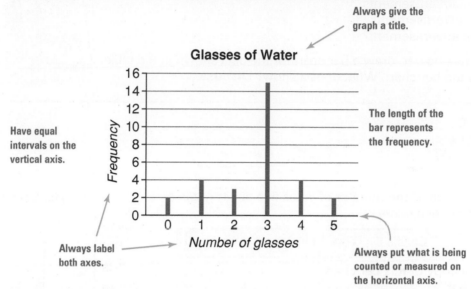

Always give the graph a title.

Have equal intervals on the vertical axis.

The length of the bar represents the frequency.

Always label both axes.

Always put what is being counted or measured on the horizontal axis.

T *Example* This table shows the favourite day of the week of pupils in year 7 at Lancewood School.

Day	Mon	Tue	Wed	Thu	Fri	Sat	Sun
Girls	10	4	4	5	7	10	4
Boys	8	7	9	9	8	9	4

We can draw a **bar chart** to show the data.

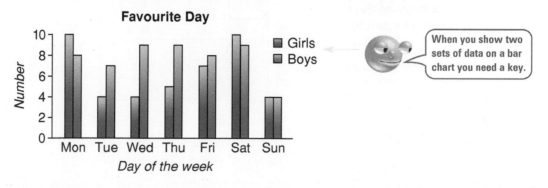

When you show two sets of data on a bar chart you need a key.

Discussion

An advertising agency wanted to know if there was any difference in where advertisements appeared in two magazines.

Page number	1	2	3	4	5	6	7	8	9	10
No. of advertisements (Her!)	8	7	6	4	0	1	1	2	0	1
No. of advertisements (Him)	2	2	1	3	4	5	3	4	5	4

This table gives the number of advertisements on the first ten pages of two magazines, Her! and Him.

How many pages in each magazine have
a two advertisements
b no advertisements
c one advertisement?

Discuss how to draw a bar chart for the data shown in the table. Draw the bar chart. What does it show? **Discuss**.

Exercise 5

 You could use a spreadsheet to draw some of these.

1 This list gives the number of votes five pupils got in a class speech contest.

Sajid 5	Cassie 7	Zeta 13	Ben 11	Claire 10

Use a copy of this bar chart.
Finish it.

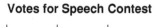

Votes for Speech Contest

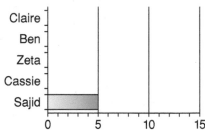

2 Sasha asked her riding friends how many shows they had entered this season.
She drew this frequency table.
Use a larger copy of the bar-line graph shown. Finish it for this data.

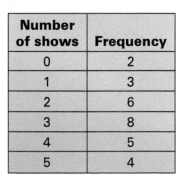 Remember to give graphs a title and label the axes.

Number of shows	Frequency
0	2
1	3
2	6
3	8
4	5
5	4

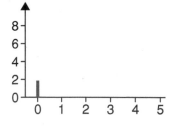

3

Number of children	Year 7	Year 11
1	3	4
2	5	6
3	8	5
4	1	2
5	3	2
6	0	1

Asha asked 20 Year 7 and twenty Year 11 pupils how many children were in their family.
This table shows her results.
Use a copy of the bar chart below. Finish it.
Shade the Year 7 and Year 11 bars differently and fill in the key.

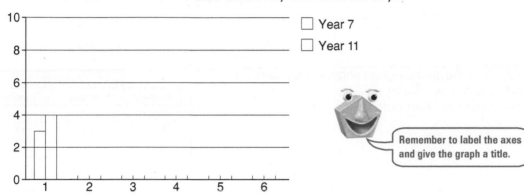

☐ Year 7
☐ Year 11

Remember to label the axes and give the graph a title.

4 Julia wrote down the number of letters in the surnames of the teachers in her school.

5 6 8 9 7 7 4 5 6 8 5 7 5
6 8 4 7 6 6 8 6 9 8 5 6 5

a Copy and fill in this frequency table.
b Draw a bar-line graph for this data.

Number of letters	4	5	6	7	8	9
Tally						
Frequency						

5 Penny asked the boys and the girls in her class which subject they liked best.
This table shows her results.

Subject liked best	Maths	Science	Tech	English	PE
Boys	4	2	5	3	1
Girls	5	4	3	1	2

Draw a bar chart for this data.
Have separate bars for boys and girls and a key.

T

Review 1 Phoebe wrote down the ages of the children who came for help at the library.

8   7   9   10   7   7   8   8   8   10   9   7
10 6   9   7   8   8   7   10 8   9   10 8

a   Copy and fill in this frequency table.

Age	6	7	8	9	10
Tally					
Frequency					

b   Use a copy of this bar-line graph. Finish it.

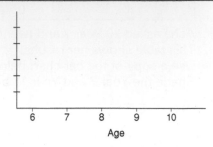

Age

Review 2 This frequency chart shows what animals were at the vets one weekend.
Use a copy of the bar chart below.
Finish it.

	Frequency	
Animal	Saturday	Sunday
Cat	13	19
Dog	16	17
Bird	1	3
Rabbit	0	2
Mouse	4	1

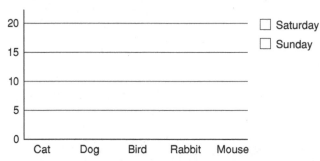

☐ Saturday
☐ Sunday

Bar charts for grouped data

The owner of a supermarket wanted to know how many items each customer bought.
He did a survey.
This frequency table shows his results.
The number of items are grouped into equal class intervals. He drew this bar chart.

Number of items	Tally	Frequency
1–5	ЖНΙ	6
6–10	ЖН ΙΙΙΙ	9
11–15	ЖН ЖН	10
16–20	ЖН ΙΙ	7
21–25	ЖН ЖН	10
26–30	ЖНΙ	6

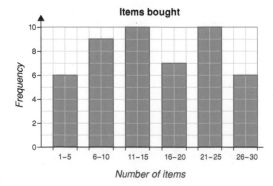

Items bought

◄──── The class interval is written under each bar.

Exercise 6

T

1 7M had a cricket competition with another class.
This table gives the number of runs made by the
students in 7M.
Use a copy of the bar chart.
Finish it.

No. of Runs	Tally	Frequency
0–4	II	2
5–9	III	3
10–14	IIII I	6
15–19	IIII IIII	9
20–24	IIII IIII	10

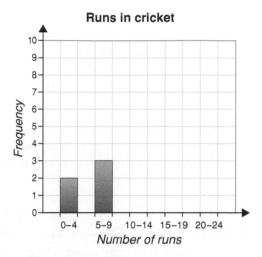

Runs in cricket

T

2 Mr Chan's class had a maths test. He wrote down the results.

| 14 | 9 | 3 | 8 | 10 | 2 | 8 | 12 | 3 | 16 | 8 | 11 | 8 | 8 | 17 | 16 |
| 2 | 7 | 13 | 6 | 12 | 9 | 10 | 10 | 6 | 4 | 17 | 6 | 2 | 11 | 7 | 19 |

Use a copy of this.

a Fill in the frequency table and finish drawing the
bar chart.

Mark	Tally	Frequency
1–4		
5–8		
9–12		
13–16		
17–20		

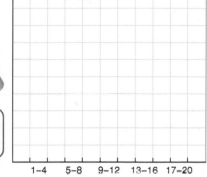

Remember to give
graphs a title and
label the axes.

b Mr. Chan's class had a second maths test.
The results are shown on the frequency
table. Finish drawing the bar chart.

Mark	Tally	Frequency
1–4	I	1
5–8	IIII	4
9–12	IIII III	8
13–16	IIII IIII	9
17–20	IIII IIII	10

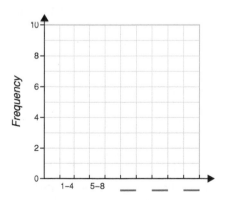

c Which test do you think Mr Chan's class found easier? Give a reason.

T

3 A health expert did a survey to find out how many pieces of fruit 23 people who came to his clinic ate in one month. These are his results.

| 5 | 26 | 7 | 31 | 37 | 25 | 6 | 10 | 8 | 24 | 36 | 33 |
| 39 | 18 | 29 | 16 | 12 | 39 | 38 | 40 | 24 | 1 | 15 | |

a Use a copy of this table. Fill it in.
b Draw a bar chart for the data.

Number of pieces	Tally	Frequency
1–10		
11–20		
21–30		
31–40		

4 Aaron did a survey on house numbers.
This list gives the house numbers of the people that Aaron knows well.

123	89	242	31	6	148	27	52	65	139
22	19	158	210	91	74	10	1	232	101
282	24	50	64	171	228	204	153	124	68
47	138	269	61	18	5	63	242	246	248

a Group this data using a frequency table. Use the class intervals 1–50, 51–100, 101–150, 151–200, ...
b Draw a bar chart for your grouped data.

T

Review 1 Some of Tammy's class entered a problem solving contest. This table shows the marks.

Marks	Tally	Frequency
1–3	IIII	4
4–6	IIIII II	7
7–9	III	3

Use a copy of this bar chart.
Finish it.

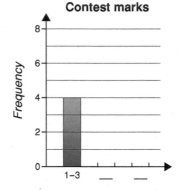

Review 2 This data shows the number of runs scored by a member of the West Indies Cricket Team in Test matches.

22	41	34	15	49	62	98	19	23	38
38	59	113	47	71	86	68	54	101	45
14	5	39	36	48	76	42	0	34	49
25	92	81	43	35	20	81	41	7	

a Group this data using the class intervals 0–19, 20–39, 40–59, ... over 100.
b Draw a bar chart for the grouped data.

Practical

Collect and organise data that can be grouped.
Graph your grouped data on a bar chart.
Remember to design a collection sheet for your data.

Ideas:
number of items bought at the school canteen or number of items of food brought to school

number of digits on the pages of this chapter

goals scored by a sports team in its games in one season

Drawing pie charts on the computer

A **pie chart** is a circle graph.
It shows how something is shared or divided.
The bigger the section of the circle, the bigger the proportion it represents.

Example This pie chart shows the proportion of kilojoules in a Big Mac Combo.

Kilojoules measure energy.

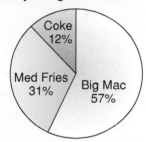

Kj in Big Mac Combo

Practical

You will need a spreadsheet programme or statistical package.

1 This frequency table shows the percentage of pupils at a school who have 0, 1, 2, 3 and 4 brothers.

Number of brothers	0	1	2	3	4
Percentage	20	40	27	10	3

Put this data into a spreadsheet or statistical package.

	A	B	C	D	E	F
1	Number of brothers	0	1	2	3	4
2	Percentage	20	40	27	10	3

Use the graph function of your spreadsheet to draw a pie chart.

2 Choose some other data that could be displayed on a pie chart.
Enter the data into the spreadsheet or statistical package.
Use the graph function to draw a pie chart.
You could enter the data from exercise 1, question 5b, page 375 or exercise 5, question 1, page 384 or collect or find your own data.

Summary of key points

 The **range** is highest data value – lowest data value.

 The **mode** is the most commonly occurring data value.

Example   For   3, 8, 8, 8, 8, 9, 15, 21, 21, 32   the mode is 8 and the range is 29.

If a frequency table has grouped data we find the **modal class**.

Example   **Age of people at Internet Cafe**

Age (years)	0–4	5–9	10–14	15–19	20–24
Frequency	2	3	19	26	21

The modal class is 15–19 because this class interval has the highest frequency.

 Mean = $\dfrac{\textbf{sum of data values}}{\textbf{number of data values}}$

Example   25,   28,   36,   24,   19,   mean = $\dfrac{25+28+36+24+19}{5}$

$$= 26 \cdot 4$$

Sometimes the data is given in a **frequency table**.

Example   **Number of scars on Tom's friends**

Number of scars	0	1	2	3	4
Frequency	5	8	4	2	1
Total number of scars	0	8	8	6	4

Multiply the number of scars by the frequency to get the totals

Mean = $\dfrac{0+8+8+6+4}{5+8+4+2+1}$ ← sum of total number of scars
← sum of frequencies

$= \dfrac{26}{20}$

$= 1 \cdot 3$

 The **median** is the middle value when a set of data is arranged in order of size. When there is an even number of values, the median is the mean of the two middle values.

Example   5, 8, 3, 7, 9, 2, 1, 10

In order of size these are 1, 2, 3, (5, 7), 8, 9, 10

median = $\dfrac{5+7}{2}$

$= 6$

There are two middle data values.

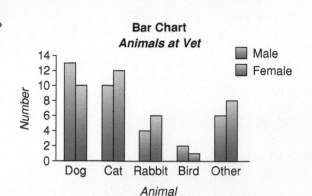

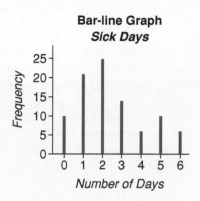

Graphs must have — what is being counted or measured on the horizontal axis
— values at equal intervals on the vertical axes
— the axes labelled
— a title.

 We can draw a **bar chart for grouped data**.
The class interval is written under each bar.

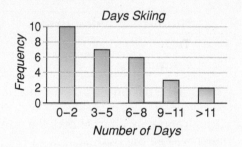

Test yourself

1 This table shows the marks that class 9L got in a test.

What is the modal group?

Mark	Tally	Frequency
31–40	II	2
41–50	IIII	4
51–60	ЖІІ II	7
61–70	ЖІІ ЖІІ III	13

2 The heights of some students are given.

158 cm   172 cm   159 cm   164 cm   167 cm

a What is the mean height of these students?
b What is the range of the heights?
c What is the median height?

3 The mean price of ten cars in a car yard was £8500.
a How much do the ten cars cost in total?
b Another two cars arrive on the yard.
Their prices are £5500 and £9400.
Calculate the new mean price of all 12 cars.
c Is it possible to work out the median price from the information given? Explain.

4 Kate got a mean of 10 and a range of 4 on her
   first six maths tests.

   | Test 10 | Test 11 | Test 9 | Test 10 | Test ? | Test ? |

   What marks did she get on her other two tests?

5 Paul got these marks in three tests: 72, 80 and 82.
   He had one more test to sit.
   He wanted an average of 80 for the four tests.
   What mark must he get in the fourth test?

6 Maureen wrote down the number of video games owned by some of the pupils in
   her class.

   | 5 | 0 | 2 | 1 | 3 | 4 | 6 | 3 | 2 | 4 | 1 | 3 |
   | 0 | 2 | 0 | 0 | 1 | 2 | 3 | 6 | 4 | 5 | 5 | 2 |

   a Copy and complete this frequency table.

Number of video games	0	1	2	3	4	5	6
Tally							
Frequency							

   b Draw a bar-line graph for this data.

7 This frequency chart shows the number of e-mails and faxes sent each day by
   Harrowdale School.

Day	Monday	Tuesday	Wednesday	Thursday	Friday
e-mails	12	16	8	9	13
faxes	3	7	1	5	7

   Draw a bar chart for this data.
   Have separate bars for e-mails and faxes and use a key.

8 Omar's class got these marks for a project.

   | 19 | 16 | 45 | 43 | 40 | 39 | 36 | 30 | 28 | 42 | 35 | 40 |
   | 32 | 38 | 41 | 48 | 27 | 18 | 29 | 38 | 42 | 26 | 41 | 35 |

   a Use a copy of this table. Fill it in.

Mark	11–20	21–30	31–40	41–50
Tally				
Frequency				

   b Draw a bar chart for this data.

17 Interpreting Graphs, Comparing Data

You need to know

Key vocabulary

**bar chart,   bar-line graph,   interpret,   mean,   median,   mode,
pie chart,   range,   survey**

 ## A Date with Data

What are some of the things you can tell from these bar charts?

Find some other published bar charts.

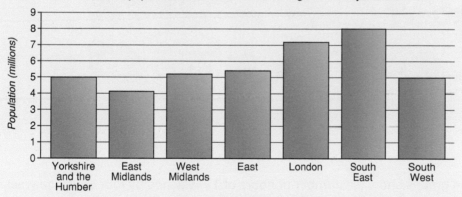

Resident populations of various areas of England in the year 2000

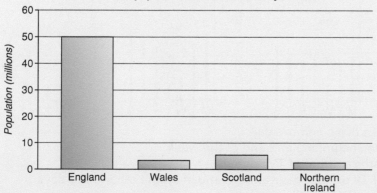

Resident populations of Britain in the year 2000

Interpreting bar charts and bar-line graphs

Sarah

Sarah used this graph to answer the questions.

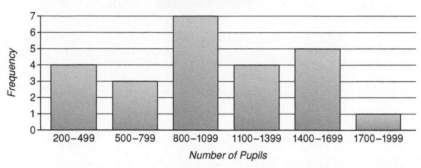

Pupils in Secondary Schools

Exercise 1

1 These bar-line graphs show the number of hours of TV watched by four boys one week.

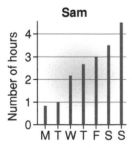

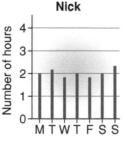

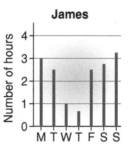

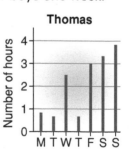

Which boys' graph matches these comments?
a I watched most TV at the beginning and end of the week.
b I watched about the same amount of TV each day.
c I watched quite a lot of TV on four days and not much on the other three days.
d Each day I watched more TV than the day before.

2

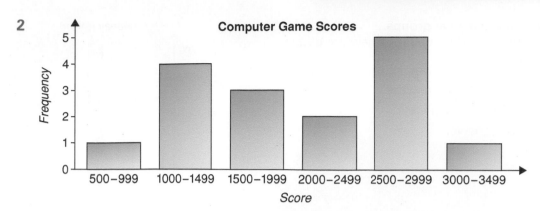

This graph shows the computer game scores for the people in a competition.
a Scores of 2500 or more won a prize. How many people won a prize?
b People with scores less than 2000 were eliminated. How many people were eliminated?
c How many people were in the competition at the start?
d A paper reported that
   'Five people in the computer games competition scored between 2700 and 3000 points.'
   Is this correct? Choose one of these answers.

   A Yes   B No   C Can't tell

3 This bar chart shows the number of tomatoes on 80 'Supertom' and 80 'Red Beauty' tomato plants.
Look at the bar chart.
Would you buy a Supertom or Red Beauty plant if you wanted lots of tomatoes?
Explain your answer.

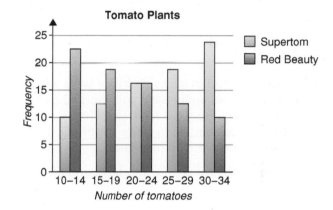

4 A red and a blue dice are each tossed 100 times.
This bar chart shows the results.
One of the dice is fair and the other unfair.
Which do you think is the fair dice? Why?

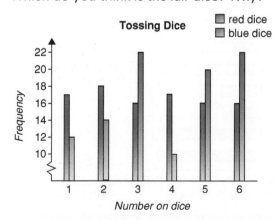

5 A school has five year groups.
Sixty pupils raised money for a new computer.
Look at the graph.
Does Year 7 have fewer pupils than year 9?
Choose one of these answers.

A Yes   **B** No   **C** Can't tell

Explain your answer.

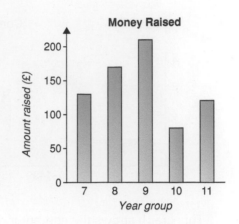

6 Here is a bar-line graph showing the average monthly temperature at 12 p.m. Christine draws a dotted line on the chart. She says 'the dotted line shows the mean for the four months'. Use the chart to explain why Christine cannot be correct.

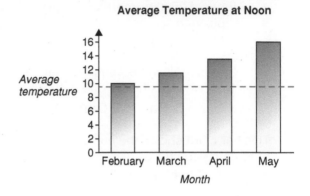

7 Rickey wanted to know if girls or boys were better at blind tasting. He surveyed 40 girls and 40 boys. He gave each pupil five different foods to taste while blindfolded. This bar chart shows his results.

Compare the results for the boys and girls.

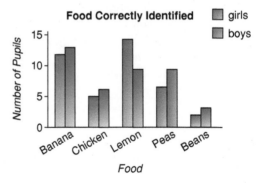

*8 In a survey people were asked what they did to keep healthy and what they would like to do in the future.
This bar chart shows the results.

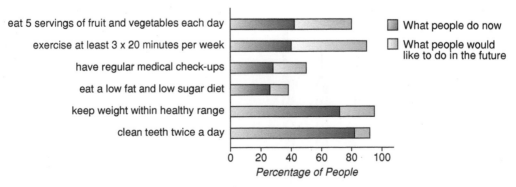

You have been asked to make an advertisement which promotes healthy lifestyles.
Choose two issues to be in your advertisement based on the information in the chart.
Explain how you chose each issue using **only** the information in the chart.

*9

Sentence Length

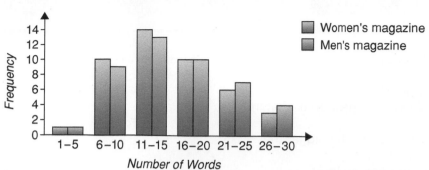

This graph shows the number of words in the first 45 sentences of two magazines.

a   Geoff says '28 out of 45 sentences in the women's magazine have fewer than 16 words. This shows that over half of the sentences in the magazine have fewer than 16 words'.
   Explain why he might be wrong.

b   Compare the number of words in the first 45 sentences of the two magazines.

Review 1   These graphs show the temperatures for three different weeks of the year. Match these statements to the bar charts.

a   It was very warm at first, then it suddenly got much colder.
b   Every day was colder than the day before.
c   Each day was hotter than the day before.

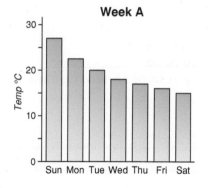

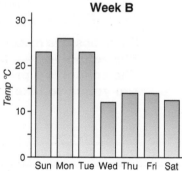

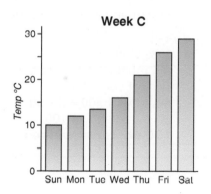

Review 2   This graph shows the amount of pocket money Julia's friends get.

a   What is the modal amount of pocket money?

b   Julia gets £1·50 pocket money.
   She asked her mother to increase it to £3. She told her mother that most of her friends got between £2 and £4.
   Is Julia's request reasonable? Give a reason.

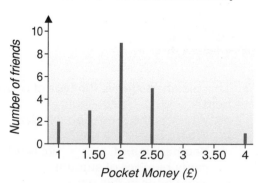

Julia's Friends Pocket Money

Interpreting pie charts

This **pie chart** shows what Sita spent her money on in July.
A pie chart is a circle graph.
It shows us the **proportion** Sita spent on each item.

The bigger the section of the circle the bigger the proportion spent on that item.
Sita spent the biggest proportion on food.
What else can you tell from this pie chart?

Spending for July

Rent
Food
Car
Going Out
Clothes

Sita spent about $\frac{1}{3}$ on food.

Sita spent about 20% on clothes.

Example   This pie chart is divided into 10 equal parts.
It shows how the students of Rayleigh School came to school.

$\frac{2}{10}$ or 20% walked.

$\frac{4}{10}$ or 40% came by bus.

$\frac{3}{10}$ or 30% came by train.

$\frac{1}{10}$ or 10% came by bike.

The tenths must add up to one whole. Why?

Method of Transport

Bike
Walk
Train
Bus

Worked Example

a There are 500 students at Rayleigh School. How many walked to school?

b Give a possible reason why fewer pupils bike.

Answer

a 20% walked.
We must find 20% of 500.
10% of 500 = 50
So 20% of 500 = 2 × 50
= 100
100 students walked.

b Some possible reasons could be:
The school is on a busy road.
The school is a primary school so most pupils are too young to bike.

Discussion

Think of some data that could be displayed on a pie chart. **Discuss**.

Exercise 2

1 This pie chart shows the type of pizza Year 9 students like best.

   a Which is the most popular pizza?

   b What goes in the gap?
Vegetarian and _____ are liked best by about the same number of students.

   c What goes in this gap?
Chicken and _____ together are liked by over half the students.

Choice of pizza

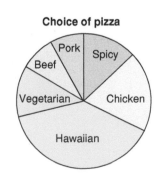

Pork
Spicy
Beef
Vegetarian
Chicken
Hawaiian

2 Cressfield School is thinking about changing the colour of the boys' jerseys.
This pie chart shows the colour 100 boys voted for.
a What proportion voted for blue?
b How many voted for blue?
c What proportion voted for red?
d How many voted for red?
e How many did not vote for red?
f What percentage voted for red?

Jersey Pie Chart

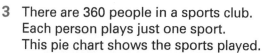

3 There are 360 people in a sports club.
Each person plays just one sport.
This pie chart shows the sports played.
a What fraction play basketball?
b What percentage play basketball?
c What percentage play netball?
d How many play hockey?
e How many play squash?
f How many play netball?

Sports played

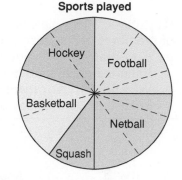

4 a There were 1200 people at a concert. How many of each age group were there?
b Why do you think there were fewer 30+ people at the concert than other age groups?

Proportion of different ages at a concert

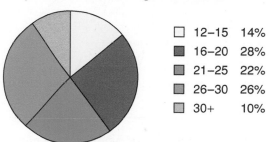

☐ 12–15   14%
■ 16–20   28%
■ 21–25   22%
■ 26–30   26%
☐ 30+   10%

5 This pie chart shows the number of men and women at a conference.
a About what percentage were women?
b About what percentage were men?
c There were 200 people at the conference. Use your percentage from **b** to work out about how many men were at the conference.
d The topic for the conference was one of these. Which do you think is the most likely? Give a reason.
Designing Racing Cars.
Back to work after raising children.

Women and Men at Conference

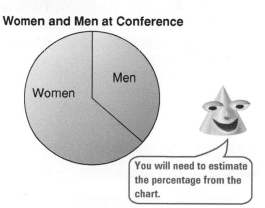

You will need to estimate the percentage from the chart.

6 These pie charts show the items sold at two cafés. Café Roma sold 5000 items altogether and Café Sun sold 2000.
a Roughly what percentage of items sold at Café Roma were muffins?
b About how many muffins did Café Roma sell?
c Jon thought the charts showed that Café Sun sold more sandwiches than Café Roma. Explain why the charts don't show this.

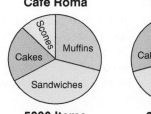

399

Review 1   This pie chart shows the amount of money spent on a holiday.

Holiday Spending

a   On which item was the most money spent?
b   What two items had over half of the money spent on them?
c   Which of these sentences are true?
   1   More money was spent on the hotel than on air travel.
   2   Less money was spent on the hotel than food, entertainment, transport and gifts taken together.
   3   More than half the money was spent on transport, entertainment and food together.

Review 2   This pie chart shows the age groups of members of a squash club.

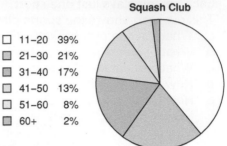

Squash Club

a   How many of each age group would there be if there are 200 members in the squash club?
b   Why do you think there are fewer people older than 60 than any other age range?

☐ 11–20 39%
☐ 21–30 21%
☐ 31–40 17%
☐ 41–50 13%
☐ 51–60 8%
☐ 60+ 2%

Review 3

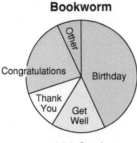

Bookworm

200 Cards

The Bookshop

600 Cards

These pie charts show what sorts of cards were in two bookshops.
Bookworm had 200 cards and The Bookshop had 600.
a   Roughly what percentage of cards at Bookworm were birthday cards?
b   Use your percentage from **a** to work out roughly how many birthday cards there were at Bookworm.
c   Max thought that the charts showed that there were more birthday cards at Bookworm than at The Bookshop.
   Explain why the charts do not show this.

More interpreting graphs

Practical

Find some published graphs.
Good places to look are newspapers, books on national statistics or the Internet.
Choose a graph to interpret. Try and choose a graph that you are familiar with like a bar chart or pie chart.
Write a short report which includes   the graph
   what the graph is about
   what you can tell from the graph

Example

This graph tells us the time that burglaries happened. We can tell from the graph that a burglary is more likely to be carried out during the evening or night than during the day; that fewer burglaries were carried out in the evening/night and morning than at other times; that more burglaries were carried out in the evening than at other times.

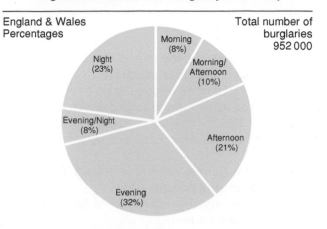

Burglaries of domestic dwellings: by time of day

England & Wales Percentages

Total number of burglaries 952 000

- Morning (8%)
- Morning/Afternoon (10%)
- Night (23%)
- Evening/Night (8%)
- Afternoon (21%)
- Evening (32%)

Source: British Crime Survey, Home Office

Comparing data

To **compare data** we use the **range** and either the **mean**, the **median** or the **mode**.

Worked Example

Brad's rugby team scored these points in matches this season.

21   24   19   22   32   22   28

Zak's rugby team scored these points in matches this season.

0   19   49   19   60   7   14

> Both teams played in the same competition this season.

a Find the mean and range of both sets of data.

b Which team should be picked to play a visiting team? Explain your choice.

Answer

a Mean for Brad's team $= \frac{21 + 24 + 19 + 22 + 32 + 22 + 28}{7}$

$= \mathbf{24}$

Range for Brad's team $= 32 - 19$

$= \mathbf{13}$

Mean for Zak's team $= \frac{0 + 19 + 49 + 19 + 60 + 7 + 14}{7}$

$= \mathbf{24}$

Range for Zak's team $= 60 - 0$

$= \mathbf{60}$

b Both teams have the same mean.

Zak's team has a much greater range. This means their play is less consistent. Sometimes they score very high points and sometimes very low points.

Brad's team has a smaller range. **This indicates their play is more consistent.**

We could choose Zak's team because there is more chance of a high score than with Brad's team or we could choose Brad's team because they are more consistent.

Exercise 3

1 This table gives the mean and range of the scores of two girls in a gymnastics competition.
Only one girl can be chosen to go to a national competition.
Use the mean and range to decide which girl you would choose.
It doesn't matter which one you choose but you must explain why you have chosen her.

	Mean	Range
Jackie	7·9	0·8
Elaine	8·1	4·3

2 Brightwater School wanted to enter a Year 7 team in the homework competition.
The same homework was given to both 7N and 7W. Each pupil was timed.

	median	**range**
7N	35 min	20 min
7W	38 min	10 min

Use the median and range to decide which class you would choose to enter.
Explain why you chose this class.

Homework Competition
Rules
1. Your class must enter 25 pupils.
2. Everyone must finish the homework and get it correct.
3. The times of all 25 pupils are added together.
4. The winner is the team with the shortest total time.

3 **A. Mann**   81   76   79   85   89   78   74   83   74   87
  B. Prebble   79   88   87   76   91   73   74   76

This data gives the golf scores of two players.
a Find the mean golf score for each player.
b Find the range of scores for each player.
c Who do you think is the better golfer? Why?

4 Ross could catch either the 83 or the 64 bus home. He caught the 83 one week and the 64 the next.
This table shows the time for the journeys.
a Find the median of the 83 bus times
b Find the median of the 64 bus times.
c Find the range of each set of data.
d Decide which bus it would be more sensible to catch. Explain why.

	M	T	W	T	F
83 Bus (time in min)	25	28	26	25	26
64 Bus (time in min)	20	34	19	40	17

5 The mean and range of the typing speeds of ten students at the beginning and end of a typing course are given.

Write two sentences comparing the results.

	Mean	Range
Beginning of course	44	20
End of course	56	21

6 The mean and range of the number of letters in 100-word samples from two newspapers are given.

Newspaper Type	Mean	Range
Tabloid	4·2	8
Sunday	4·0	15

Write two sentences comparing the results.

7 Ryan did an experiment to see which was his better catching hand. He threw five cubes in the air 100 times and wrote down how many of these he caught each time.

	Mode	Range
Left hand	3	3
Right hand	2	5

Which do you think is Ryan's better catching hand? Give a reason.

Review 1 This table shows the rainfall each month in two different parts of the world.

	J	F	M	A	M	J	J	A	S	O	N	D
Rainfall at A (mm)	500	610	650	410	150	20	0	0	50	470	594	626
Rainfall at B (mm)	340	348	352	350	350	344	352	347	349	348	340	356

a Find the mean and range for the rainfall at A and for the rainfall at B.
b Use the mean and range to choose in which place you would rather live. It doesn't matter which place you choose but you must use the mean and range to explain why you would rather live there.

Review 2 The median and range of the time taken (in minutes) to travel to school by bus and by car over a two week period are given.

	Median	Range
By car	15·6	5·3
By bus	13·8	9·7

Write two sentences comparing the results.

Practical

Collect two sets of data that you can compare.
Find the mean, median or mode of the data and the range.
Use these to compare the data.

Suggested data:
- midday temperatures at two different places each day for a week
- hours of sunshine, in two different areas, each day for 20 days
- number of students in all classes at two different schools
- tries scored, in matches last season, by two rugby teams

- height (to the nearest centimetre) of students in your class and another class
- price of cars for sale at two different car dealers
- test results of two groups of students
- number of goals scored by first and second division teams throughout the season
- number of hours of TV watched by two different classes or by two different year groups or by two different ages each day for a month
- number of fast food meals eaten each week by two different age groups

Surveys

Investigation

Census

What is a census?
Who organises a census?
When was the last full census in Britain?
How often is a full census taken?
Are there different sorts of census?
How is the data gathered?
What data is gathered?
What is the information used for?
Who is able to use the information?
Investigate these and other questions to do with a census.

You could use the Internet.

Remember
Before you carry out a survey you need to think about these.

- What you want to find out.
- What you might find out.
- What data you need to collect.
- How you will collect the data and who from.
- When and where you will collect the data.
- How you will display the data.
- How you will interpret the data.

Practical

A 1 Think of a question you want answered or a statement you want to test.
   2 Plan a survey. Think about all of the points given above.
     Try to choose something that interests you.

You could use some published data which answers your question.

Suggestions

Questions

- Where are more rubbish bins needed around the school?
- What are the television viewing habits of Year 7 pupils?
- Which sport do people like watching most on TV? Why?
- What is the most common month for people in this class to have a birthday?

- How old are the grandparents of students in this class?
- Which foods should the canteen stop or start selling?
- Which road rules do people obey least?
- What things do people do to look after their health (or the environment) and what might they do in the future?
- How much pocket money do Year 7 pupils get?
- What are the music likes and dislikes of Year 7 pupils?
- What types of video are watched by Year 7 pupils and how often do they watch videos?
- How many calculations can Year 7 pupils do in 10 minutes?
- What type of product is advertised most on TV? Does it depend on the time of year?
- How will the population of different countries change over the next 50 years?
- Do magazines or newspapers use words or sentences of different length?

Statements to test

More boys are left-handed than girls.

Adults' pulse rates are slower than 5 to 16 year olds' pulse rates.

3 Carry out the survey.
4 Display the results on a table and graph.
5 Write some conclusions.
This means, write some sentences about what you found out.
Make sure your conclusions relate to the questions you asked or the statements you tested.
6 Write about any difficulties you had and how you solved these difficulties.

> You could use the mean, median or mode and the range to help interpret the data.

B Find some published data which is given in table form.
Here is an example.

Adults living with their parents: by gender and age

England			Percentages
	1977–78	1991	1998–99
Males			
20–24	52	50	56
25–29	19	19	24
30–34	9	9	11
Females			
20–24	31	32	38
25–29	9	9	11
30–34	3	5	4

Source: National Dwelling and Household Survey of English Housing, Department of the Environment, Transport and the Regions, Labour Force Survey, Office for National Statistics

The Internet is a good place to find up-to-date data.
Try the website http://www.statistics.gov.uk
Use your data to draw some graphs.
Write a short report which includes your graphs and what you found out.

Summary of key points

 A **Interpreting bar and bar-line graphs**

We often use a graph to help us **interpret data**.

Example   This bar chart tells us that Year 10 students watch more television than Year 7 students. The amount of TV watched varies each week for both year groups.

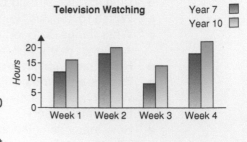

 B A **pie chart** is a circle graph.

It tells us the proportion in each category.

Example   This pie chart tells us the country of birth of Year 7 students at a school.

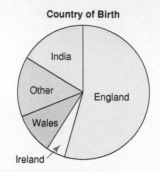

 C To **compare data** we can use the **range** and one of the **mean**, **median** or **mode**.

Example   Melissa and Yolande both wanted to be chosen to represent the school in the National cross country competition.

The mean and range of their last ten races are

	Mean	Range
Melissa	1 hour 46·62 mm	25 minutes
Yolande	1 hour 47·25 mm	4 minutes

We could choose Melissa because her mean time is better or Yolande because her times are more consistent.

Test yourself

1 This bar chart shows the activities done on Saturday and Sunday at a sports centre. **A**

  a One day the weather was warm and on the other day it was cool. Which day do you think was warm? Explain.

  b Which of these is the most likely reason nobody did aerobics on Sunday?

  A Everyone slept in.
  B The centre didn't have aerobics classes on Sunday.
  C An earthquake destroyed the aerobics room.

  c The sports centre is thinking of closing the exercise room and squash courts on either Saturday or Sunday. Which day do you think they should close them? Why?

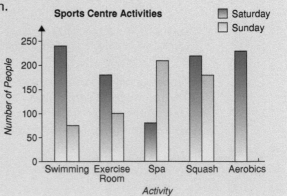

2 A tennis club has four committees. Each gave a donation towards new club rooms.
The bar chart shows the total amount donated by each committee.

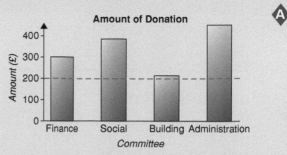

Amount of Donation

a Does the administration committee have the most members?
Choose one of these answers.

  A Yes   **B** No   **C** Can't tell

b The chairperson said the mean donation given by all the committees was £200.
How can you tell from the graph that this is wrong?

3 Kirsty's drama school is putting on a play for the students' families.
This pie chart shows the ages of the students' brothers and sisters.

Age of Siblings

▨	0–4	31%
▨	5–9	36%
☐	10–14	20%
▮	15+	13%

a There are 200 brothers and sisters altogether. How many of each age group are there?

b Do you think it would be better to start the play at 6 p.m. or 8 p.m.? Explain your answer.

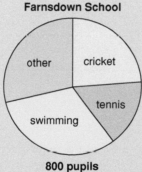

4

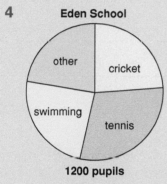

Eden School

1200 pupils

Farnsdown School

800 pupils

Erin was trying to organise a summer sports competition between Eden School and Farnsdown School. She said 'Cricket will be easy to organise because both schools have the same number of cricketers'.
Explain why Erin is wrong.

5 Pete and Kishan both want to represent the school in the public speaking competition.
The mean and range of the scores they have achieved in the last ten speeches is given.

	Mean	Range
Pete	73	15
Kishan	73	35

Who would you choose to represent the school? Use the mean and range to justify your choice.

18 Probability

······ **Key vocabulary** ····································

certain, chance, dice, doubt, equally likely, even chance, fair, fifty-fifty chance, good chance, impossible, likelihood, likely, no chance, outcome, poor chance, possible, probability, probability scale, probable, random, risk, spin, spinner, uncertain, unfair, unlikely

Can I Play?

Some games are said to be pure chance.
Think of some.
Some games are said to be pure skill.
Think of some.
Some games are unfair.
Is this game fair or unfair?

MOVE FOUR

START	1	2	3	4	5	6	7	8	9	10	11	12
26	25	24	23	22	21	20	19	18	17	16	15	14	13
27	28	29	30	31	32	33	34	35	36	37	38	STOP

Find a partner, a dice and two counters.
Decide who is player A and who is player B.
Take turns to move your counters.

Rules for moving Player A throws the dice and moves the number shown.
Player B moves 4 squares every time.

A game is fair if all players have an equal chance of winning.

Make up a game using dice, counters, spinners or cards that is unfair.

How likely

Discussion

- Peter wants to go rock climbing.
  There is a 'risk' he will be injured.
  How great is the risk? **Discuss**.
  Discuss the risk or chance of getting injured in these sports.
  Use words such as *likely*, *high risk*, *good chance*, *unlikely*,
  poor chance, *low risk*, *uncertain*, *doubt*, *probable*, ...

  cricket   horse-riding   rugby   swimming   skiing
  skating   badminton   netball   yachting

  Which do you think has the highest injury risk? **Discuss**.

- Is there more chance of having a road accident at certain times of the day? **Discuss**.
  What about at certain times of the year?
  You might be able to find statistics on times of road accidents to help your discussion.
  The Internet could be a good place to start.

- In some countries there is a higher chance of dying before you are 70 than in other countries.
  Is this true? **Discuss**.
  You could find out if this is true using the library, Internet, CD ROM, ...

- What is the chance of these happening in England? **Discuss**.
  an earthquake   a cyclone   a meteorite crashing
  Which do you think is most likely to happen? **Discuss**.

Remember

Some events are **certain**, some are **impossible**, some are
likely to happen, some are **unlikely** to happen and some
events have an **even chance** of happening.
It is certain that next year will have 52 weeks and impossible that
it will have 13 months.
It is likely I will go to sleep tonight and unlikely that I will dream
of skiing.
There is an even chance of getting a head when you toss a coin.

Some events are **more likely** to happen than others.

Example   These shapes are put in a bag.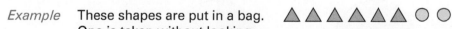
               One is taken without looking.
               It is more likely to be a triangle than a circle because there are more triangles
               in the bag.

Some events have an **equal chance** of happening.

Example   These shapes are put in a bag. △ △ △ △
               One is taken without looking.   ○ ○ ○ ○
               There is an equal chance of it being a triangle or a circle.

There are the same
number of circles
as triangles.

Exercise 1

1 Choose a word from the box to describe the likelihood of these events.
   a You will eat something between now and bedtime.
   b Next week will have eight days.
   c The day after Tuesday next week will be Wednesday.
   d You will see the Prime Minister tomorrow.
   e You will walk into your house through the front door after school.
   f You will have take-aways for tea every night next week.

> certain
> likely
> unlikely
> impossible

2 Charlotte puts these shapes in a bag.

   a She takes one without looking.
     Which shape is she most likely to get?
   b Which shape has she the same chance of getting as a circle? Explain why.

3 All of these spinners are spun at the same time.
   a Which one is most likely to stop on   i red   ii blue?
   b Which one has an equally likely chance of stopping on red as on yellow?

   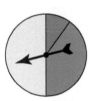

   Spinner A   Spinner B   Spinner C   Spinner D

4 A class is going to play three games.
   The class is divided into two groups, yellow and blue.
   A spinner is spun.
   If it stops on yellow, the yellow group gets a point.
   If it stops on blue, the blue group gets a point.

   Spinner for game 1   Spinner for game 2   Spinner for game 3

   a Which game is yellow most likely to win? Why?
   b Which game is blue most likely to win? Why?
   c Which game is it impossible for yellow to win?
   d Which game is it equally likely that yellow or blue will win?
   e Explain why games 2 and 3 are unfair.

5

The numbers 1 to 10 are written on 10 cards. The cards are shuffled and then put face down in a line.

a The first card is turned over. It is a 6.
Is the next card to be turned over likely to be higher or lower than this? Why?

b The next card turned over is a 3.
Is the next card likely to be higher or lower than this? Why?

6 Gwen chose one of these cards without looking.

a Which shape is she most likely to get?

b Which shape is she least likely to get?

c Gwen mixed the cards up and then put them face down.
She turned one over.

Is the *next* card she turns over more likely to be

 or ? Explain why.

d Gwen wanted to put just and cards in a pile so that it was more likely she

would choose a than a .

What cards could she put in the pile?

Review 1 All of these spinners are spun at the same time.
a Which one has the greatest chance of stopping on red?
b Which has the greatest chance of stopping on green?

Spinner A   Spinner B   Spinner C   Spinner D

Review 2 Anton, Fiona and Kirsty play a game of **TOSS 6**. The first to toss 10 sixes wins.

Anton's cube has 5, 6, 6, 3, 4, 2, on it.
Fiona's cube has 4, 4, 4, 6, 6, 6 on it.
Kirsty's cube has 6, 3, 4, 1, 1, 2 on it.
Is this a fair game?

How could you make this game fair?

Review 3

a Nina takes a card without looking.
   Is she more likely to get a triangle or a circle?
b Is Nina card more likely to have an odd or an even number on it?
c Sam mixes the cards up and turns them face down.

He turns one over. It has a ③ on it. He turns another card over.
Is it more likely he will get a number smaller than 3 or a number bigger than 3? Explain.

 Ups and Downs – a game for two players

You will need ten cards numbered 1 to 10.
You could use playing cards and use the ace as 1.

To play

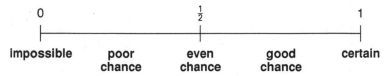

- Shuffle the cards.
- Place them face down in a line.
- Turn over the first card.
- Player 1 predicts whether the next card will be higher or lower than the first card.
- The card is turned over. Player 1 gets 1 point if the prediction was correct.
- Take turns to predict if the next card will be higher or lower than the card just turned over. Stop when all the cards are turned over.
- Play the game four times, taking turns to start.
- The player with the most points wins.

Probability scale

Probability is a way of measuring the chance of a particular outcome.
We can show probabilities on a **probability scale**.
Impossible is at one end and **certain** is at the other.

```
    0                    ½                    1
    ├────────┼──────────┼──────────┼─────────┤
impossible  poor       even        good     certain
            chance     chance      chance
```

The probability of an event that is **certain** to happen is **1**.
The probability of an event that is **impossible** is **0**.
The probability of all other events is between 0 and 1.
The more likely an event is to happen, the closer the probability is to 1.

Discussion

Bobby had a pack of 52 playing cards with no Jokers.

If I take one card without looking, there is a one-in-two chance it will be red.

Bobby

A pack of playing cards has 4 suits, diamonds, hearts, clubs and spades. There are 13 cards in each suit.

Is Bobby right? **Discuss**.
What fraction is this?
What fraction would you use to describe the chance of the card being a diamond? **Discuss**.
What about a spade?

0   1

Where would you put the fractions on this probability scale? **Discuss**.

Exercise 2

1 Use a copy of this probability scale.

0                            $\frac{1}{2}$ 1
                             ↑
                             a

Show where each of these events might be on the scale.
The first one is done.
a   You will get a tail when you toss a coin.
b   Your teacher will live for 200 years.
c   Next year will be 1954.
d   There will be 12 months in the year next year
e   The red marble will land in box A.
f   I will get a red counter when I take a counter from
    the bag without looking.
*g   I will get a 6 when I toss a dice.

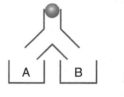

Review   Use a copy of this probability scale.

0                            $\frac{1}{2}$ 1

impossible   certain

Put an arrow on the scale to show the probability of these.
a   the spinner shown stopping on red
b   the spinner shown stopping on green
c   the next baby born being a boy
d   a black bead being taken, without looking, from this bag

413

Outcomes

If we toss a coin, we can get a head or a tail.
These are called **outcomes**.

Example   The possible outcomes of throwing a dice are: 1, 2, 3, 4, 5 or 6.
Each of these outcomes has the same chance of happening.
We call these **equally likely outcomes**.

Exercise 3

1   Bob spun this spinner.
    a   Copy and finish this list of possible outcomes.
        10, 15, __, __, __, __
    b   Do each of these outcomes have the same chance of happening?
        Explain.

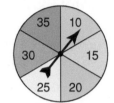

2   Alex was invited to four friends' houses one Saturday.
    She couldn't decide which to go to.
    She put the names in a hat.
        Sarah   Amelia   Charlotte   Kate
    She pulled one out without looking.
    a   Copy and finish this list of possible outcomes.
            Sarah ____, ____, ____
    b   Do each of the friends have the same chance of being chosen?

3   James put some counters in a bag.
    He put in 4 green, 4 red, 4 yellow, 4 blue and 4 black counters.
    He took one without looking.
    a   Copy and finish this list of possible outcomes for the colour.
            green, red, ...
    b   Does each colour have an equal chance of being chosen? Explain.

4   Ryan put some counters in a bag.
    He put in 7 green, 8 red, 1 yellow, 2 blue, and 1 black counter.
    He took one without looking.
    a   Copy and finish this list of possible outcomes for the colour
            green, ...
    b   Are these equally likely outcomes? Explain.

5   A multi-pack of crisps has 4 salt and vinegar, 3 ready salted, 2 barbecue and 3 sour cream
    and chives.
    Kanta takes one without looking.
    a   Write down the list of possible outcomes.
    b   Are these equally likely outcomes? Explain.

Review   The letters of the word WINNER are put in a box.

One is chosen without looking.
a   Write down the list of possible outcomes
b   Are these equally likely outcomes? Explain.

Calculating probability

We can **calculate the probability** of events which have equally likely outcomes.

Example   Harry buys three tickets in a raffle. 100 tickets are sold altogether. Each ticket is equally likely to be chosen.
Harry has 3 out of 100 chances of winning first prize.
The probability is $\frac{3}{100}$ or 3% or 0·03.

For **equally likely outcomes,**

$$\text{probability of an event} = \frac{\textbf{number of favourable outcomes}}{\textbf{number of possible outcomes}}$$

For the example above
probability of Harry winning = $\frac{3}{100}$ ← number of ways Harry could have a winning ticket.
← total number of tickets that could be chosen

Worked Example
The letters of the word CHANCE are put in a bag.
Hein took one without looking.
What is the probability it is the letter   **a** E   **b** C?

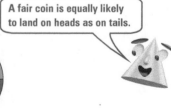

Answer
a There is one way the letter E can be chosen.
The number of favourable outcomes is 1.
The total number of possible outcomes is 6.
So the probability is $\frac{1}{6}$.
b There are two ways the letter C can be chosen.
The number of favourable outcomes is 2.
The total number of possible outcomes is 6.
So the probability is $\frac{2}{6}$ or $\frac{1}{3}$.

Hein didn't look when she chose a letter. We could have said 'Hein chose a letter at random'.

Choosing at **random** means every item has the same chance of being chosen.

Exercise 4	**Only use a calculator if you need to.**

1 What is the probability of getting a tail when a fair coin is tossed?

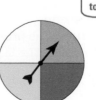

A fair coin is equally likely to land on heads as on tails.

2 The arrow is spun.
   a What is the probability it will stop on red?
   b What is the probability it will stop on grey?

3 Robyn can't make up her mind what to do one Saturday.
She puts these four pieces of paper in a box.
She takes one without looking.
   a Write down all the possible outcomes.
   b What is the probability Robyn's piece of paper will have 'bike ride' on it?

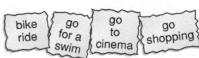

4 Sami has 15 socks in his drawer, four of which are blue. He pulls out a sock at random.
   What is the probability that the sock he has pulled out is blue?

5 Sirah puts 4 purple and 6 black balls in a bag.
   She takes one without looking.
   a What is the probability she takes a purple one?
   b Marie thinks there is a 60% chance of getting a black ball.
   Is she right? Explain.

6 A letter of the word IMAGINATION is picked at random.
   What is the probability of getting
   a an M   **b** an N   **c** an A   **d** an I?

7 There are 50 pieces of chalk in a box. All of them are white. Mrs Patel chooses one piece at
   random. What is the probability that this piece of chalk is
   a white   **b** yellow?

8 Colin spins this spinner.
   What is the probability it will stop on
   a purple   **b** white
   c blue   **d** red?

9 The letters of the alphabet are put in a bag.
   One letter is chosen at random. What is the probability it is
   a the letter T   **b** a vowel   **c** a consonant?

10 This table shows the colours of 100 tickets sold in a raffle.
    All the tickets were put in a hat.
    One ticket was pulled out without looking.
    The arrow on this probability scale shows the probability that
    the ticket pulled out was pink.

green	20
pink	60
blue	5
yellow	15
Total	100

    0   ↑   1
    impossible   pink   certain

    Use a copy of this line.
    a Put an arrow on the line to show the probability that the ticket pulled out was green.
    Explain why you put the arrow here.
    b Put an arrow on the line to show the probability that the ticket pulled out was blue.
    c Put an arrow on the line to show the probability that the ticket pulled out was yellow.

11 These discs were put in a bag.
    Kim took one without looking.
    This probability scale shows the probability of her getting each colour.

    0   ↑   ↑   ↑   1
             red green   orange

    Draw a probability scale to show the probability of taking each colour if these discs were
    put in a bag.
    a 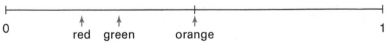   **b**

12 A fair dice is rolled.
What is the probability of rolling

    a a 6
    b an odd number
    c a number greater than 4
    d a factor of 8
    e a prime number
    f zero
    g 7
    h a number between 0 and 7?

Mark these probabilities on a probability scale.

1, 2, 3, 4, 5 and 6 are all equally likely events on a fair dice.

13 Ben and Sarah are playing a word game.
Letters are chosen at random from a bag.
There are 4 letters left in the bag.

Ben needs the T to make a word and Sarah needs the A or the E.
What is the probability that the next letter chosen will be
a the letter Ben needs
b one of the letters Sarah needs?
One of the letters is chosen at random by Ben's sister.
She tells them it is **not** a T.
What is the probability that the letter Ben's sister chose is
c the one Ben needs
d one of the letters Sarah needs?

*14 Each person at a concert was given a coloured ticket.
The number of tickets of each colour given out were:

green	120	orange	100	blue	250
purple	280	yellow	180	red	70

What is the probability that a person picked at random has
a a green ticket
b a blue or purple ticket
c neither an orange nor a yellow ticket?

1000 tickets were given out in total.

*15 Five cards numbered 1 to 5 are shuffled.
One is chosen at random.
    a What is the probability it will be a prime number?
    b Repeat **a** for six cards numbered 1 to 6.
    c Repeat **a** for seven cards numbered 1 to 7.
    d Draw a bar-line graph showing the probability of getting a prime number on one card
        chosen at random from 5 up to 15 cards numbered 1, 2, 3, 4, 5, ... 15.

Hint: You will need to change the fractions into decimals to 2 d.p.

*16 Benjamin has a dice with rectangular faces.
Would the probability of getting a 6 be the same as on a cube-shaped dice?

Link to shape and space.

Review 1   Frank comes home late and must open the front door in the dark.
His key ring has five keys on it.
What is the probability that he chooses the right key on the first try?

Review 2   There are 16 counters in a bag. Five of them are black. Miranda is blindfolded
when she chooses a counter from the bag.
What is the probability that Miranda chooses a black counter?

Review 3   What is the probability that your maths teacher's birthday is on the 30th of
February?

Review 4 Free gifts were given to customers at a shop.
Customers were equally likely to get one of four things.

Jenni wants the perfume. Cass wants the notebook or the rose.
What is the probability that
a Jenni will get the perfume   b Cass will get the notebook or the rose?

Review 5 20 cards are numbered 1 to 20.
One card is chosen at random.
What is the probability it will be
a a prime number   b a non-prime number   c a number from 1 to 9   d an odd number?

Probability from experiments

Discussion

Roseanne made a biased dice.
She wanted to know the probability of throwing a 6 with her dice.

Can she use this formula to find the probability of throwing a 6?

$$\text{probability} = \frac{\text{number of successful outcomes}}{\text{number of possible outcomes}}.$$

Discuss.

> A biased dice is not fair.
> Each number does not
> have an equal chance of
> being thrown.

Often we cannot calculate the probability of an event happening. Instead, we can sometimes
get an **estimate of the probability** by doing an **experiment**.

Example If we want to know the probability of someone catching a ball we need to do an
experiment to find out an estimate.

Example Owen tossed a biased dice 100 times. He got a four 47 times.
From this he estimated the probability of getting a four as

$$\text{probability of getting a 4} = \frac{\text{number of times 4 was tossed}}{\text{total number of tosses}}$$

$$= \tfrac{47}{100}$$

He said 'The probability of getting a 4 on my dice is just under $\tfrac{1}{2}$'.

If Owen wanted a more accurate estimate he would need to toss the dice more times.

Practical

A You will need four different cards and a copy of the table below.

1 Shuffle the cards. Turn them face down.
Take one card without looking.
Before you look at it write down which card you think it is.
If you are right put a tick in the 1st column of your table.
If you are wrong put a cross.
Keep the card in your hand.
Repeat this until you have picked up all four cards.

For the 2nd card put a tick or cross in the second column.

2 Repeat all of **1** 20 times. Use a table like this to record your results.

Guesses

	1st	2nd	3rd	4th
1				
2				
3				
4				
⋮				
20				

What chance is there of being right on the 1st guess?
Is there no chance, a poor chance, an even chance, a good chance or a certain chance?
What about the 2nd, 3rd and 4th guesses? Justify your answers.

B You will need a bag and a mixture of coloured counters.

1 Ask someone to put ten of the counters in the bag without you seeing.

2 Without looking, take a counter out of the bag, write down its colour, then put it back.

3 Repeat **2** 50 times. Use a tally chart like this to record your results.

Colour	Tally	Frequency

Which colour counter do you think there is most of in the bag? Explain.
Which colour do you think there is least of in the bag? Explain.
Estimate the probability of each colour counter.
Use the formula

$$\text{probability of} \_\_\_ = \frac{\text{number of times colour pulled from bag}}{50}.$$

Now empty the bag.
Are your estimated probabilities about right?

C **You will need** some thin card and some Blu-Tack.

Make a biased dice like this.

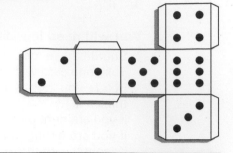

1 Use a larger copy of this net and cut it out.

2 Tape a small piece of Blu-tak to the inside of one face.

3 Tape your dice together.

Toss your dice 100 times.
Record the results in a tally chart.
Did you get the results you expected?
Explain why or why not.
Use your results to estimate the probability of getting each number on your dice.
Compare these probabilities with those you would expect to get using a fair dice.

Number	Tally	Frequency
1		
2		
3		
4		
5		
6		

Investigation

Coin Flips

Someone once told me that if a coin was placed tails up on the thumb nail, then flicked into the air, it was likely to land tails up!
Investigate.

Comparing calculated probability with experimental probability

Discussion

There is a 50% chance of getting a head when a coin is tossed.

Does this mean you would get exactly five heads if you tossed a coin 10 times?

Would you get exactly 50 heads if you tossed it 100 times?

What about 1000 times? **Discuss**.

Practical

A **You will need** a coin.

1 Toss a coin 10 times.
How many heads did you get?

The calculated probability of getting a head is $\frac{1}{2}$.

2 Toss a coin 100 times.
Record your results in a tally chart.
How many heads did you get?
Is the probability of getting a head closer to $\frac{1}{2}$ in **1** or **2**?

B **You will need** a fair spinner.
You could make one using a piece of circular card and a pencil.

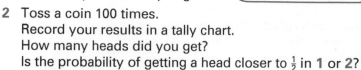

Spin the spinner 100 times.
Record the colour on a tally chart like this one.
Did the spinner stop on each colour the same number of times?
Compare your results with the results of others.

Colour	Tally	Frequency

C Toss a dice 60 times. Record the results in a tally chart.
Draw a bar chart to show your results.
Are the results what you would expect?
If you repeated this experiment would you expect to get the same results?
What results would you expect if you tossed the dice 600 times?

T *** D** **You will need** a computer software package that simulates tossing a coin or a dice.

You could use the random (RND) function on a spreadsheet.

1 Simulate tossing a coin 50 times.
How many heads do you get? How many heads would you expect to get based on a calculated probability?
Repeat the simulation.
Did you get the same result?
Simulate tossing a coin 1000 times, then 2000 times, then 3000 ...
What do you notice?

2 Simulate rolling a dice 60 times. How many 3s would you expect to get based on a calculated probability?
Compare your results with this.
What if you simulate tossing the dice 600 times, 6000 times, ...
Compare your results with the results of others.
Did you all get the same results? Explain why or why not.
Add your results for rolling a dice 60 times together with nine other people's results.
How many heads were tossed altogether out of 600 tosses?
What do you notice?
Add the results of the whole class together.
What proportion of heads were tossed?
What do you notice?

Handling Data

Exercise 5

1 Brian spins a fair coin four times. He gets a head three times.
   Brian says that if he spins it another four times, he'll get three heads again.
   Explain why Brian is wrong.

2 Olivia, Ben and Joshua each rolled a different
   dice 360 times. Only one of the dice was fair.
   Whose was it? Explain your answer.

Number	Olivia	Ben	Joshua
1	27	58	141
2	69	62	52
3	78	63	56
4	43	57	53
5	76	56	53
6	67	64	5

3 Mr Peters gave his class this homework.
       Roll a dice 100 times. Record your results on a chart like this.

Number on dice	1	2	3	4	5	6
Number of times rolled						

   Two students handed in exactly the same results.
   Mr Peters asked if one had copied the other's results without rolling the dice.
   Explain why he might have thought one had copied.

4 There are 100 sweets in a box. Eric takes a sweet without looking inside the box. He writes
   down what sort it is and then **puts it back**. He does this 100 times.
   This chart shows Eric's results.

toffee	20
mint	38
jelly	14
choco	25
caramel	3

   a Eric thought there must be exactly 20 toffees in the box.
      Explain why he is wrong.
   b What is the smallest number of caramels that could be in the box?
   c Is it possible there is any other sort of sweet in the box?
      Explain.
   d Eric starts again and does the same thing another 100 times.
      Will his chart have exactly the same numbers on it?
   e Eric's friend takes a sweet. What sort is he most likely to get?

Review Helal, Michelle and Millie each rolled a fair dice 300 times.
Here are the results they wrote down.
Only one of them had written down the results correctly. Which one?

Number	Helal	Michelle	Millie
1	52	50	25
2	47	50	32
3	58	50	28
4	46	50	89
5	46	50	63
6	51	50	63

Boxes – a game for a group

Draw a 'box' like this.

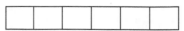

Choose a leader.
The leader calls out six digits from 0 to 9, each one different.
As each digit is called, write it in one section of your 'box'.
The winner is the person who has made the largest possible number.
The winner becomes the leader for the next round.

Remember place value on page 14.

Summary of key points

 We can describe the probability of an event happening using one of these words.

certain   likely   even chance   unlikely   impossible

Some events are **more likely** to happen than others.

Example   You are more likely to get a heart than a club if you take one of these cards without looking.

 Probability is a way of measuring the chance or likelihood of a particular outcome.
We can show probabilities on a probability scale.

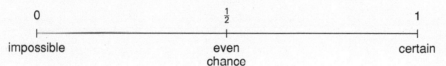

All probabilities lie from 0 and 1.

 This spinner is spun.
It could stop on red, green or yellow.
These are called the possible **outcomes**.
Equally likely outcomes have an equal chance of happening.

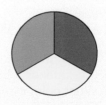

Example   The spinner above is divided into three equal sections of red, green and yellow.
Stopping on red, stopping on green, stopping on yellow are equally likely outcomes.

D For equally likely outcomes

$$\text{probability of an event} = \frac{\text{number of favourable outcomes}}{\text{number of possible outcomes}}$$

Example The names of 5 boys and 3 girls were put in a box.

Tyrone took one at random.

Probability of a girl = $\frac{3}{5}$ ← 3 ways of getting a girl
← total number of possible outcomes

E Sometimes we cannot calculate the probability, but we can **estimate** it by **doing an experiment**.

Example If we wanted to know the probability that a car at a particular junction will turn right we would need to do an experiment.

We could find out how many, out of say 1000 cars, turned right at the intersection.

From this we could estimate the probability from

$$\text{probability} = \frac{\text{number of cars that turned right}}{1000}$$

Example Robert found by experiment that he kicked a rugby ball over the post 61 times out of 100.

He estimated the probability of his kicking the ball over the post as $\frac{61}{100}$ or about 0·6.

When we **repeat an experiment** the **results are often slightly different**.

Example Josh tossed a coin 20 times and got 11 heads.

He tossed the same coin another 20 times and got 8 heads.

The more trials Josh did each time, the more similar the results would be.

Test yourself

1 Madu puts 5 purple beads and 1 pink bead in a bag.
Madu takes a bead without looking.
What colour bead is she more likely to get?
Explain why.

2 Draw a probability scale.
Put an arrow on the scale to show the probability of these.
   a getting an odd number when you throw a dice
   b the spinner shown stopping on purple
   c the spinner shown stopping on blue
   d getting a 3 when you throw a dice

3 Ian put these shapes in a bag.

Sarah pulled one out without looking. Write down all the possible outcomes.

4 A shop sells red, green, purple, yellow and orange lollipops. They are in a jar.
There are the same number of each colour.
a   Clive chooses one at random. What is the probability it is orange?
b   Sheryl likes green and red ones. She chooses one at random.
   What is the probability it is a colour she likes?
c   Matt likes all the colours. What is the probability he will get a colour he likes?
d   Put the probabilities you found in **a**, **b** and **c** on a probability scale.
e   Sue chooses one lollipop.
   The arrow on this scale shows the probability that Sue gets a colour she likes.

How many colours does Sue like?

5 Ellie had these ten number cards.

Her friend took one card without looking.
What is the probability the card will have
a   an odd number   b   an even number
c   a number less than 10   d   the number 3
e   a number less than 16   f   a number greater than 6?

6 Bill put eight purple counters and eight black counters in a bag.

He took a counter without looking and then **put it back**.
He did this 16 times.
Bill took out purple counters 10 times and black counters 6 times.
Bill said 'I must have made a mistake and put 10 purple and 6 black counters in the bag.'
Explain why Bill is wrong.

7 Raffle tickets are numbered 1 to 300.
Lena buys five tickets and gets the numbers 8, 9, 10, 11, and 12.
Freddie buys five tickets and gets the numbers 18, 29, 182, 207, 234.
One ticket is chosen at random.
Is Lena or Freddie more likely to have the winning number? Explain your answer.

*8 There are six counters in a box.
The probability of taking a green counter out of the box is 0·5.
A green counter is taken out of the box and put to one side.
Jason now takes a counter from the box at random.
What is the probability it is green?

Test yourself answers

Chapter 1 page 29

1 a 8 hundreds or 800 b 8 ones or 8 c 8 tenths or 0·8 d 8 hundredths or 0·08 e 8 thousandths or 0·008
2 870·5
3 a One hundred and sixty-eight point zero three b Eight point nine six four
  c Ninety-six point zero eight six d Four and two hundred and seventy-three thousandths
4 a 89 062·07 b 11·64 c 402·039 5 a 8·8 b 13·96 c 3·99 d 4·99
6 a 48 000 b 9·6 c 5·4 d 0·083 e 57 f 89 g 520 h 0·00637 i 0·052 7 £3·65
8 a False b False c True d True 9 a 4·074, 4·27, 4·7, 4·72 b 0·06, 0·078, 0·689, 0·798, 0·85, 0·869
10 a 18·5 b 10 c 7·5 d 0·45 e 0·85 f 1·95 g 6·85 h 8·45
11 a Rob b Aled c Jake d Samuel
12 a 870 cm b 900 cm c 8 m, 9 m, 9 m, 8 m d 8·3 m, 8·7 m, 8·9 m, 8·1 m 13 8

Chapter 2 page 56

1 ⁻6, ⁻4, ⁻1, 0, 1, 3 2 ⁻3·5 3 11 °C
4 a i ⁻3 ii ⁻1 and ⁻3 or 2 and ⁻6 iii ⁻6 iv ⁻1 and 3 or ⁻3 and 1, 1 and 5, ⁻2 and 2, ⁻6 and ⁻2
  b ⁻6 + ⁻3 + ⁻2 = ⁻11 c 5 − ⁻6 = 11
5 a 600, 640 b 108, 117, 216, 387, 600, 702 c 108, 216, 600, 640 d 108, 216, 600, 702
  e 216, 600, 640 f 108, 117, 216, 387, 702
6 Possible answers are: 1 is a square number, a triangular number and a factor of all numbers. 3 is a prime number, a triangular number and a factor of 6, 4 is a square number, a multiple of 2 and a factor of 20, 8 is a cube number, a multiple of 2 and a factor of 16.
7 a 136, 163, 316, 361, 631 b 316
8 a 12 b 8 c 1 and 72, 2 and 36, 3 and 24, 4 and 18, 6 and 12, 8 and 9
9 5 and 19 or 7 and 17 or 11 and 13
10 10 + **5** = **15** 5 is the next counting number. The 5th triangular number is 15.
11 a 8 b 5 c 9 d 16 e 3 f 36 12 a 16 b 10 c 49 d 7 e 0·25 f 0·6
13 a 46·24 b 841 c 2·9 d 6·3 14 a 4·6 b 5·7 c 10·4 15 a 3600 b 40 000 c 60 d 400
16 11, 17, 37, 71, 73, 79, 97

Chapter 3 page 76

1 a 21 b 70 2 a 46 b 50 c 24 d 100 e 117 f 225 g 429 h 32 i 4816
3 a 1·7 b 1·9 c 9·8 d 1·7 e 23·6 f 11·6 g 5·1 h 13·2
4

14	21	16
19	17	15
18	13	20

5 a 60 b 130 c 148 d 410 e 450 f 39 g 32 h 800 i 3600 j 902 k 420
  l 130 m 1800 n 925
6 a 2·6 b 21·5 c 39·2 d 2·4 e 0·7 f 2·4 g 5·4 h 0·2 i 0·6 j 0·7 k 56·1
  l 129 m 265 n 61·1
7 a 11 b 388 c 485 8 a £231 b £15·60 c £0·40
9 a 11 b 14 c 3 d 18 e 11 f 60 g 3 h 4 i 50 j 24 k 4 10 a C b B
11 Possible answers are: a 500 + 200 = 700 or 450 + 200 = 650 b 3 × 5 = 15 or 3·5 × 5 = 17·5
  c 1000 − 400 = 600 or 950 − 350 = 600.
12 a 3 + 8 − 1 = 10 or 4 + 8 − 2 = 10 or 5 + 8 − 3 = 10 or 6 + 8 − 4 = 10 or 7 + 8 − 5 = 10 or 4 + 7 − 1 = 10 or
    5 + 7 − 2 = 10 or 6 + 7 − 3 = 10 or 5 + 6 − 1 = 10 or 4 + 6 − 0 = 10 or 2 + 8 − 0 = 10 or 3 + 7 − 0 = 10
  b 48 × 2 = 96 or 12 × 8 = 96 c One possible answer is 56 + 78 = 134.

Chapter 4 page 102

1 a 2312 b 85 082 c 16·26 d 24·64 e 42·509 f 9412 g 820 h 1·36 i 88·82
  j 295·06 k 796·81
2 7500 m 3 7·14 tonne

4 a 1904 b 4368 c 15 525 d 23 052 e 50 641 f 44 184 g 17 366 h 33 558
i 19 363 j 49 494 k 19 346
5 4472 **6** a 32·2 b 91·5 c 15·12 d 33·81 e 82·1 f 109·44 g 1141·6 **7** £38·30 **8** 66·35 m
9 a 56 b 89 c 235 d 56 e 59 f 41 g 31 **10** 18 **11** a 0·8 b 2·3 c 14·6 d 21·3 e 23·56
12 a £20·48 b £4·52
13 There are many possible answers. One is
a key ⑨①④⊖⑤⑥⊜ b key ⑨③⑨⊖⑨③⊜
14 a Incorrect. All of the amounts are less than £15 and $6 \times £15 = £90$ so the total should be well below £90.
b Incorrect. 52 is about 50 and 204 is about 200. $200 \div 50 = 4$.
15 a $35\,112 \div 42 = 836$ or $35\,112 \div 836 = 42$ b $192 \times 1·9 = 364·8$
c $28·08 - 19·72 = 8·36$ or $28·08 - 8·36 = 19·72$
16 Possible answers are: a $(600 \times 27) + (3 \times 27) = 16\,281$ b $(40 \times 59) + (2 \times 59) = 2478$ or $(42 \times 60) - 42 = 2478$
c $(632 \div 2) \div 2 = 158$
17 a $423 - 156 + 56 = 323$ b $589 + 86 - 185 = 490$ **18** 18 **19** a 5550 b 10·13 c 2·2 d 4·8
20 a 7 days 18 hours b 6 minutes 13 seconds

Chapter 5 page 124

1 A and D **2** a $\frac{51}{100}$ b $\frac{49}{1000}$ c $\frac{43}{60}$ **3** $\frac{1}{6}$
4 a $\frac{3}{5} = \frac{6}{10} = \frac{9}{15} = \frac{12}{20}$ b $\frac{5}{8} = \frac{10}{16} = \frac{15}{24} = \frac{20}{32} = \frac{25}{40}$ c $\frac{4}{9} = \frac{8}{18} = \frac{12}{27} = \frac{20}{45} = \frac{24}{54}$ d $\frac{2}{7} = \frac{4}{14} = \frac{6}{21} = \frac{8}{28} = \frac{10}{35}$
5 There are many possible answers. One possible answer is:
a $\frac{4}{6}, \frac{6}{9}, \frac{8}{12}, \frac{10}{15}$ b $\frac{6}{8}, \frac{9}{12}, \frac{12}{16}, \frac{15}{20}$ c $\frac{2}{16}, \frac{3}{24}, \frac{4}{32}, \frac{5}{40}$ d $\frac{3}{4}, \frac{6}{8}, \frac{9}{12}, \frac{12}{16}$ e $\frac{5}{9}, \frac{10}{18}, \frac{15}{27}, \frac{25}{45}$
6 a $\frac{1}{2}$ b $\frac{1}{4}$ c $\frac{1}{3}$ d $\frac{1}{3}$ e $\frac{2}{3}$ f $\frac{3}{5}$ g $\frac{4}{5}$ h $\frac{4}{1}$ i $\frac{2}{3}$ j $\frac{3}{1}$ **7** a $\frac{1}{3}$ b $\frac{3}{4}$ **8** a $3\frac{3}{7}$ b $7\frac{1}{4}$ c $7\frac{2}{5}$ d $5\frac{3}{8}$ e $8\frac{1}{2}$
9 a 21 b 18 c 14 **10** a $\frac{5}{4}$ b $\frac{13}{5}$ c $\frac{39}{8}$ d $\frac{32}{9}$ e $\frac{47}{7}$
11 a $\frac{1}{2}$ b $\frac{1}{10}$ c $\frac{3}{5}$ d $\frac{43}{100}$ e $\frac{73}{100}$ f $\frac{6}{25}$ g $\frac{7}{20}$ h $\frac{47}{50}$ i $\frac{47}{50}$ j $\frac{7}{25}$ k $\frac{9}{25}$ **12** a 0·2 b 0·57 c 0·35 d 1·8
13 a $\frac{3}{4}, \frac{2}{5}, \frac{3}{8}, \frac{1}{4}, \frac{1}{6}$ b $1\frac{7}{8}, 1\frac{4}{5}, 1\frac{3}{4}, 1\frac{5}{8}$ **14** a $\frac{6}{7}$ b $\frac{3}{8}$ c $1\frac{2}{5}$ d $\frac{1}{4}$ e $\frac{1}{2}$ f 1 g $\frac{6}{13}$ **15** a $\frac{1}{2}$ b $\frac{3}{8}$ c $1\frac{1}{2}$ d $\frac{1}{8}$
16 a £42 b 9 m c 20 ℓ d 15 km e £16 f 35 cm g 9 m h 140 g i £4·50 j 24 m k 5 m
17 a 50 minutes b 6 **18** a 8 b 4 c 12 d 21 e 30 f $\frac{2}{5}$ g $\frac{10}{3}$ or $3\frac{1}{3}$ h $\frac{20}{3}$ or $6\frac{2}{3}$ i $10\frac{1}{2}$ j 20 k 63

Chapter 6 page 137

1 a $\frac{7}{10}$ b $\frac{3}{4}$ c $\frac{7}{20}$ d $\frac{31}{50}$ e $1\frac{3}{4}$ **2** a 0·7 b 0·75 c 0·35 d 0·62 e 1·75 **3** a $\frac{2}{5}$ b 0·4
4 a 25% b 90% c 80% d 65% e 300% f 180% **5** a 63% b 84% c 5% d 11%
6 a $\frac{1}{5}$ b 20% c 0·2 **7** a 40% b 30% c 45% **8** a $\frac{37}{100}$ b 0·37
9 a Any answer between 45% and 55% is correct. b Any answer between 85% and 95% is correct.
c Any answer between 20% and 30% is correct.
10 a 6 b 10 c 54 d 6 e 1·8 f 0·5 g 6 h 902 **11** 24
12 a £6·65 b 8·16 ℓ c 34·2 m d £1387·76 e £0·52 f £4·38 to the nearest p. **13** C
14 a 136 cm b 169·2 cm **15** Mary

Chapter 7 page 150

1 a £10·40 b £15·60 c £20·80 **2** 31 : 4 **3** a 9 : 14 b 14 : 9 **4** 2 : 1 **5** a 12 b 3 c 20
6 a 3 : 2 b 60% c 0·4 **7** a $\frac{13}{20}$ b 35% c 7 : 13 **8** 15 litres **9** 8 **10** Natasha 300 g, Sister 400 g
11 35 **12** Victoria

Chapter 8 page 186

1 a $3y$ b $4b$ c c^2 d $4(x + 3)$ e $2(a + 7)$ f $6(y - 4)$ g p^2 h $3b^2$ i $(5y)^2$ or $25y^2$
2 a $n - 4$ b $7n$ c $2n + 3$ d $2(n + 3)$ e $(3n)^2$
3 a $4x$ b $11a$ c $5b$ d $15m$ e $7n$ f $11p$ g $5q$ h $14p + 4$ i $5y + 9$ j $5x + 4$ k $6m + 3$ l $5a + 6b$
m $2x + 4y$ n $5p + q$ o $9n$
4 $8x + 32$ **5** a 5 b $4n$ c $4x$ d 30 e 15 f $\frac{3a}{2}$ **6** a $5x + 15$ b $7n + 28$ c $8a + 32$ d $12x + 24$
7 a D b G c H d F e C f A g B h E
8 a 6 b 1 c 9 d 30 e 2 f 6 g 0 h 16 i 12 j 16 k 6 l 32
9 a $y + x = 6$, $6 - y = x$, $6 - x = y$
b There are many possible answers. One is $x = 4$, $y = 2$. x and y must always add to 6.
10 i a £6 b £9 c £8·50 d £5·50 ii $C = \frac{A}{2} + 1$ **11** 6 **12** a £340 b £600 c £437·5
13 a $n + 2$ b $n - 3$ c $3n$ d $\frac{n}{5}$ e $4n - 3$
14 a $n - 5 = 7$ b $4n = 32$ c $\frac{n}{6} = 4$ d $3n = 450$ e $2n + 4 = 20$
15 a 12 b 13 c 10 d 36 e 5 f 9 g 8
16 a $4n + 9 = 37$; $n = 7$ b $5n - 4 = 51$; $n = 11$
c $9x = 180°$; $x = 20°$ d $4n - 1 = 127$; $n = 32$
17 $n + n + 4 = 78$ $n = \frac{74}{2}$ $n = 37$ ←

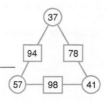

Chapter 9 page 218

1 **a** 4, 6·5, 9, 11·5, 14, 16·5   **b** 1, ⁻2, ⁻5, ⁻8, ⁻11, ⁻14   **2** **a** D   **b** A   **c** B   **d** C

3 2·8, 3·2, 3·6 **first term** 0·8 **rule** add 0·4 or each term is a multiple of 0·4

4

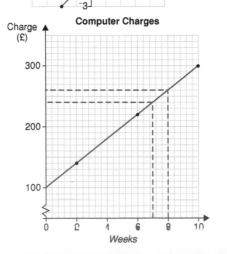

5 **a** 23, 20, 17   **b** 1000, 200 and 8   **c** 1, 2, 4, 8

6 **a** There are many possible answers. Two are 0, 5, 10, 15, 20, ... and 14, 19, 24, 29, 34, 39, ...
   b No. Every second time you add 5 you must get an even number.

7 **a** 6, 7, 8, 9, 10   **b** ⁻1, 1, 3, 5, 7   **c** 18, 16, 14, 12, 10   **8** **a** 20   **b** 27   **c** ⁻10

9

Term number	1	2	3	4	5	6	...	n
Term	6	12	18	24	30	36	...	$6n$

10 **a** There are 9 dots in this diagram.

   b The first diagram has 3 dots and 2 dots are added for each additional diagram.
   c No because 24 is an even number and all of the diagrams have an odd number of dots.
   d One possible answer is: Shape n has n dots along the bottom and $n + 1$ dots along the top. The nth shape will have $n + n + 1$ dots.   **e** $2n + 1$

11 The nth term is 3 times n plus 2   **12** **a** 16, 19, 21, 28   **b** 4, 3, 9, 12   **c** 19, 28, 13, 10   **d** 16, 32, 10, 26

13

x	1	2	3	4	5
y	9	11	13	15	17

14 **a** $y = 3x + 4$ **b** $y = (x - 3) \times 4$ or $y = 4(x - 3)$ **15** **a** $x \rightarrow \frac{x}{4} + 3$ **b** $x \rightarrow (x + 2) \times 5$ or $x \rightarrow 5(x + 2)$

16 **a** $x \rightarrow \boxed{\text{multiply by 2}} \rightarrow \boxed{\text{subtract 4}} \rightarrow y$ **b** $x \rightarrow \boxed{\text{subtract 4}} \rightarrow \boxed{\text{multiply by 2}} \rightarrow y$

   c No, because the machines are in a different order. Gemma's function machine would give output of 16. Ricky's would give output of 12.

17 **a** C   **b** D   **c** A   **d** B

18 **a** Add 4   **b** Divide by 3   **c** Multiply by 3, add 2   **d** Add 1, multiply by 3 **or** multiply by 3, add 3

19 **a** 6, 10   **b** 30, 24   **20** **a** 17   **b** 5   **21** Multiply by 2, subtract 7 in this order

Chapter 10 page 240

1 ⁻7

2 **a**

x	1	3	5
y	4	12	20

b

x	⁻1	2	5
y	⁻5	⁻2	1

c

x	0	1	5
y	9	8	4

3 **a**

x	⁻2	0	3
y	⁻3	⁻1	2

b (⁻2, ⁻3), (0, ⁻1), (3, 2)   **c**

   d Yes
   e There are many possible answers. Two are (5, 4) and (⁻10, ⁻11).

4 **a** Any number greater than 2 is correct.
   b $y = x + 3$ **c** C

5 **a** and **c**

6 **a**

Weeks	2	6	10
Charge (£)	140	220	300

   c ⧦ means the numbers on the axis do not start at zero.
   d Go up from 9 weeks to the graph. From there go across to the vertical axis.
      The charge for 9 weeks is £280.
   e Go across from £240 to the graph. From there go down to the horizontal axis.
      She hired the computer for 7 weeks.

b

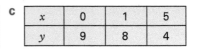

Chapter 11 page 278

1 a ∠Q b ∠M   2 a ∠PQR or ∠RQP or PQ̂R or RQ̂P   b ∠LMN or ∠NML or LM̂N or NM̂L
3 a a — acute, b — reflex, c — reflex, d — straight, e — obtuse, f — acute
   c a — 27°, b — 270°, c — 238°, d — 180°, e — 120°, f — 60°
5 a AD and BE, AB and DE   b AC and CB, AB and BE, AD and DE, BE and ED, AB and AD
6 A rectangle   7 a 129° b 61° c 86° d 75° e 70° f $l = 31°$, $m = 38°$ g 73° h 65° i $e = 73°$, $f = 57°$

Chapter 12 page 300

1 Possible answers are: a △PQR or △PRQ b △ABC or △ACB c △LMN or △LNM   2 a p b c c m
3 a True b True c False   4 a 3 b 4 c 2 d 0 e 1 f 1
5 a An equilateral triangle b Rhombus   6 a $p = 130°$ b $q = 140°$, $r = 40°$ c $s = 126°$, $t = 54°$
7 $x = 130°$, $y = 65°$, $z = 50°$
8 No, because if it has two lines of symmetry, all three angles must be equal which means it would be an equilateral triangle with 3 lines of symmetry.
9 A square   10 A right angle, 60°, 30°   11 a 68° b 80°   12 3
13

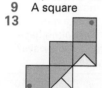
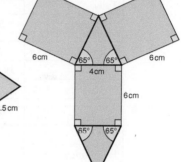

14 Possible answers are given. The nets are not full size.

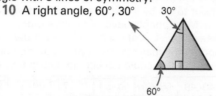
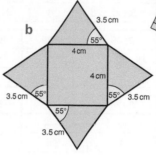

Chapter 13 page 321

1 b 4 c There are many answers. Two are (0, 3) and (3, 1) or (⁻2, 0) and (1, ⁻2). Yes it is possible to make a square because only two points are given. We can plot two more points at (⁻1, 1) and (2, ⁻1) to make a square. d second
2 Use a mirror to check your answers   3 Arrowhead or delta
4 P and P′ are not the same distance from the mirror line S.
5 a b   6

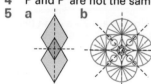

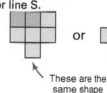

or

or ... or ... or

These are the same shape

7 a

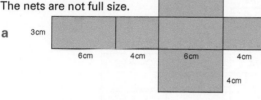

8 b and d   9 a 2 b 4 c 3 d 3
10 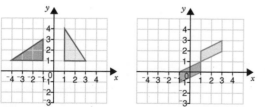   11 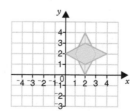   12

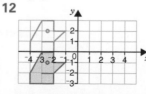
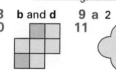

13 a 5 units left and 3 units up b 4 units right and 6 units down   14 B and C

Chapter 14 page 355

1 **a** 32 **b** 2800 **c** 360 **d** 680 **e** 7600 **f** 0·47 **g** 0·86 **h** 6·852 **i** 7·52 **j** 5·2 **k** 0·864 **l** 60
m 0·0585 **n** 3860 **o** 620   2 **a** 16 **b** 4 **c** 35 **d** 22 **e** 50 miles or 80 km
3 **a** A is at 2·2 lb, B is at 3·4 lb C is at 1·8 lb   **b** A is at 0·34 Newtons, B is at 0·23 Newtons, C is at 0·27 Newtons
  c A is at about 0·5 kg, B is at about 1·8 kg, C is at about 3·2 kg   4 About 1 kg   5 **a** A **b** B
6 Possible answers are: **a** Use kitchen scales to measure in g   **b** Use a ruler to measure in cm by making a
mark at the ends of your foot and measuring the distance between the marks.   **c** Use a measuring jug to
measure in mℓ by filling the egg cup with water and then pouring it into a measuring jug.
7 **a** 36 cm² **b** 1 m² **c** 160 cm²   8 **a** 46 m² **b** 23   9 **a** 544 cm² **b** Blue 352 cm², red 192 cm²
10 200 cm²

Chapter 15 page 372

1 **a** Possible answers are: watching TV or videos is the most popular activity, doing homework is the least
popular activity.   **b** Sheet 2, because it is easier to tell by looking at it which activity is most popular.
2 **a** C **b** B **c** A   3 Possible answers are: **a** Most people have takeaways more than once a month.
Friday night is the most popular night for takeaways.   **b** Data needed for each person; how often have
takeaways and which day of the week   **c** About how many times each month do you have takeaways?
Which night/day do you most often have them?   **d** He could collect the data in tally charts.

4 **a**

Number of books	0–19	20–29	30–39	40–49	50–59	60–69	70–79	80–89	90–99	100–109	110–119
Tally	|||	|||	||||	||						|||					
Frequency	3	3	4	2	1	0	1	3	1	1	1

b

Number of books	0–49	50–99	100–149
Tally	卌 卌 ||	卌		||
Frequency	12	6	2

c

Number of books	0–24	25–49	50–74	75–99	Over 99
Tally	||||	卌 ||	||	||||	||
Frequency	4	7	2	4	2

d The 0–24, 25–49, ... because we can see the patterns in the data best. Using the intervals 0–19, 20–29, ...
gives too many intervals and doesn't show up the patterns. Using the intervals 0–49, 50–99 gives too few
intervals and doesn't show up any patterns.

Chapter 16 page 391

1 61–70   2 **a** 164 cm **b** 14 cm **c** 164 cm   3 **a** £85 000 **b** £8325 **c** No, because we can't tell
what price the 10 cars are individually. We can't find the middle value.   4 8 and 12   5 86

6 **a**

Number of video games	0	1	2	3	4	5	6
Tally	||||	|||	卌	||||	|||	|||	||
Frequency	4	3	5	4	3	3	2

b

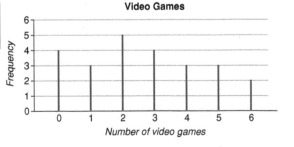

7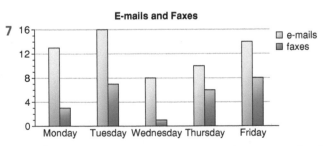

8 **a**

Mark	11–20	21–30	31–40	41–50
Tally	|||	卌	卌 ||||	卌 ||
Frequency	3	5	9	7

b

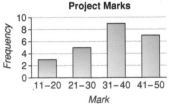

Chapter 17 page 406

1 **a** Saturday because more people went swimming and fewer had a spa. This is what you would expect on a warm day. **b** B **c** Sunday because fewer people use them on a Sunday.

2 **a** C **b** Because $200 is below the tops of all of the bars so it cannot be the mean.

3 **a** 0–4, 62; 5–9, 72; 10–14, 40; 15+, 26 **b** 6 p.m. because most of the brothers and sisters are under 10 and 8 p.m. would be too late for young children.

4 The charts show the proportion who play each summer sport and not the actual numbers. There are 1200 pupils at Eden School and only 800 at Farnsdown. Eden School will have more cricketers.

5 One possible answer is: I would choose Pete because he has the same mean as Kishan but his range is much smaller which means he is more likely to get a mark fairly close to 73. Kishan has a larger range so he might get a mark much higher than 73 but he also might get a mark much lower than 73.

Chapter 18 page 424

1 A purple bead because there are more of them

2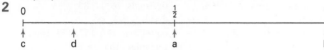

3 Triangle, circle, square, hexagon

4 **a** $\frac{1}{5}$ **b** $\frac{2}{5}$ **c** 1 **d**
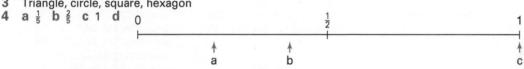

  e The probability given on the scale is $\frac{6}{10}$ or $\frac{3}{5}$. Sue likes three of the colours.

5 **a** $\frac{6}{10}$ or $\frac{3}{5}$ **b** $\frac{4}{10}$ or $\frac{2}{5}$ **c** $\frac{7}{10}$ **d** 0 **e** 1 **f** $\frac{6}{10}$ or $\frac{3}{5}$

6 Although each colour has an equal chance of being selected it does not mean they will be selected equally. Bill might have taken out some counters more than once and others not at all.

7 They have the same chance of winning because they both have 5 tickets.

8 $\frac{2}{5}$

Index

Index